KT-436-895

KNOTS AND SURFACES

Knots and Surfaces

N. D. Gilbert

Department of Mathematical Sciences,
University of Durham.

T. Porter

School of Mathematics,
University College of North Wales, Bangor

OXFORD NEW YORK TOKYO
OXFORD UNIVERSITY PRESS
1994

Oxford University Press, Walton Street, Oxford OX2 6DP

Oxford New York
Athens Auckland Bangkok Bombay
Calcutta Cap Town Dar es Salamm Delhi
Florence Hong Kong Istanbul Karachi
Kuala Lumpur Madras Madrid Melbourne
Mexico City Nairobi Paris Singapore
Taipei Tokyo Toronto

and associated companies in
Berlin Ibadan

Oxford is a trade mark of Oxford University Press

Published in the United States
by Oxford University Press Inc., New York

A catalogue record for this book is available from the British Library

Library of Congress Cataloging in Publication Data
Gilbert, N.D.
Knots and surfaces / N.D. Gilbert, T. Porter.
1. Knot theory. 2. Surfaces. I. Porter, T. (Timothy), 1947–
II. Title.
QA612.2.G55 1994 514'.224—dc20 94–5209

ISBN 0 19 853397 7

Typeset by Pure Tech Corporation, Pondicherry, India
Printed in Great Britain by St Edmundsbury Press,
Bury St Edmunds, Suffolk

Preface

The fascination exerted by interlaced patterns and knotted forms is evident in their use in decorative and symbolic art across the centuries. Knotwork is a distinguishing feature of Celtic art, and the intricate beauty of decorated stonework and illuminated manuscripts such as the Lindisfarne Gospels, the Book of Durrow, and the Book of Kells manifests the technical and creative mastery over knotted forms achieved by Celtic artists before AD 800.

The mathematical theory of knots has a comparatively short history, properly beginning with some remarks of C.F. Gauss and being advanced by Victorian pioneers inspired by Lord Kelvin's vortex theory of atomic structure. Whilst the vortex theory fell by the wayside, knot theory became a flourishing branch of pure mathematics. Recent developments have revealed unexpected connections with the theoretical physics of quantum field theory, re-uniting knot theory with physics. One of the aims of this book is to provide the background material so as to make accessible the ideas and discoveries of this challenging and expanding area of mathematical research.

Our main theme is of course the mathematical theory of knots, and especially its interaction with, and enrichment by, the theory of surfaces and of group presentations. We begin with a study of knots based on pictures of interlacing, and discuss two complementary problems. Firstly, what is the precise nature of the interlacing that we wish to describe? More precisely, when should different pictures be regarded as showing the same interlacing? Secondly, how can we classify the different interlaced forms represented by the pictures? Indeed, the fundamental problem of classifying knots, of organizing them according to their properties and characteristics, is a guiding theme of the entire book. There is a wealth of raw material for this process: the knots themselves! A few will be familiar from the practical problems of tying shoelaces, parcel string, mooring and rigging lines, safety ropes, and so on. Most interlacing patterns do not produce knots of practical use, but their decorative appeal remains, and their mathematical interest does not depend on their practical value. We have plenty of knots to work with: for example, it is a fact that there are precisely 12 965 different knots that can be represented by interlacing patterns in which the string crosses itself at most 13 times.

The pictures of interlacing represent spatial relationships in three-dimensional space, and to pursue this aspect of knot theory we use topology, the study of shape and continuity. Topological methods provide new tools to aid our understanding of knotted forms and connect the study of knots with the second major subject of the book: the study and classification of surfaces. The immediate vicinity of any point of a surface looks just like two-dimensional space, and we want to understand what the possibilities might be for the shape of the entire surface. Here we have a second topological classification problem, and one with a complete solution whose essentials were known in the nineteenth century. The classification theorem is a masterpiece of pure mathematics in action. A subtle problem is broken down into intermediate steps, the essential information is recorded by simple invariants, and the invariants can then be organized to produce a list of all the possible objects under scrutiny.

Knots, surfaces, and the connections between them lead us naturally into other subject areas, principally to graphs and to group presentations. These subjects have their own motivating questions, their own deep problems, and their own fascinations and appeal. We try to draw out the connecting strands, the exchange of ideas, concepts, and methods. We hope that the reader will become happily entangled in the web that is woven by the theory of knots and surfaces.

Durham N.D.G
Bangor T.P.
Oct 1993

Acknowledgements

This book is based on a third-year undergraduate course given at the University College of North Wales, Bangor. We are grateful to our colleagues who have been involved with the course at various times, and in particular to Ronnie Brown, Peter Cromwell, Hugh Morton, and Tom Thickstun for discussions, advice, and the provision of some material. We are especially grateful to the large number of students that have responded to our enthusiasm for the subject by electing to take knot theory as an option course at Bangor, and have criticized, complained, and commended at various points. The first draft of this book was based on lecture notes of improbable neatness taken by Ms Caren Williams in the academic year 1987–8.

The first author carried out most of his share of the preparation of this book whilst a lecturer in mathematics at Heriot-Watt University, and thanks his former colleagues for the congenial working environment that they created. He is also most grateful to Robin Nicholls for her assistance in preparing the typescript for the publisher.

Contents

1

Knots, links, and diagrams

Interlaced or knotted patterns and forms are encountered in many settings. Their occurrence may be essential, as in the practical applications of knot tying, or inconvenient, as with a knotted shoelace or inextricably entangled garden hose. Interlacing motifs are the basis for much Celtic and Islamic art, and language is enriched by many figures of speech referring to knots. Our subject is how mathematics gets into knots and our aim is to knit a coherent theory of knots from the ravelled strands of our cultural and practical fascination.

1.1 Knot and link diagrams

Instructions on the skills of knotting are commonly given by means of pictures of the kind shown in Fig. 1.1.

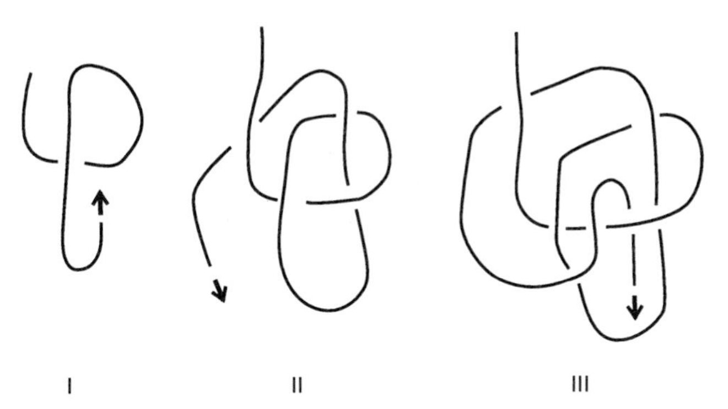

Fig. 1.1

Some attempt may be made to render the appearance of the strands rope-like, as shown in Fig. 1.2, but the significant information conveyed by the picture is the pattern of interlacing of the

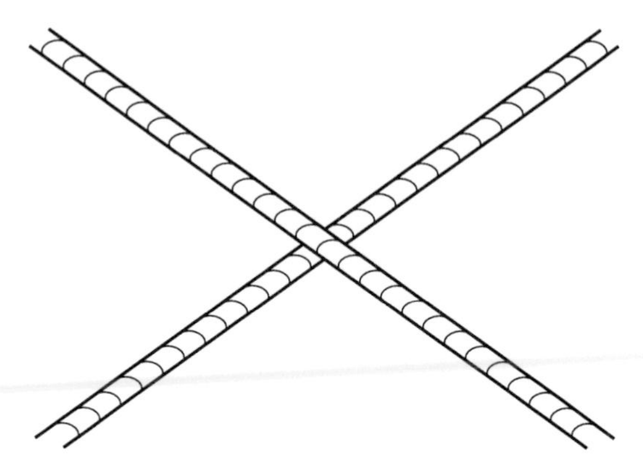

Fig. 1.2

strands: a break in a strand indicates that the broken strand passes under the unbroken one. Such pictures will be the object of study in this chapter and we shall see to what extent we can develop a pictorial theory of knottedness, for both *knots*, in which a single strand is interlaced, and for *links*, in which more than one strand is involved.

We call the pictures *knot* or *link diagrams*; no formal definition will be given and none seems necessary at this point. Figure 1.3 contains some examples, together with the names commonly given to the knots and links represented.

There are a number of observations to be made, and some terminology to be established. In tying a knot in a piece of string, we

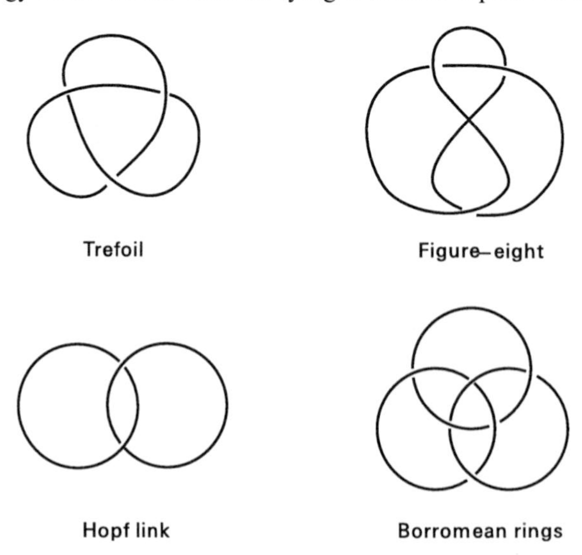

Fig. 1.3

are left with two loose ends. In a knot diagram (Fig. 1.4) these loose ends are shown spliced together, and this removes any need to treat the loose ends in a special way. The splicing also fixes the knottedness in the diagram; there is no chance of the loose ends moving so as to undo the knot! To sum up then, a link diagram encodes a way of interlacing one or more circles in three dimensions.

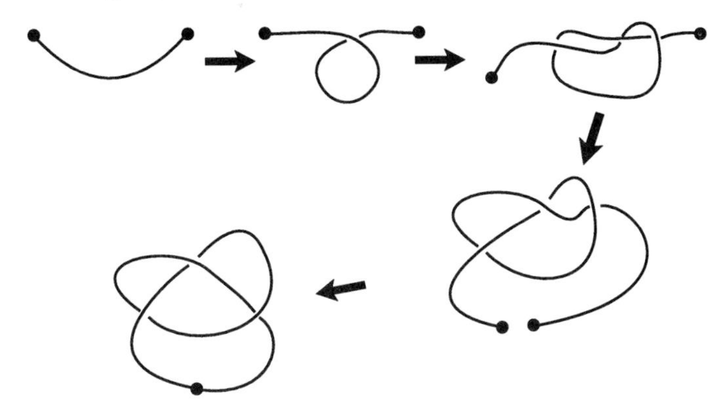

Fig. 1.4

A continuous segment in a diagram is an *arc*, and arcs meet at *crossings*. The unbroken arc at a crossing is the *overpass*, and the remaining two arcs form the *underpass*. All knot and link diagrams will have only finitely many crossings, and arcs may only meet at crossings. The arcs in a diagram are divided into a number of separate classes, called *components*, corresponding intuitively to the different pieces of string used in the physical knot or link being modelled. The unknot, the trefoil and the figure-eight have one component, and so are *knots*, whilst the Hopf link has two components, and the Borromean rings have three. By definition, a *knot* diagram has *exactly* one component. A *link* diagram may have any finite number of components, and we shall use link as the general term, using knot only when we wish to specialize to the case of one component. An *orientation* of a link diagram is a choice of direction on each component, indicated on the diagram by arrows as shown in Fig. 1.5.

In an oriented diagram (Fig. 1.6) we distinguish two types of crossing and every crossing is of one type or the other. One method of remembering the allocation of signs is to imagine an approach to the crossing along the underpass in the direction of the orientation: if the overpass orientation runs from left to right, the crossing is positive, and if from right to left then the crossing is negative. The *writhe* of an oriented diagram is the sum of the signs of all its crossings: we denote the writhe of a diagram D by $w(D)$.

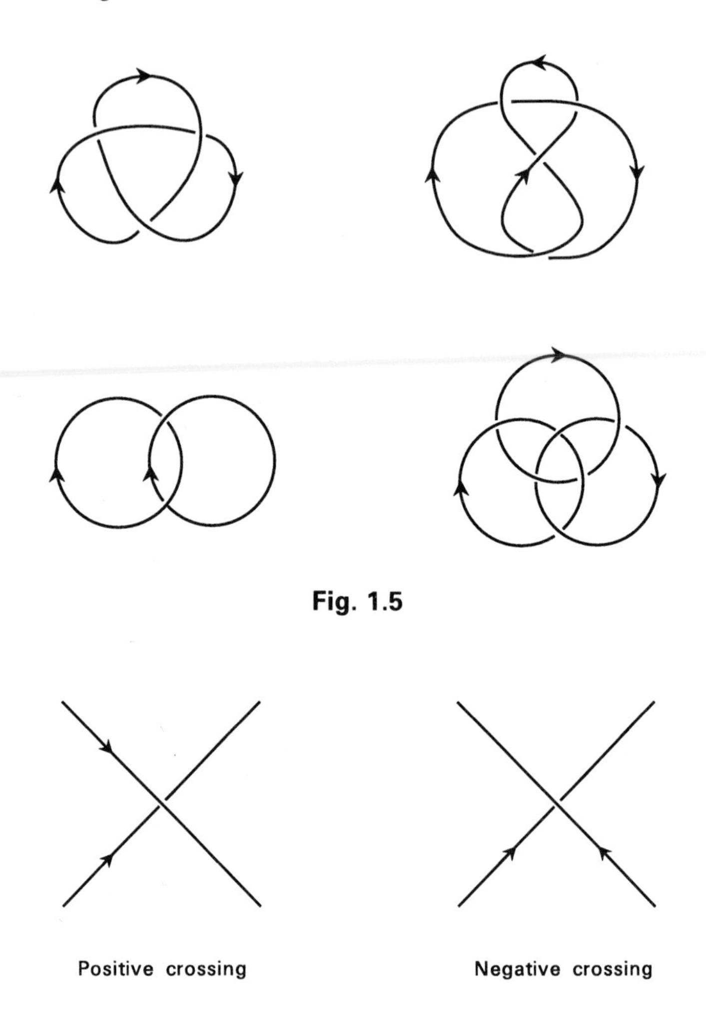

Fig. 1.5

Fig. 1.6

Examples

A variation of the writhe is useful for the study of links. Given an oriented link diagram D, with components $C_1, \ldots, C_m$ define the *linking number of* C_i *with* C_j, where C_i and C_j are distinct components of D, to be one-half the sum of the signs of the crossings of C_i with C_j, denoted $lk(C_i, C_j)$. The *linking number of* D is then the sum of the linking numbers of all pairs of components:

$$lk(D) = \sum_{1 \leq i \leq j \leq m} lk(C_i, C_j).$$

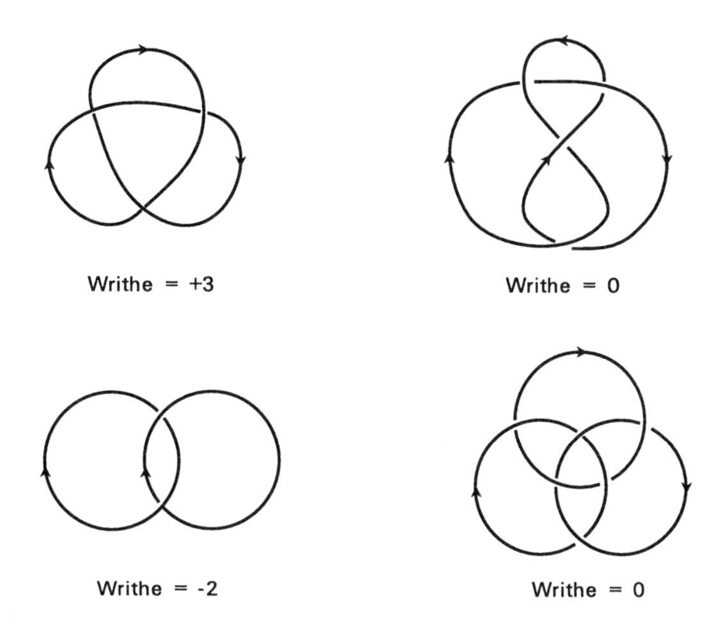

Fig. 1.7

1.2 Isotopy and the Reidemeister moves

We now introduce the idea of *isotopy*, an equivalence relation between diagrams that relates their study more closely to our intuitive notions of interlacing, and, thanks to a deep theorem of Reidemeister, also to the study of knotted curves in three-dimensional space. We shall meet Reidemeister's theorem in Chapter 3, in which we look at the topological approach to knot theory.

The *Reidemeister moves* are the changes shown in Fig. 1.8 to link diagrams. It is assumed that the diagram is otherwise unchanged, and that no other part of the diagram appears at the crossings changed by the Reidemeister moves. We shall use a similar convention for the manipulation of diagrams in Chapter 2, and only draw that part of a diagram at which the manipulation occurs. Reidemeister moves first occur in the book by Reidemeister (1932).

We also consider a class of diagram changes, which for convenience we label **R0**, though they are not Reidemeister moves. Two diagrams D and D' differ by a move of type **R0** if there is a continuous deformation of the plane that carries D to D' without changing any crossings. Moves of type **R0** correspond to pulling and stretching a knot without affecting its interlacing (Fig. 1.9).

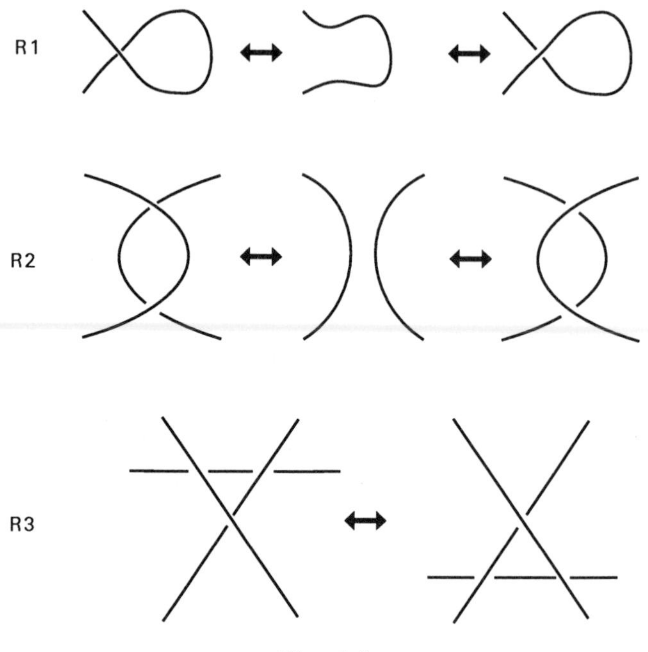

Fig. 1.8

Link diagrams D and D' are *isotopic* if D can be transformed into D' by some succession of moves **R0**, **R1**, **R2**, and **R3**, and *regularly isotopic* if the transformation can be achieved *without* use of the move **R1**. It is clear that isotopy is an equivalence relation on link diagrams, and the same is true of regular isotopy. An equivalence

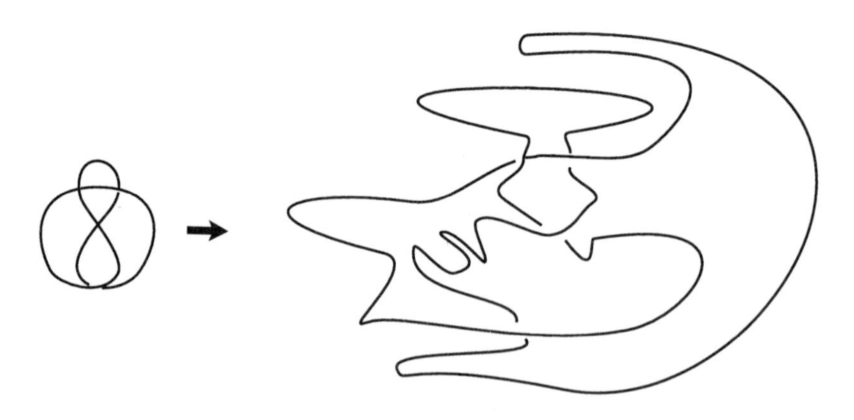

Fig. 1.9

class of diagrams modulo isotopy is an *isotopy class*, and evidently diagrams in the same class represent the same essential pattern of interlacing. An isotopy class of link diagrams will be our model for an interlaced form in three dimensions; the correctness of this model is in fact assured by Reidemeister's theorem mentioned already, that deformations of interlaced circles in three-dimensional space can be mirrored exactly by successions of Reidemeister moves on diagrams. We shall call isotopy classes of link diagrams by the name customarily given to the associated link.

What are the problems that pictorial link theory must face? Ultimately we might hope for a classification of all possible knots and links, in the form of a list of isotopy classes and a means of deciding into which class a given link diagram belongs. A more fundamental problem is to decide if two given link diagrams are isotopic. How can we decide if a given link diagram is in fact isotopic to an m-component unlink? What sort of properties should we seek, in order to establish some means of comparison and differentiation of diagrams? Plainly, a property useful for such a purpose must be *isotopy invariant*: if it holds for a diagram D then it must hold for all diagrams isotopic to D. We can verify isotopy invariance by examining the effects of the Reidemeister moves (and the plane deformations **R0**).

The obvious first classification of links is by the number of components: none of the moves **R0**, **R1**, **R2**, **R3** alters the number of components of a diagram. Hence the number of components of a link diagram is an isotopy invariant. We deduce that the Hopf link is not isotopic to the Borromean rings!

To go further, let us consider linking numbers. A Reidemeister move that involves at least two components of a link (and must therefore be an example of **R2** or **R3**) leaves all linking numbers of an oriented diagram unchanged. A move **R2** removes or introduces two crossings of opposite sign, whilst a move **R3** leaves the number of crossings and their signs unaltered. Since any two components of an unlink have zero linking number, the Hopf link is not isotopic to the two-component unlink, for any assignment of orientations to the components of the Hopf link results in a non-zero linking number. Check that this is so! However, for the Borromean rings, the linking number of any two components is zero. Furthermore, linking number cannot help us to study the isotopy of knots, and the writhe of a diagram is not an isotopy invariant. Indeed, we can alter the writhe at will by moves of type **R1**. We now go on to consider some isotopy invariants particularly adapted to the study of knots.

1.3 3-colouring

A knot diagram is *3-colourable* if we can assign colours to its arcs such that

3C1: each arc is assigned one colour;

3C2: exactly three colours are used in the assignment;

3C3: at each crossing, either all the arcs have the same colour, or arcs of all three colours meet.

Examples

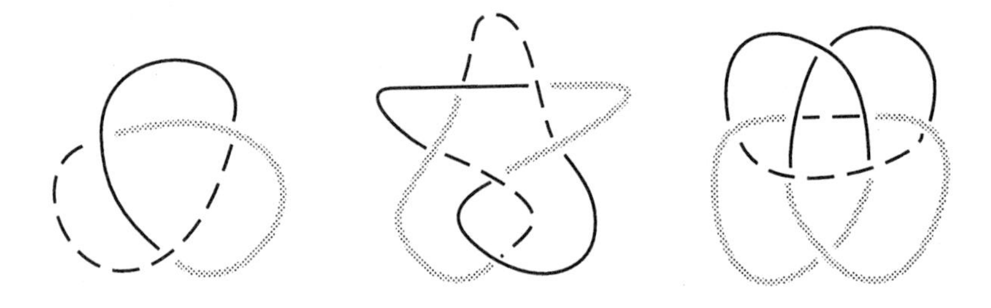

Fig. 1.10

1.1 Theorem. *3-colourability is an invariant of isotopy type.*

Proof. A deformation of type **R0** changes no crossing, so cannot change the property of 3-colourability. For the Reidemeister moves, we show that the colouring at any crossing changed by the move can be amended if necessary so that the resulting diagram is again 3-coloured. In Figs 1.11 and 1.12, we show only that part of the knot diagram changed by the Reidemeister moves.

R1

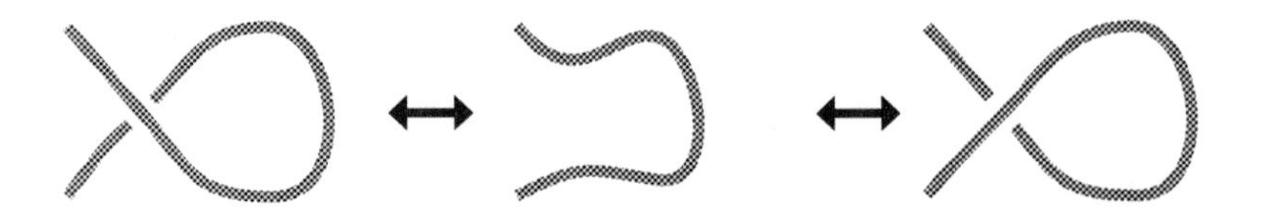

Fig. 1.11

R2

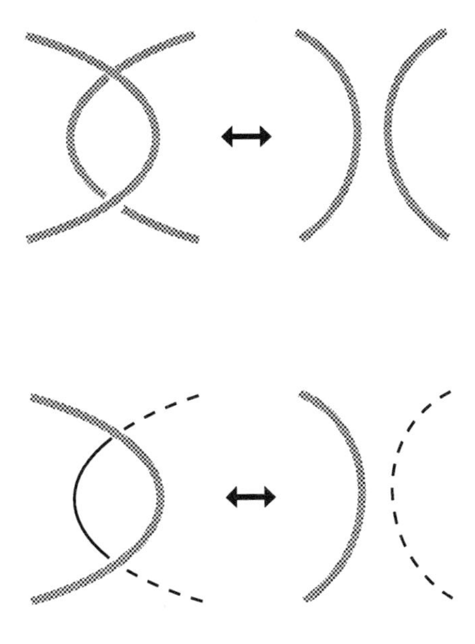

Fig. 1.12

R3

There are many cases to check: try some! You should bear in mind the following facts: to give a complete proof for the case **R3**, you need to be exhaustive; that is, to have checked all the possibilities. However, permutation of the colours will not affect the problem, but you are not allowed to change the colours on arcs that leave the diagram, for such changes might effect 3-coloured configurations at other crossings. ■

Since the unknot is *not* 3-colourable (why not?) the 3-coloured diagrams above are not isotopic to the unknot; so knots exist! We

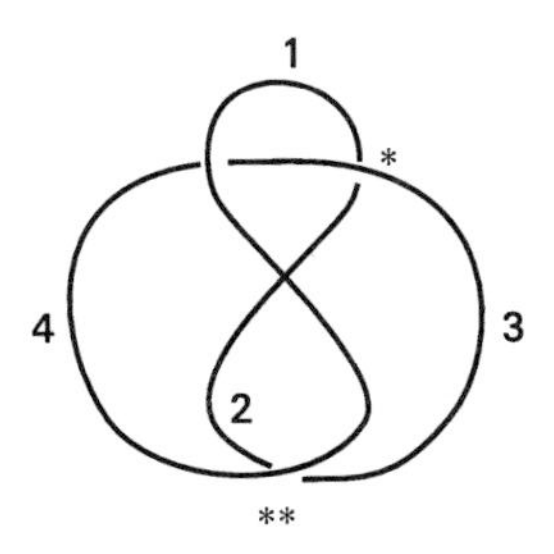

Fig. 1.13

can also show that certain knots are different, that is not isotopic to one another. For example, the figure-eight knot in Fig. 1.13 is not 3-colourable.

If arc 1 were coloured red, and arc 2 red as well, then arc 3 would have to be red to satisfy **3C3**, and arc 4 is then also forced to be red. Then the whole diagram is red, which contravenes **3C2**. If instead we try arc 1 red and arc 2 white, then arc 3 must be blue (at *) and arc 4 red (at **), but this forces the other two crossings to contravene **3C2**.

Exercises 1.3

1. Which of the knots in Fig. 1.14 are 3-colourable?

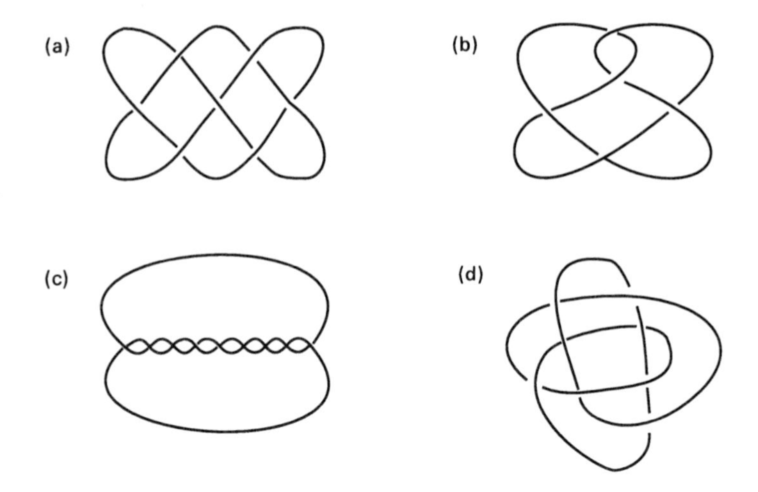

Fig. 1.14

2. Can 3-colourability be used to study links with $m \geqslant 2$ components?
3. The notion of 3-colourability generalizes to n-colourability. Let $\mathbb{Z}_n$ denote the integers modulo n. An *n-colouring* of a knot diagram is an assignment to each arc of an element of $\mathbb{Z}_n$ in such a way that, at each crossing, the sum of the values assigned on the underpass arcs is twice that on the overpass, and such that at least two elements of $\mathbb{Z}_n$ are assigned. Is this notion really a generalization of 3-colourability? Investigate the two meanings of the term 3-colourable and see that they do coincide. Find several examples of 5-colourable knots. Prove that n-colourability is an isotopy invariant.

1.4 Numerical invariants

Properties of knot diagrams that are defined solely in terms of the isotopy class to which the diagram belongs will certainly be isotopy invariant. In considering such properties, remember that we shall call an isotopy class of diagrams simply a *knot*, and if we wish to particularize, to use names of knots such as the trefoil or figure-eight for the isotopy class containing the named diagram. This matches our intention to use isotopy classes as the models of interlaced form.

Crossing number

Let K be a knot. The *crossing number* $c(K)$ of K is the minimum number of crossings in a diagram in the isotopy class K. The crossing number is therefore the number of crossings in the simplest picture of a knot. A diagram of a knot K with $c(K)$ crossings is a *minimal* diagram.

Examples

$$c(\text{unknot}) = 0; \quad c(\text{trefoil}) = 3; \quad c(\text{figure-eight}) = 4.$$

The lists of knots given in Burde and Zieschang (1985) and Kauffman (1987a) are arranged according to crossing number: a minimal diagram is given for each knot with crossing number at most 9. Choices have to be made of one mirror image or the other, a topic to which we shall return shortly. The crossing number provides a means of arranging data on knots in a systematic form according to an intuitively reasonable measure of complexity. However, given some arbitrary diagram, the crossing number of the knot that it represents may be hard to determine.

Unknotting number

A *crossing change* in a diagram exchanges the overpass and underpass at a crossing as shown in Fig. 1.15.

Crossing changes *will* alter the isotopy type of a diagram.

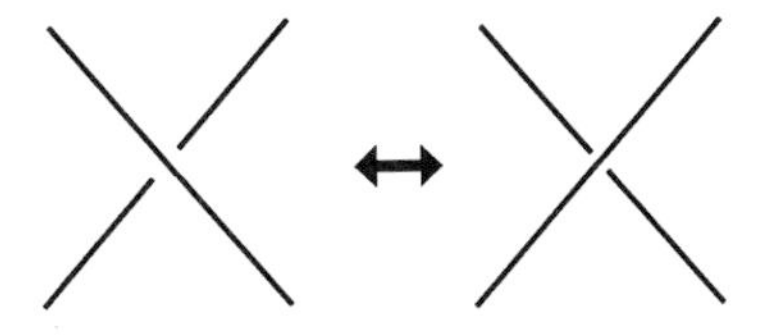

Fig. 1.15

1.2 Lemma. *Let D be a diagram with c crossings. Then changing at most $c/2$ crossings of D produces a diagram of the unknot.*

Note. You will probably find it useful, before reading the following proof, to try out a few examples of using crossing changes to produce unknot diagrams. Try to find a procedure for doing this systematically, and then compare your procedure with the one suggested in the proof. You should certainly test the proof on some examples of your own devising.

Proof. Travel once around the diagram D from any starting point: at each first encounter with a crossing, note whether you are on the overpass or the underpass. Count the final tally of overpasses and underpasses and change all the crossings contributing to the *smaller* count: this can involve at most $c/2$ changes. By construction, at each first encounter with a crossing in the new diagram, you arrive consistently on the overpass or on the underpass, so that no interlacing ever occurs and the new diagram is an unknot. ∎

The *unknotting number* $u(K)$of a knot K is the smallest number of crossing changes required to obtain the unknot from some diagram of the knot. The unknotting number is hard to calculate, and is still unknown for some knots with crossing number 9. The essential difficulty is illustrated by the following.

1.3 Theorem. (Bleiler 1984) *The unknotting number of a knot does not necessarily occur in a minimal diagram.*

Proof. Let K be the knot with the diagram shown in Fig. 1.16.

It can be shown that this diagram is minimal and is unknotted by three crossing changes. However, no two crossing changes will unknot it. Now the diagram in Fig. 1.17 is isotopic to the previous

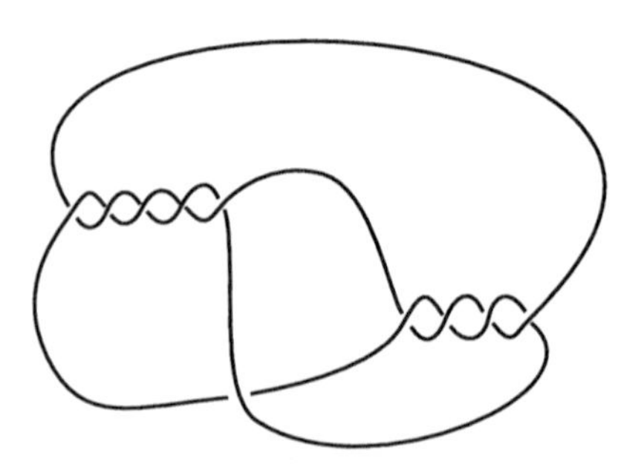

Fig. 1.16

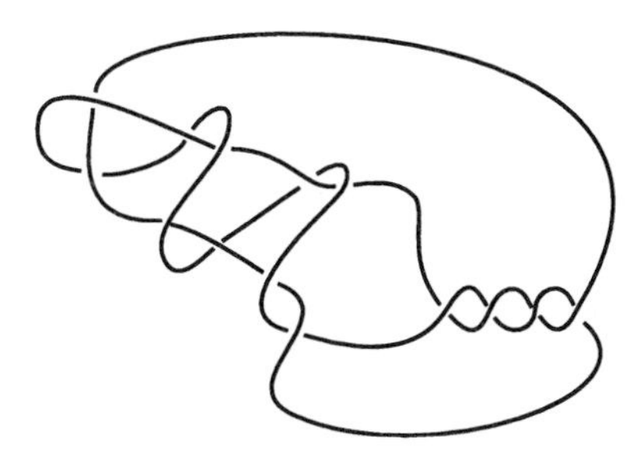

Fig. 1.17

one. It has 14 crossings, so is not minimal, but is unknotted by two crossing changes. ■

Note. In fact $u(K) = 2$. See Bleiler's (1984) paper for more details.

Bridge number

A *bridge* in a knot diagram is an arc that is the overpass in at least one crossing.

Example

Fig. 1.18

The *bridge number* $b(K)$ of a knot K is the minimum number of bridges occurring in a diagram of the knot. By convention the unknot has bridge number 1.

Example

The standard diagram of the trefoil has three bridges as shown in Fig. 1.19.

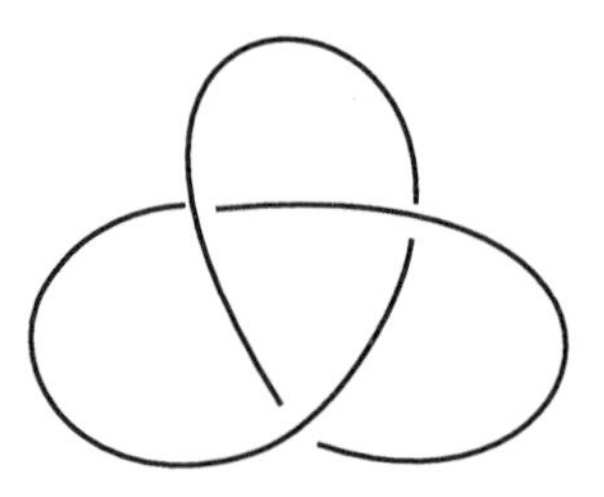

Fig. 1.19

However, the diagram of the trefoil (Fig. 1.20) has only two bridges, so $b(\text{trefoil}) \leqslant 2$.

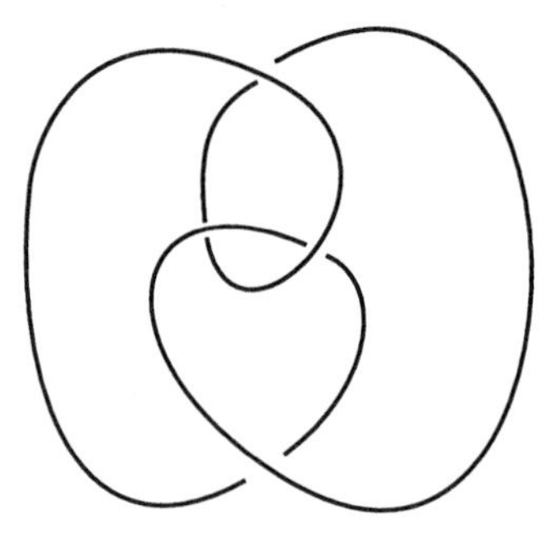

Fig. 1.20

1.4 Proposition. *A knot K has bridge number* 1 *if and only if it is the unknot.*

Proof. It is our convention that the unknot has bridge number 1. So suppose that K is a knot with $b(K) = 1$ and let D be a diagram of K with one bridge and the smallest possible number of crossings, m say. If $m \leqslant 2$ then K is the unknot, so we may assume that $m \geqslant 3$. No arc of the diagram, except the bridge, is an overpass at any crossing. Therefore, there must occur two adjacent crossings at which the same arc is the underpass, and a Reidemeister move **R2** is applicable, which reduces the number of crossings by two. This produces either a diagram of the unknot, in which case the original diagram was the unknot, or else a 1-bridge diagram with a number of crossings smaller than m, which contradicts the choice of m. ■

It now follows that the trefoil has bridge number 2 (why?). The 2-bridge knots have been classified by H. Schubert: see his original paper (Schubert 1956) or the account in Burde and Zieschang (1985).

Exercises 1.4

1. Can you improve the bound on the number of crossing changes needed to unknot a diagram? Try some examples.
2. Construct an isotopy between the two diagrams in the proof of Theorem 1.3.
3. Calculate the unknotting number of the knots shown in Fig. 1.21.

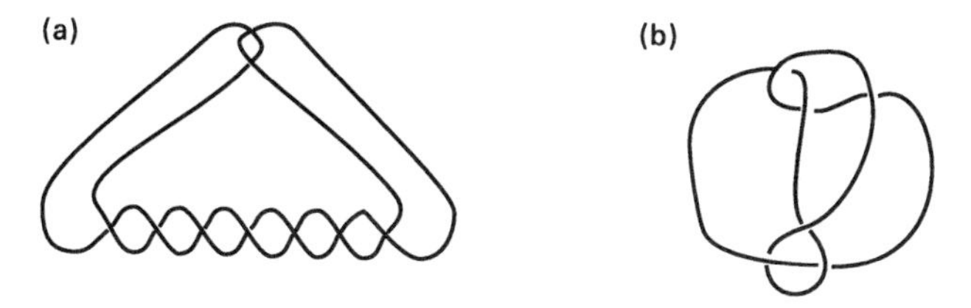

Fig. 1.21

4. What is the unknotting number of the knot with the diagram shown in Fig. 1.22.

Fig. 1.22

5. Construct an isotopy between the two trefoil diagrams in Fig. 1.19 and Fig. 1.20.

1.5 Chiral and invertible knots

Need the mirror image D' of a knot diagram D be isotopic to D? Consideration of the trefoil in Fig. 1.23 indicates that the answer is no.

No amount of persuasive manipulation of a string tied as on the left seems to transform it to being tied as on the right; it appears that

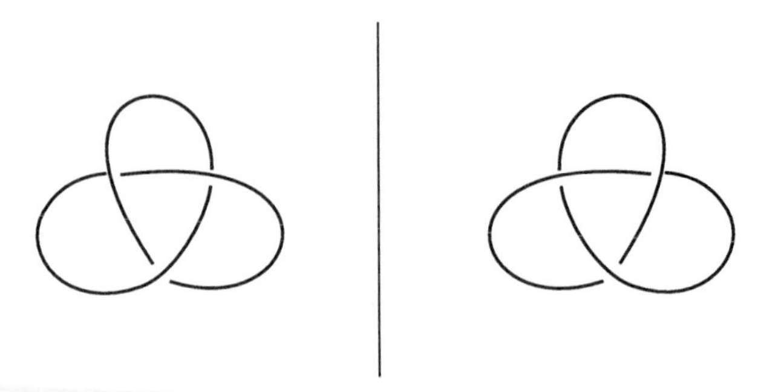

Fig. 1.23

the diagrams lie in distinct isotopy classes. This is in fact the case, though we do not have the means to prove it as yet. However, the following manipulations show that the figure-eight and its mirror image are isotopic (Fig. 1.24).

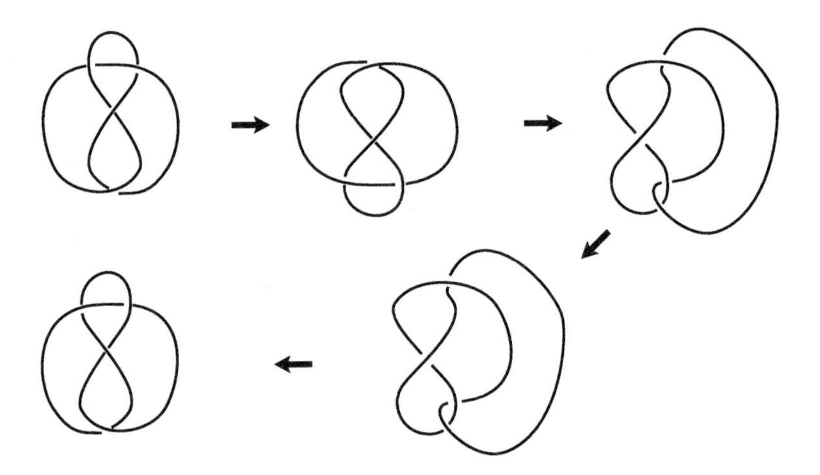

Fig. 1.24

The effect of a mirror reflection on a single crossing is to perform a crossing change as shown in Fig. 1.25.

To obtain D' from D we can simply change *all* the crossings in D, for a mirror image of D is obtained by changing all the crossings and then rotating the entire diagram through 180° about the line representing the mirror. This last operation does not affect the interlaced form, so does not change the isotopy class of the diagram (Fig. 1.26).

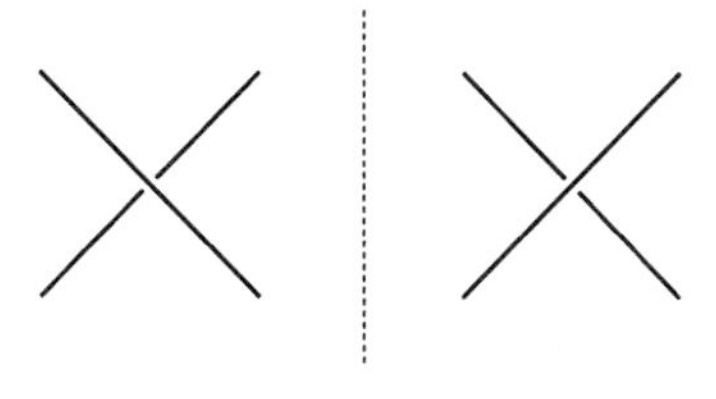

Fig. 1.25

A knot that is isotopic to its mirror image is said to be *achiral*, so that the figure-eight is achiral. Otherwise, the knot is *chiral* and we shall see in the next chapter that the trefoil is indeed chiral; the polynomial invariants of Chapter 2 give a ready means of deciding chirality for many knots.

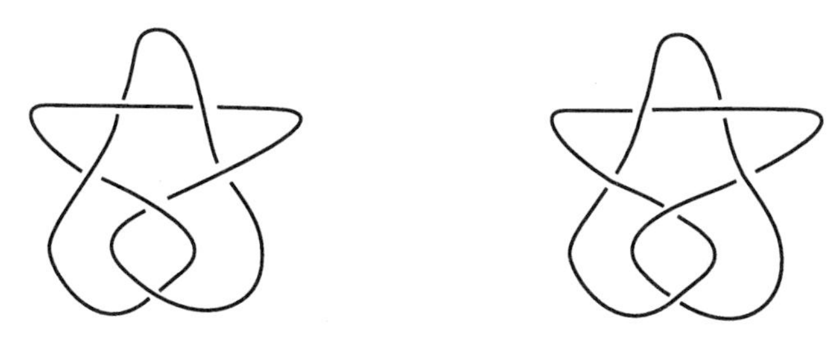

Fig. 1.26

If D is an oriented diagram, let $\bar{D}$ be the diagram with the orientation reversed. Are D and $\bar{D}$ necessarily isotopic? Here we require the isotopy to match up the orientations. An oriented knot that is isotopic to itself with the orientation reversed is *invertible*. The trefoil is invertible (Fig. 1.27).

Fig. 1.27

Non-invertible knots were not known to exist until 1964, when H.F. Trotter found an infinite family of them (Trotter 1964). One is shown in Fig. 1.28.

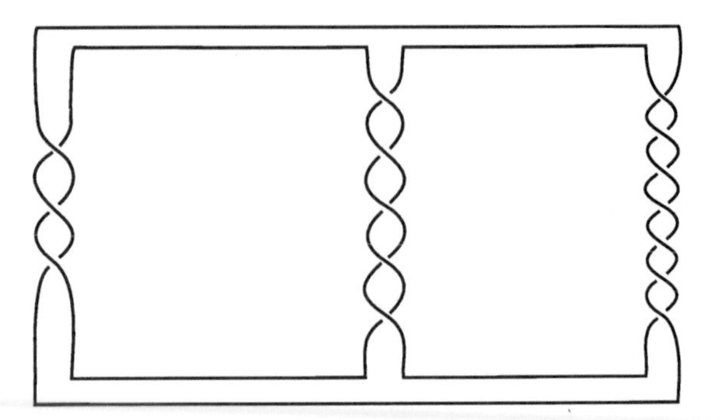

Fig. 1.28

We close this chapter with a definition that will be crucial in the next. A knot diagram is *alternating* if you meet crossings alternately at overpasses and underpasses as you travel around the diagram. Not every knot possesses an alternating diagram: the first examples that do not have crossing number 8. See the list in Kauffman (1987*a*)

Exercises 1.5

1. Find a sequence of Reidemeister moves carrying the figure-eight to its mirror image.
2. Show that the figure-eight and its mirror image are regularly isotopic.
3. Simply flipping over shows that the two diagrams in Fig. 1.27. are indeed isotopic. Give a sequence of Reidemeister moves that effects the isotopy.

2
Knot and link polynomials

Polynomial invariants of knots and links have played an important part in knot theory since the work of J.W. Alexander (1928). The appearance in 1984 of a new polynomial invariant, found by V.F.R. Jones, led to an upsurge in research activity and to a number of striking advances in our knowledge of knots and links and their representation by diagrams, as well as to new interactions between knot theory, geometry, and quantum physics. In recognition of his achievements, Jones was awarded a Fields Medal (the equivalent in mathematics of a Nobel Prize) at the 1990 International Congress of Mathematicians. Jones discovered his polynomial as an offshoot of work in a quite different area—the theory of operator algebras. He tells the story in Jones (1985). Subsequently, Alexander's and Jones' polynomials were generalized in a two-variable polynomial; this generalization was accomplished virtually simultaneously by several groups of mathematicians, using a variety of approaches.

Jones' polynomial may also be derived from some results due to L.H. Kauffman. Some of this work is phrased in the terminology of statistical physics, which provides a relevant theoretical parallel; see Kauffman (1991). The place of Jones' polynomial and its generalizations within the framework of pre-existing knot theory remains a matter of much interest and no little mystery.

We begin this chapter with Kauffman's work and examine his *state model* polynomial, and the Jones polynomial that is derived from it. We then move on to the two-variable generalization, and finally look at Alexander's original treatment of his polynomial.

2.1 State models and the Jones polynomial

The title of this section is also that of Kauffman's paper explaining his construction (Kauffman 1987*b*).

Let D and D' be (unoriented) link diagrams. Recall from Chapter 1 the notion of regular isotopy: D and D' are *regularly isotopic* if they differ by a sequence of Reidemeister moves of types **R2** and **R3**; the move **R1** that introduces or removes a loop is not allowed.

We use U to denote a diagram of the unknot with no crossings, and if D is a link diagram, then we will use DU to denote a diagram with one extra unknotted component that introduces no extra crossings. The *bracket polynomial* of a diagram D is the Laurent polynomial $\langle D\rangle$ in one variable A defined by means of the following rules:

BP1: $\langle U\rangle = 1$

BP2: $\langle DU\rangle = -(A^2 + A^{-2})\langle D\rangle$

BP3: $\langle \times \rangle = A\langle \asymp \rangle + A^{-1}\langle)(\rangle$

In rule **BP3** we use the convention from Chapter 1 of drawing only that part of a knot diagram that is changed in the intended manipulation. **BP3** is a shorthand for the following computational procedure. Take any diagram D and choose any crossing of D: delete the crossing and consider the now separated strands (Fig. 2.1). Imagine travelling along each of the two strands in the remains of the overpass towards the former crossing point: join these strands to those from the underpass that lie to your left, to produce a new diagram L. If instead you join the strands from the overpass to those from the underpass that lie on your right, you obtain a new diagram R. The rule **BP3** now says that $\langle D\rangle = A\langle L\rangle + A^{-1}\langle R\rangle$. The removal of crossings and the rejoining of strands are *not* Reidemeister moves; a little experimentation now, and a wealth of experience to come in the next few pages, will certainly convince you that the isotopy type of diagrams will be changed by these manipulations.

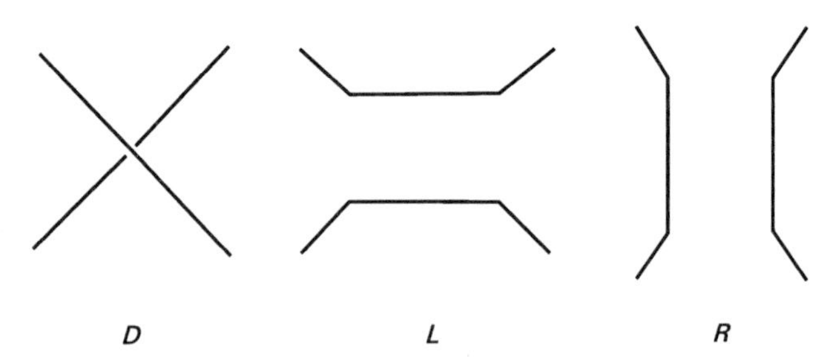

Fig. 2.1

Note that **BP3** then implies the rule

BP3′: $\langle \times \rangle = A\langle)(\rangle + A^{-1}\langle \asymp \rangle$

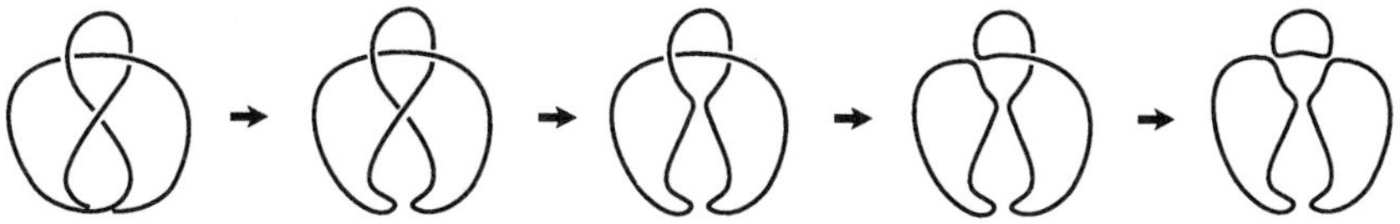

Fig. 2.2

Some care is needed in applying these rules to the computation of bracket polynomials. It is evident that **BP1** and **BP2** together determine the bracket polynomial for any diagram of an unlink with no crossings, whilst **BP3** enables the calculation of $\langle D\rangle$ for any diagram D by successive reduction of the number of crossings. Starting with a diagram D we can produce a polynomial $\langle D\rangle$. To do this we have to change the diagram by successive reduction of the number of crossings. For example, with the figure-eight knot we could reduce as shown in Fig. 2.2 or as in Fig. 2.3.

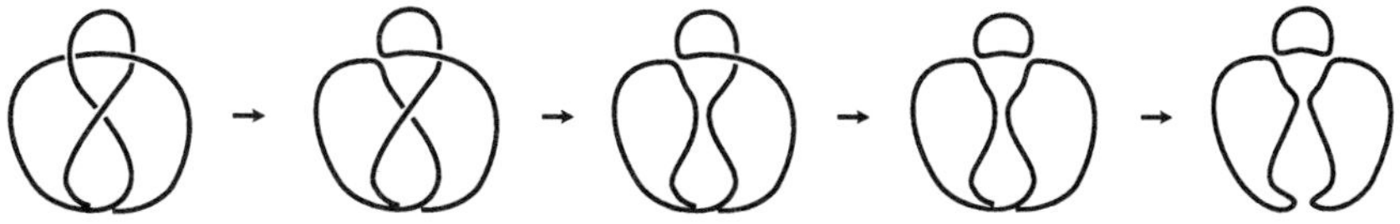

Fig. 2.3

This raises the problem of deciding if the end result might depend on the route taken. We shall see that $\langle D\rangle$ does not depend on the order in which the crossings in D are removed; for the moment, we assume this fact and go on to derive some of the properties of the bracket polynomial that underlie its theoretical significance and facilitate its computation.

2.1 Lemma. *The bracket polynomial is an invariant of regular isotopy; that is, applications of Reidemeister moves* **R2** *and* **R3** *to a diagram leave its bracket polynomial unchanged.*

Proof. For move **R2** we proceed as follows, making full use of the device of showing only that portion of a diagram at which changes are taking place.

$$\langle\ \rangle = A\langle\ \rangle + A^{-1}\langle\ \rangle \qquad \textbf{(BP3)}$$

$$= A\,(A\langle\ \rangle + A^{-1}\langle\ \rangle) + A^{-1}(A\langle\ \rangle + A^{-1}\langle\ \rangle) \qquad \textbf{(BP3')}$$

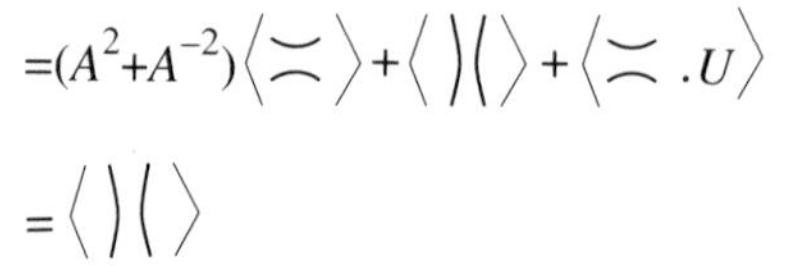

$$=(A^2+A^{-2})\langle \asymp \rangle + \langle)(\rangle + \langle \asymp .U \rangle$$

$$= \langle)(\rangle \qquad \textbf{(BP2)}$$

Invariance under move **R2** now implies invariance under move **R3**:

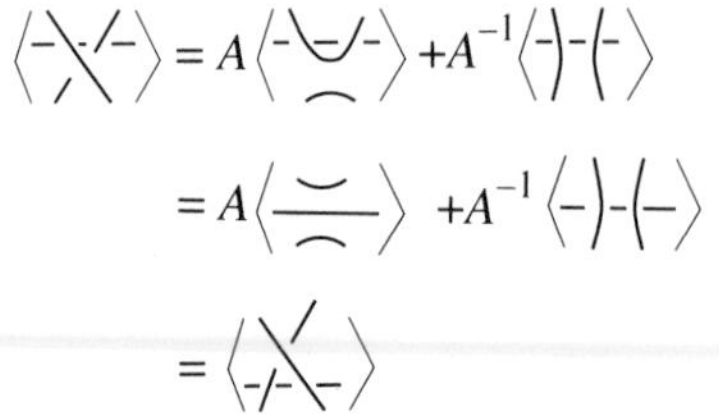

$$\langle \; \rangle = A\langle \; \rangle + A^{-1}\langle \; \rangle \qquad \textbf{(BP3)}$$

$$= A\langle \; \rangle + A^{-1}\langle \; \rangle \qquad \textbf{(R2)}$$

$$= \langle \; \rangle$$

by symmetry. ■

2.2 Lemma. *The bracket polynomial is not an isotopy invariant. Reidemeister move* **R1** *produces the following changes to bracket polynomials*:

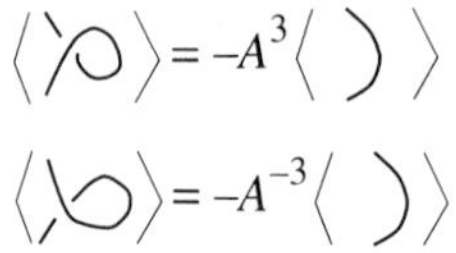

$$\langle \; \rangle = -A^3\langle \;) \; \rangle$$

$$\langle \; \rangle = -A^{-3}\langle \;) \; \rangle$$

Proof. We have

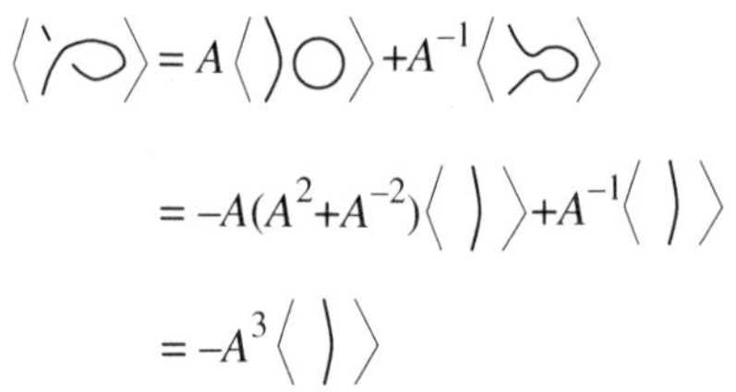

$$\langle \; \rangle = A\langle)\bigcirc \rangle + A^{-1}\langle \; \rangle \qquad \textbf{(BP3)}$$

$$= -A(A^2+A^{-2})\langle \;) \; \rangle + A^{-1}\langle \;) \; \rangle \qquad \textbf{(BP2)}$$

$$= -A^3\langle \;) \; \rangle$$

The other case is left to the reader. ■

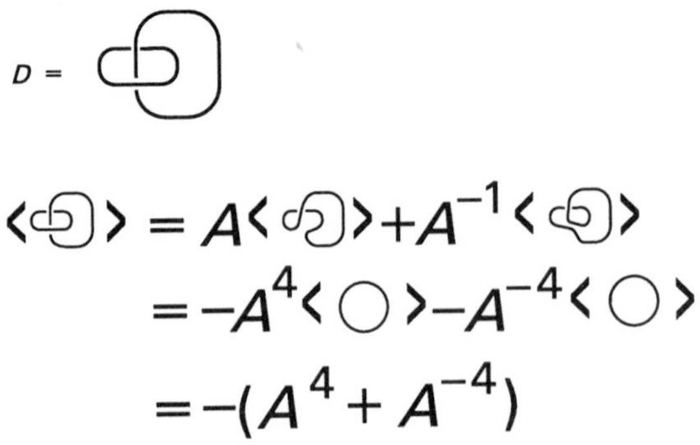

$D =$

$$\langle \; \rangle = A\langle \; \rangle + A^{-1}\langle \; \rangle$$

$$= -A^4\langle \bigcirc \rangle - A^{-4}\langle \bigcirc \rangle$$

$$= -(A^4 + A^{-4})$$

Fig. 2.4 (a)

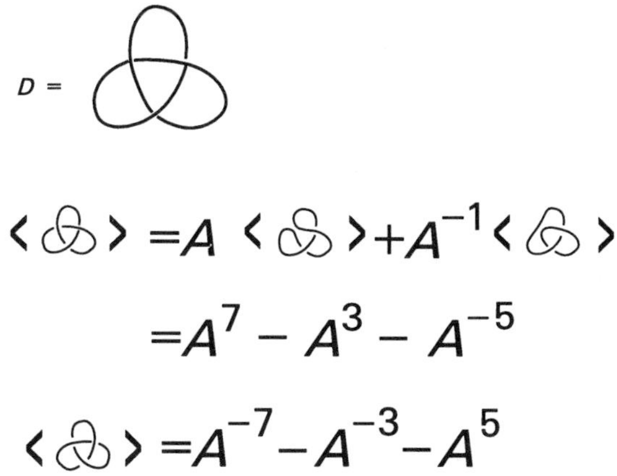

$D =$

$$\langle\ \rangle = A\,\langle\ \rangle + A^{-1}\langle\ \rangle$$
$$= A^{7} - A^{3} - A^{-5}$$
$$\langle\ \rangle = A^{-7} - A^{-3} - A^{5}$$

Fig. 2.4 (b)

We now proceed to some illustrative computations of bracket polynomials in Figs 24(a)–(c).

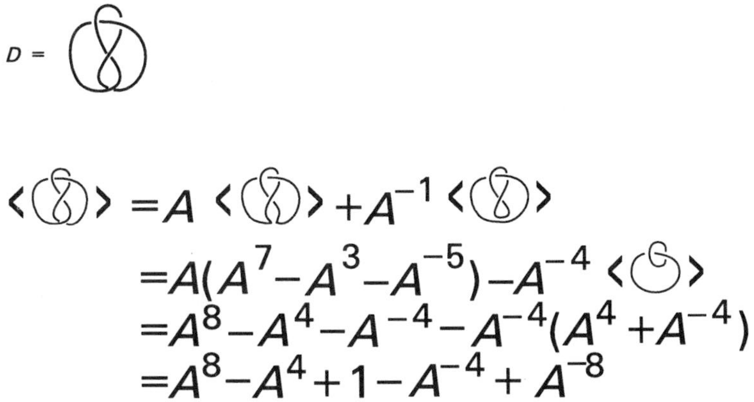

$D =$

$$\langle\ \rangle = A\,\langle\ \rangle + A^{-1}\langle\ \rangle$$
$$= A(A^{7} - A^{3} - A^{-5}) - A^{-4}\langle\ \rangle$$
$$= A^{8} - A^{4} - A^{-4} - A^{-4}(A^{4} + A^{-4})$$
$$= A^{8} - A^{4} + 1 - A^{-4} + A^{-8}$$

Fig. 2.4 (c)

We now turn to the effect on the computation of a bracket polynomial of a change in the ordering of the crossings and their removal. We use Kauffman's idea of a *state* for a knot diagram.

For each crossing in a diagram D there are two ways to eliminate it, as we discussed in relation to the rule **BP3** above: either as shown in Fig. 2.5(a) or as shown in Fig. 2.5(b).

Having made such a choice, we attach to each crossing a *splitting marker* that connects the two regions of the diagram that will be

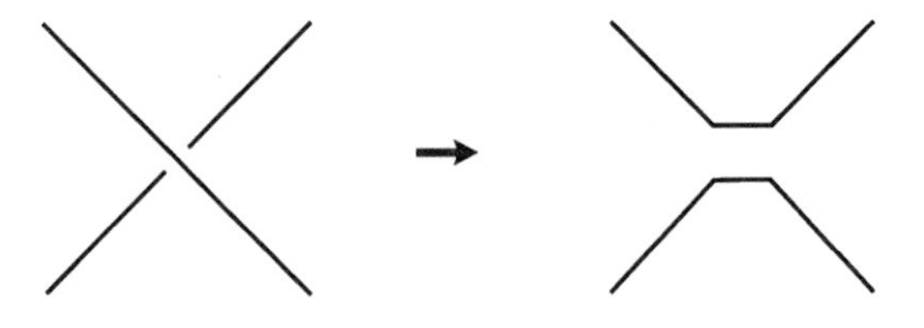

Fig. 2.5 (a)

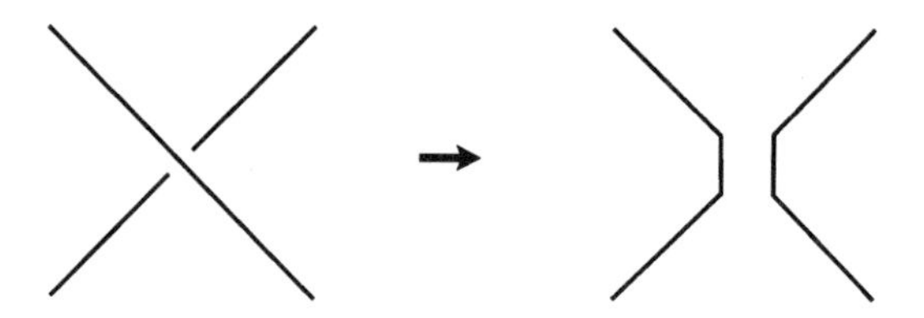

Fig. 2.5 (b)

joined by removal of the crossing: either as shown in Fig. 2.6.(a) or as shown in Fig. 2.6(b).

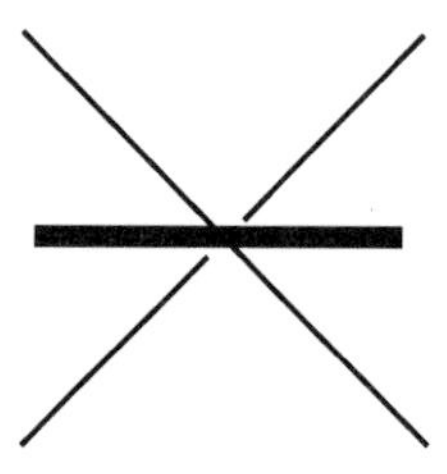

Fig. 2.6 (a)

Once a splitting marker has been attached to every crossing in D we have a *state* of D; thus a diagram with c crossings has a total of 2^c possible states. If we now remove *all* the crossings in accordance with the splitting markers in a given state, we obtain a diagram of an unlink with the number of components in the unlink depending on the state. The bracket polynomial of the unlink diagram is determined by **BP1** and **BP2**: it is $(-(A^2 + A^{-2}))^{k-1}$, where k is the number of components.

Fig. 2.6 (b)

Now consider the computation of $\langle D\rangle$ for a diagram D. Repeated use of **BP3** results in an expression for $\langle D\rangle$ as a sum of terms $\pm A^m\langle U\ldots U\rangle$, where $U\ldots U$ is a diagram of an unlink, with exactly one such term for each state of D. *This* expression is uniquely determined by D, and $\langle D\rangle$ is therefore independent of the route chosen for its computation. We can be a little more precise and obtain a formula for $\langle D\rangle$ as a *sum over its states.*

Consider the four regions adjacent to a crossing in D (Fig. 2.7): the two regions that lie between the overpass and an underpass, in that order, when circling the crossing anticlockwise, we shall call *positive*; the other two regions we shall call *negative*. Observe that in the earlier discussion of the elimination of crossings, in order to apply **BP3** the diagram L was obtained from D by rejoining the separated strands so as to connect the two positive regions, and the diagram R was obtained by connecting the two negative regions.

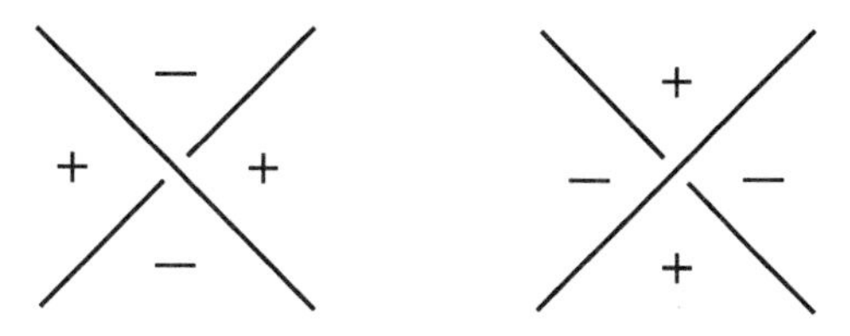

Fig. 2.7

A splitting marker connects regions of the same sign, and in **BP3** the bracket $\langle L\rangle$ is multiplied by A and the bracket $\langle R\rangle$ by A^{-1}. Let S be a state of the diagram D and let $p(S)$, $n(S)$ be respectively the number of splitting markers in S joining positive and negative regions. Now define $\langle D\,|\,S\rangle = A^{p(S)-n(S)}$. We write $|S|$ for the number of components in the diagram of unlinks obtained from D by removing all crossings in accordance with the state S: this diagram receives coefficient $\langle D|S\rangle$ in the expression for D. We can summarize the above remarks in a formula for D.

2.3 Proposition. *Let D be a link diagram. Then*

$$\langle D\rangle = \sum_S \langle D|S\rangle \left[-(A^2 + A^{-2})\right]^{|S|-1}$$

where the sum is taken over all states S of the diagram D. ∎

The formula just given does not provide a practical method for the evaluation of $\langle D\rangle$ but it is of theoretical interest because of its resemblance to the form of expressions found in statistical physics. It is a pleasing and surprising nexus between two strands of mathematical thought, and an excellent place to seek further information is Kauffman's book (Kauffman 1991).

At first sight—and certainly according to the spirit of Chapter 1—the bracket polynomial is not much use in the study of knots and links, since it is not an isotopy invariant. Happily, this deficiency is easily remedied to produce an isotopy invariant for *oriented* links.

Let L be an oriented link and let D be an oriented diagram of L with writhe $w(D)$. Recall from Chapter 1 that the writhe is not an isotopy invariant; however, it is an invariant of regular isotopy, as may easily be checked by examining the effect of Reidemeister moves **R2** and **R3**. Define the Laurent polynomial $f[L]$ by

$$f[L] = (-A)^{-3w(D)}\langle D\rangle$$

where $\langle D\rangle$ is the bracket polynomial of the diagram obtained from D by simply forgetting its orientation. The notation already suggests that $f[L]$ is determined by L; that is, it is an isotopy invariant.

2.4 Theorem. *The polynomial $f[L]$ is an isotopy invariant.*

Proof. Since $w(D)$ and $\langle D\rangle$ are invariants of regular isotopy, so is $f[L]$. Therefore, we need only to check the effect of Reidemeister move **R1**. One application of this move either introduces or removes a crossing and so changes $w(D)$ to $w(D) \pm 1$; the term $(-A)^{-3w(L)}$in $f[L]$ contributes an additional factor $-A^{\pm 3}$ which is exactly cancelled by the factor contributed by $\langle D\rangle$ and given by Lemma 2.2. ∎

Combining our calculations of bracket polynomials and writhes, we have the following computations of Kauffman's polynomial $f[L]$:

(1) $L =$ *left-hand Hopf link* (Fig. 2.8)

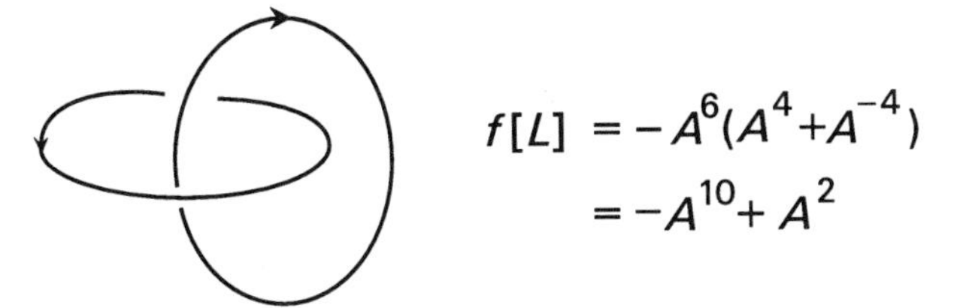

Fig. 2.8

(2) $L =$ *left-hand trefoil* (Fig. 2.9)

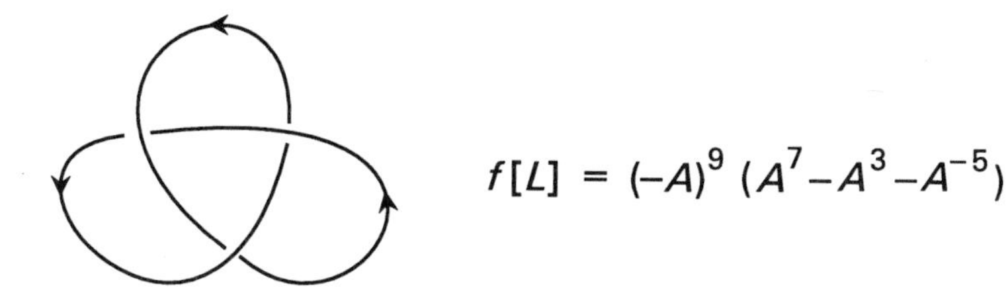

Fig. 2.9

(3) $L =$ *right-hand trefoil* (Fig. 2.10)

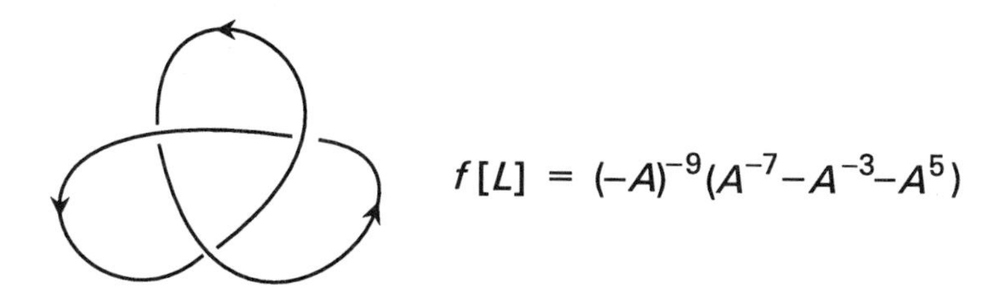

Fig. 2.10

(4) $L =$ *figure-eight knot* (Fig. 2.11)

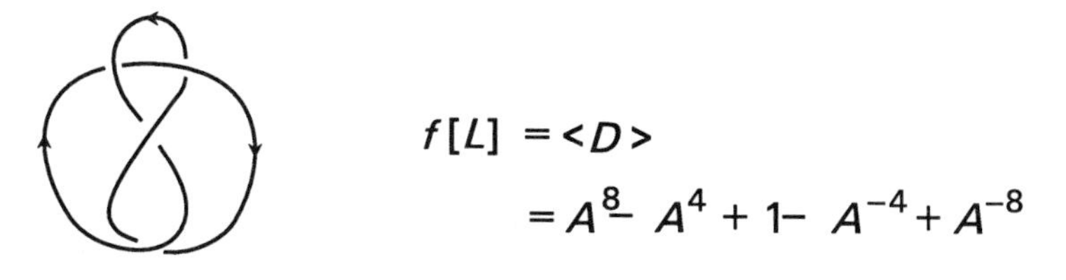

Fig. 2.11

Observe that the Kauffman polynomial for the right-hand trefoil is obtained from that for the left-hand trefoil by interchanging A and A^{-1}. This observation prompts the question: is it always the case that the Kauffman polynomial of a mirror image is obtained by interchanging A and A^{-1}? If the answer is yes then achiral knots must have Kauffman polynomials that are symmetric in A and A^{-1}, and this is the case for the figure-eight knot. It is therefore reasonable to propose, as our first formal property of the Kauffman polynomial, the following result.

2.5. Proposition. *Let L be a link diagram and let $L^!$ be its mirror image. Then $f[L^!]$ is obtained from $f[L]$ by interchanging A and A^{-1}. Moreover, $\langle L^! \rangle$ is also obtained from $\langle L \rangle$ in this way.*

Proof. We know from Chapter 1 that a diagram of $L^!$ is obtained from one of L by reversing all the crossings, and a computation of $\langle L^! \rangle$ from this diagram proceeds exactly as for $\langle L \rangle$ except that the roles of A and A^{-1} are interchanged. Since it is also the case that $w(L^!) = -w(L)$, the result now follows. ■

Exercises 2.1

1. Show that the two diagrams in Fig. 2.12 are regularly isotopic.

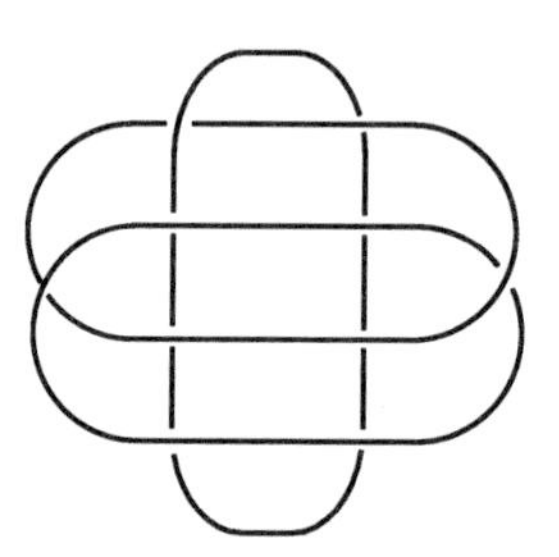
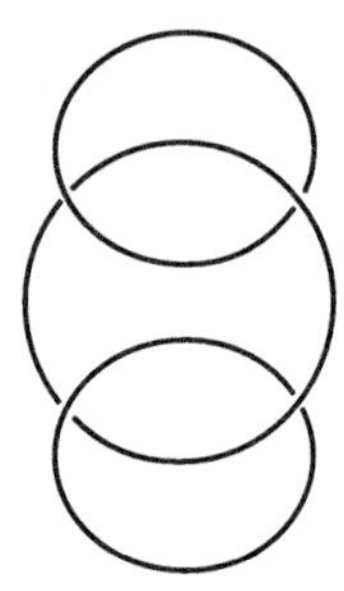

Fig. 2.12

2. Show that the two diagrams in Fig. 2.13 are not regularly isotopic.
3. Compute $\langle D \rangle$ directly from the formula in Proposition 2.3 for the diagrams D in Fig. 2.14.
4. Compute $f[L]$ for L equal to the links in Fig. 2.15.

2.2 The Jones polynomial

The Jones polynomial and the Kauffman polynomial are really the same, apart from a change of variable, so that the Jones polynomial

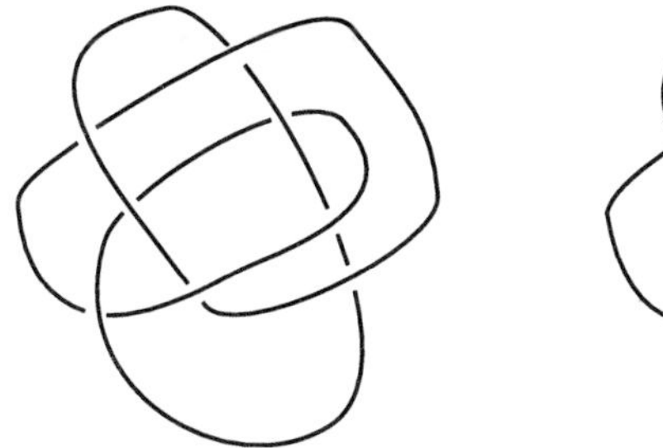

Fig. 2.13

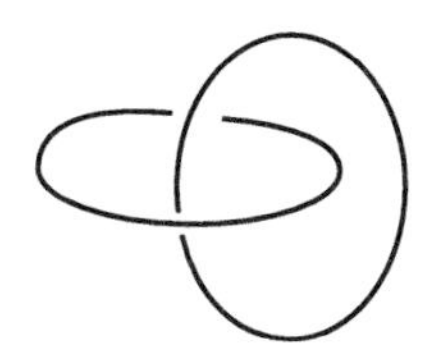
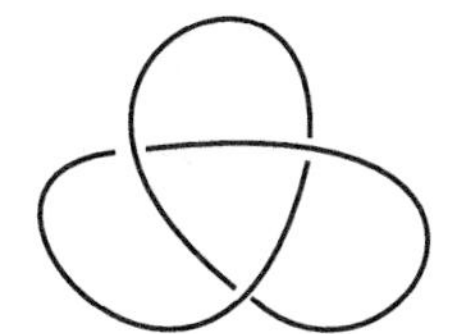

Fig. 2.14

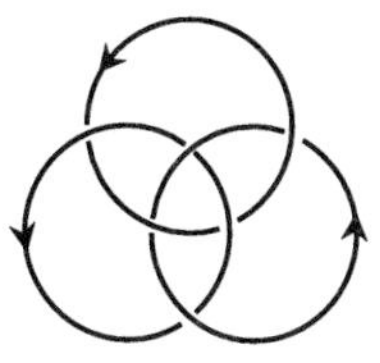

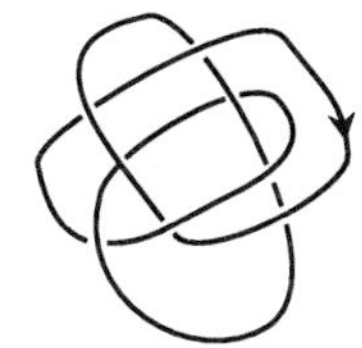

Fig. 2.15

is an isotopy invariant of oriented links. The Jones polynomial of an oriented link L is denoted V_L and the customary choice of variable is t. The formula connecting V_L and $f[L]$ is then

$$V_L(t) = f[L](t^{-1/4}).$$

The change of variable implies that the Jones polynomial can be defined by the following two rules:

JP1: $V_U(t) = 1$, where U is the oriented unknot;

JP2: $t^{-1}V(t) - tV(t) = (t^{1/2} - t^{-1/2})V(t)$.

In **JP2** we show only that part of the link at which changes occur: so **JP2** relates the Jones polynomials of three links, identical except at one crossing, at which we find the configurations shown in Fig. 2.16.

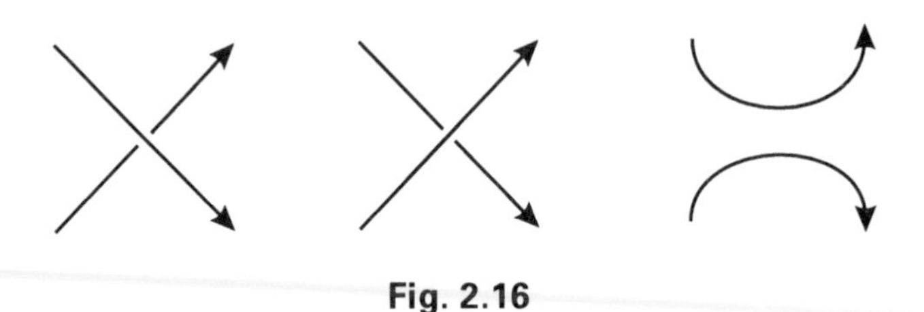

Fig. 2.16

Exercises 2.2

1. Show that $V_L(t) = f[L](t^{-1/4})$ does satisfy **JP1** and **JP2**.
2. Use **JP1** and **JP2** to compute the Jones polynomial for the links in Exercise 2.1.4.

2.3 Applications of the Kauffman polynomial

Kauffman's approach to the Jones polynomial enables relatively simple proofs to be given of some of its formal properties; one such property is the *reversing result*. It tackles the question: how is the Kauffman or Jones polynomial of an oriented link affected if the orientations of some of the components are reversed?

Exercises 2.3

(Exercises are placed at strategic points throughout this section, and are numbered concurrently within it.)

1. If K is an oriented knot, and $\bar{K}$ is the reverse of K, show that $f[K] = f[\bar{K}]$.
2. If L is a two-component oriented link, and the orientation of just one component is reversed to form a new oriented link L', how is $f[L]$ related to $f[L']$? Experiment with some two-component links and conjecture a formula relating $f[L]$ and $f[L']$. Try to prove that your conjecture is correct. [**Hint**. Your formula should involve the linking number of the two components.]
3. Consider the general setting of a c-component oriented link L and a link L' obtained from L by reversing the orientation of r components of L. How are $f[L]$ and $f[L']$ related?

One of the most striking applications of the Kauffman polynomial concerns its relationship to the crossing number of a knot. The result may be summed up as follows: *any two reduced alternating diagrams of a link L must have the same number of crossings.* The meaning of the term *reduced* will be explained in due course. Not every link has an alternating diagram, but the result asserts that for those that do, the crossing number in a reduced diagram is an isotopy invariant. Proofs of the result have been given in Kauffman (1987*b*), Murasugi (1987), and Thistlethwaite (1987). We shall follow Kauffman, as we have done so far, but strongly recommend that the reader examine the alternatives. We have need of one result whose proof is deferred until Chapter 7, since it is most naturally seen as a special case of Euler's formula for planar graphs; see Section 7.2.

Firstly we shall define *reduced* diagrams. An *isthmus* in a link diagram is a crossing at which only three regions of the diagram meet, as in the examples in Fig. 2.17.

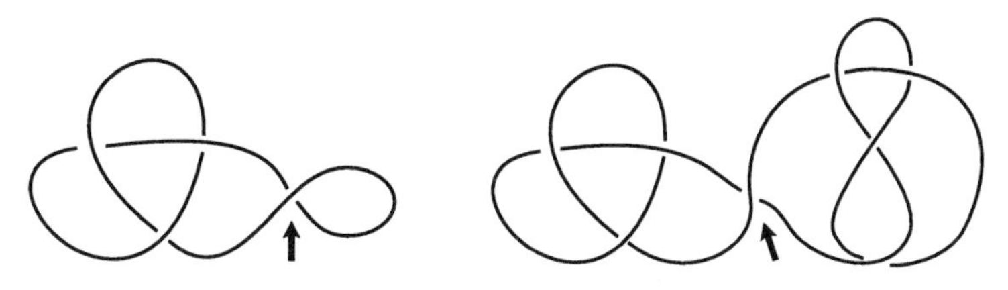

Fig. 2.17

A diagram is *reduced* if it does not contain an isthmus, and is *connected* if it is not the disjoint union of two smaller diagrams. The special case of Euler's formula that we need is the following.

2.6 Lemma. *Let D be a connected link diagram with c crossings. Then D divides the plane into c + 2 regions.* ■

We need one more piece of terminology before proceeding to the results. The *span* of a bracket polynomial $\langle D \rangle$ is the difference between the maximum and minimum powers of A occurring in it; recall that $\langle D \rangle$ may contain both negative and positive powers of A. As an example, $-A^5 - A^{-3} + A^{-7}$ has span $5 - (-7) = 12$. Clearly the definition of span extends to any Laurent polynomial in one variable.

Exercise 2.3

4. If D is a diagram of link L, what is the relationship between the span of $\langle D \rangle$ and the span of $f[L]$?

We now state the main theorem on crossing numbers and the Kauffman polynomial, and follow it by a more technical proposition that contains the core of the argument. We then show how the theorem is derived from the proposition, and give the proof of the proposition assisted by a series of exercises.

2.7 Theorem. *If a link L possesses a connected, reduced, alternating diagram D with c crossings then the span of $\langle D\rangle$ is $4c$. Hence any two connected, reduced, alternating diagrams of a given link have the same number of crossings.*

The proposition uses the notion of positive and negative regions in a link diagram, introduced earlier in the chapter. The sign of a region is determined at a crossing to which it is adjacent, and in general the allocation of signs will not be consistent for the same region at different crossings. However, in an *alternating* diagram, the allocation of signs *is* consistent, so that every region can be given an unambiguous sign.

2.8 Proposition. *Let D be a connected, reduced, alternating link diagram with c crossings, p positive regions, and n negative regions. Then the term in $\langle D\rangle$ of maximum degree is $(-1)^{n-1}A^{c+2n-2}$ and the term in $\langle D\rangle$ of minimum degree is $(-1)^{p-1}A^{-c-2p+2}$.*

Proof of the theorem. From the proposition we see that the span of $\langle D\rangle$ is $c + 2n - 2 - (-c - 2p + 2) = 2c + 2(n + p) - 4$. Now $n + p$ is the total number of regions in D, which is also equal to $c + 2$. Therefore the span of $\langle D\rangle$ is $2c + 2(c + 2) - 4 = 4c$. ■

Proof of the proposition. Let S be the state of D with all the splitting markers joining positive regions, so that $p(S) = c$ and $n(S) = 0$. Then $|S| = n$ and $\langle D|S\rangle = A^c$ and we see that S contributes a term $(-1)^{n-1}A^c(A^2 + A^{-2})^{n-1}$ to the expansion of $\langle D\rangle$ as a sum over its states, and so a term $(-1)^{n-1}A^{c+2n-2}$ to $\langle D\rangle$. We wish to show that all other states contribute terms of smaller degree.

Exercise 2.3

5 Consider a state S_1 obtained from S by changing exactly one splitting marker. Show that $|S_1| < |S|$. [**Hint**. Use the fact that D is reduced.]

Now consider an arbitrary state S'. There exists a sequence of states

$$S = S_0, S_1, S_2, \ldots, S_{m-1}, S_m = S'$$

such that S_{i+1} is obtained from S_i by changing exactly one splitting marker. Therefore $p(S_{i+1}) = p(S_i) - 1$ and $n(S_{i+1}) = n(S_i) + 1$, and it follows that $\langle D \mid S_{i+1} \rangle = A^{-2} \langle D \mid S_i \rangle$.

Exercises 2.3

6. Show that $|S_{i+1}| = |S_i| \pm 1$, and deduce that the largest degree of a term in $\langle D \rangle$ contributed by S_{i+1} is less than or equal to the largest degree of a term contributed by S_i.
7. Combine Exercises **2.3.5** and **2.3.6** to show that S' contributes terms to $\langle D \rangle$ of degree smaller than $c + 2n - 2$.

   This completes the proof of the proposition for the term of maximum degree.

8. Prove the statement in the proposition for the term of minimum degree. ■

We now consider a generalization of Theorem 2.7 observed independently by Murasugi and by Thistlethwaite; however, we continue to follow Kauffman's account of the result, which we state at once.

2.9 Theorem. *If a link L possesses a connected diagram D with c crossings then the span of the bracket polynomial ⟨D⟩ is at most 4c.*

This theorem follows from a result that Kauffman calls *the dual state lemma*. Given a state S of a connected diagram D, the dual state $\hat{S}$ is obtained from S by changing *all* the splitting markers.

2.10 Dual state lemma. *Let S be a state of a connected diagram D and let $\hat{S}$ be the dual state. Suppose that D has r regions. Then* $|S| + |\hat{S}| \leqslant r$.

Proof. Once again, some parts of the proof are left to you.

Exercise 2.3

9. Verify the dual state lemma for diagrams with at most two crossings.

Now suppose that D has c crossings: select one crossing and split D at that crossing to obtain diagrams L and R.

Exercise 2.3

10. Use the connectivity of D to show that at least one of L and R is connected.

We may assume that it is L that is connected. The splitting to form L is done according to a splitting marker in one of the states S or

$\hat{S}$; by interchanging these states if necessary, we may assume that the splitting marker is in S. The other splitting markers in S now give a state S' of L. Note that $|S'| = |S|$.

Exercise 2.3

11. Show that L has $r - 1$ regions. Use induction on c to deduce that $|S'| + |\hat{S}'| \leqslant r - 1$.
12. Show that $|\hat{S}'| + 1 \geqslant |S|$.
13. Complete the proof of the dual state lemma. ■

Proof of Theorem 2.9. Again we consider the state S of D with all the splitting markers joining positive regions, so that $p(S) = c$ and $n(S) = 0$.

The proof of Proposition 2.8 shows that S contributes a term $A^c[-(A^2 + A^{-2})]^{|S|-1}$ to $\langle D \rangle$, and so a term of maximum degree $c + 2|S| - 2$, and that no other state can contribute a term of higher degree to $\langle D \rangle$. Similarly $\hat{S}$ contributes the term of minimum degree $-c - 2|\hat{S}| + 2$ Therefore the span of $\langle D \rangle$ is $2c + 2(|S| + |\hat{S}|) - 4$ and so by the dual state lemma, the span of $\langle D \rangle$ is at most $2c + 2r - 4$, where D has r regions. Now the fact that $r = c + 2$ completes the proof. ■

2.4 The oriented or Homfly polynomial

The appearance of the Jones polynomial revealed an unexpected interplay between different fields of mathematics, and opened up the possibility of advances in knot theory using the new results and techniques. There were formal similarities between the defining formulae for the Jones polynomial and that given for the Alexander polynomial by J.H. Conway in his work on the enumeration of knots and links (Conway 1969). These similarities indicated that the Alexander polynomial and the Jones polynomial were special cases of a more general construction. A new polynomial invariant for knots and links, generalizing both the Alexander and the Jones polynomial, was found virtually simultaneously by four sets of mathematicians: P.J. Freyd and D.N. Yetter, J. Hoste, W.B.R. Lickorish, and K.C. Millett, and A. Ocneanu. Their work was published as a co-authored paper (Freyd *et al.* 1985) and the polynomial was dubbed the H.O.M.F.L.Y.

Let L be an oriented link. The Homfly polynomial for L, denoted $P(L)(\ell, m)$, is the unique two-variable Laurent polynomial defined by

HP1: $P(L)$ is an isotopy invariant;

HP2: $P(U) = 1$;

HP3: if L_+, L_-, and L_0 are three oriented links identical except at one crossing, at which we see the configurations as shown in Fig. 2.18, then $\ell P(L_+) + \ell^{-1}P(L_-) + mP(L_0) = 0$.

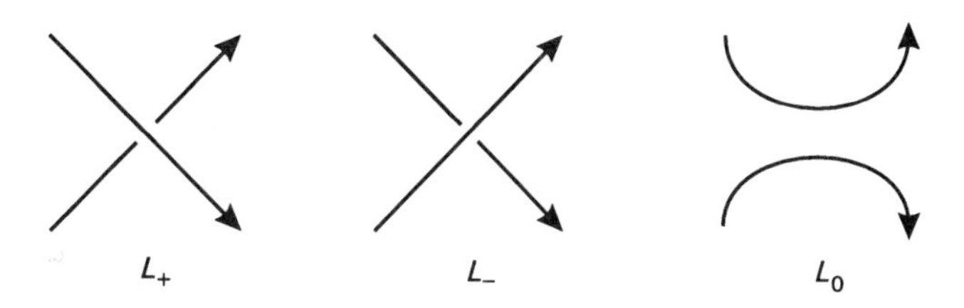

Fig. 2.18

HP1 allows the computation of $P(L)$ from *any* diagram of L, whilst **HP2** and **HP3** give a practical procedure for carrying out such a computation, by crossing changes and the elimination of crossings in a diagram. Let us consider some examples. Observe that the orientation of L is crucial, for it tells us how to eliminate crossings in an application of **HP3**.

(1) The c-component unlink (Fig. 2.19)

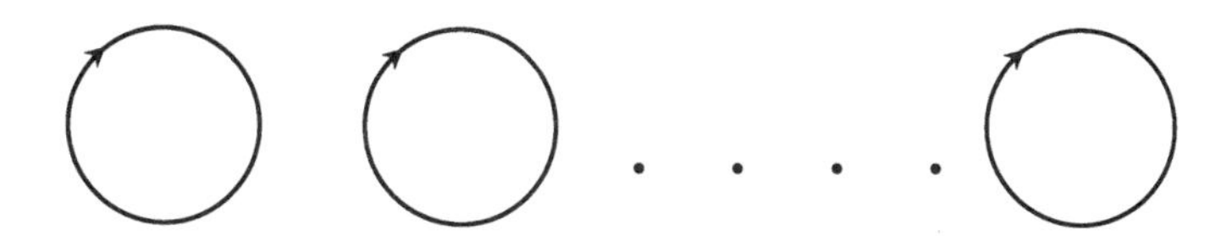

Fig. 2.19

We denote the c-component unlink by U^c. Since $U^1 = U$, **HP1** tells us that $P(U^1) = 1$. Now consider U^2. We begin with a diagram of U^2 with no crossings, and take it as our L_0 (Fig. 2.20). Then we introduce a crossing to give L_+ and L_-.

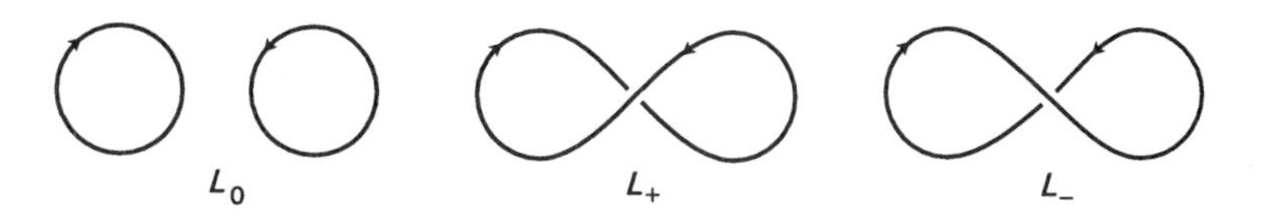

Fig. 2.20

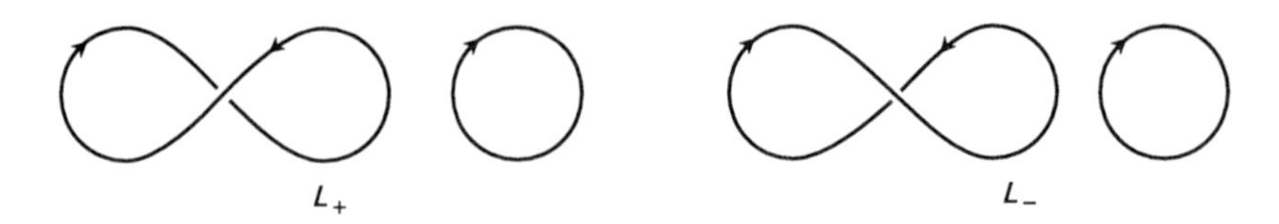

Fig. 2.21

Each of L_+, L_- is now an unknot, and so an application of **HP3** gives $\ell + \ell^{-1} + mP(U^2) = 0$, and it follows at once that $P(U^2) = -m^{-1}(\ell + \ell^{-1})$. What happens for U^3? We follow the same procedure, taking as L_0 a three-component unlink with no crossings, and introducing a crossing to form L_+ and L_- (Fig. 2.21); these are two-component unlinks.

We have $P(L_+) = -m^{-1}(\ell + \ell^{-1}) = P(L_-)$ and we want to calculate $P(L_0)$ **HP3** gives $\ell[-m^{-1}(\ell + \ell^{-1})] + \ell^{-1}[-m^{-1}(\ell + \ell^{-1})] + mP(L_0) = 0$ and we see that $P(U^3) = P(L_0) = [-m^{-1}(\ell + \ell^{-1})]^2$. Write $\mu = -m^{-1}$ $(\ell + \ell^{-1})$. We leave it to you as an exercise to complete an inductive proof that $P(U^c) = \mu^{c-1}$.

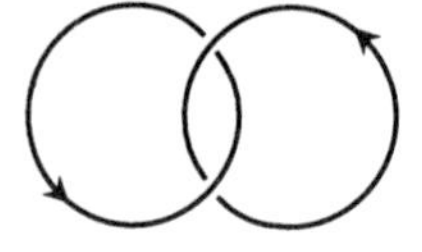

Fig. 2.22

(2) The left-hand Hopf link (Fig. 2.22)

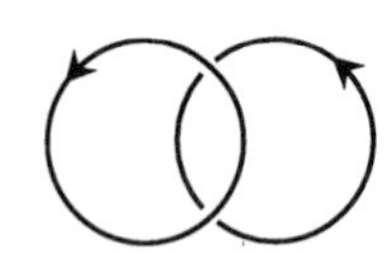

Fig. 2.23

We operate on the top crossing, which has negative sign (do you agree?), so that we take the given diagram as L_- to obtain the triple shown in Fig. 2.23.

We know that $P(L_+) = \mu$ and that $P(L_0) = 1$. An application of **HP3** gives the equation $\ell\mu + \ell^{-1}P(L_-) + m = 0$, from which it follows that

$$P(L_-) = -\ell(m + \ell\mu) = -\ell m + \ell^3 m^{-1} + \ell m^{-1}.$$

Some formal properties of the Homfly polynomial follow easily from the definition. As for the Kauffman polynomial, we may easily compute the Homfly polynomials of reverses and mirror images.

2.11 Proposition. *Let L be an oriented link, with reverse $\bar{L}$ and mirror image $L^!$. Then $P(L)(\ell, m) = P(\bar{L})(\ell, m) = P(L^!)(\ell^{-1}, m)$.*

Proof. Reversing orientation on all components of L leaves the signs of crossings unchanged, so that a computation of $P(\bar{L})$ is identical to one of $P(L)$. Taking a mirror image changes the sign of every crossing, so that a computation of $P(L^!)$ is the same as one of $P(L)$ with ℓ and ℓ^{-1} interchanged. ■

Computer-assisted calculations of Homfly polynomials, the most comprehensive of which cover all the 12 965 knots with crossing number at most 13, invite speculation on two key questions about the Homfly polynomial:

1. Can we characterize the two-variable Laurent polynomials that arise as Homfly polynomials of links?
2. Is there a non-trivial link L with $P(L) = 1$?

We have yet to establish that the computational procedure implicit in **HP1**, **HP2**, and **HP3** does give a unique result. The details of this fact can be found in Lickorish and Millett (1987). Their proof is by induction on the number of crossings in a link diagram; the argument is necessarily more subtle than for the bracket polynomial, since the use of **HP3** does not guarantee a reduction in the number of crossings of diagrams. The inductive hypothesis is as follows.

Let $\mathscr{L}_n$ be the set of all link diagrams with at most n crossings, with an ordering chosen for the components and a starting point selected on each component. We assume that for every $K \in \mathscr{L}_{n-1}$, a polynomial $P(K)(\ell, m)$ has been defined, such that $P(K)(\ell, m) = P(K')(\ell, m)$ if K and K' differ only in a choice of ordering of components and of starting points on components, that $P(K)$ is unchanged by Reidemeister moves carried out within $\mathscr{L}_{n-1}$, that it satisfies **HP2** and **HP3**, and that in addition, if $U^c \in \mathscr{L}_{n-1}$ is a projection of a c-component unlink, then $P(U^c)(\ell, m) = \mu^{c-1}$.

This is a bit of a mouthful, so as an aid to digestion it is recommended that you check that the induction starts successfully on $\mathscr{L}_0$. Note that $P(U^c)$ is then *defined* by the inductive hypothesis.

The definition of P is now extended to $L \in \mathscr{L}_n$ by applying crossing changes to the components of D in order, beginning on each component at the specified starting point, and carried out so that the first encounter with each crossing is made along an underpass, as in the proof of Lemma 1.2. The result is an unlink αL with c components, and $P(\alpha L)(\ell, m)$ is defined as μ^{c-1}. Now at each crossing change, the inductive hypothesis is used to compute $P(K_0)(\ell, m)$ for each K_0 that arises, and **HP3** permits the calculation of a polynomial defined as $P(L)(\ell, m)$. With this detailed inductive definition of $P(L)$ to hand, it is now a matter of showing that it satisfies the requirements of the inductive hypothesis on $\mathscr{L}_n$. This verification is accomplished in Lickorish and Millett (1987), to which we refer the reader for further details.

Exercises 2.4

1. Verify that if L is the right-hand Hopf link (Fig. 2.24) then $P(L) = -\ell^{-1}m + \ell^{-3}m^{-1} + \ell^{-1}m^{-1}$.

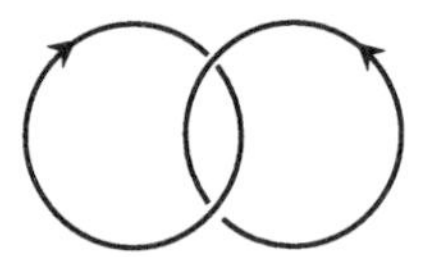

Fig. 2.24

2. Use the triple of diagrams in Fig. 2.25 to show that the Homfly polynomial of the left-hand trefoil is $\ell^2m^2 - 2\ell^2 - \ell^4$.

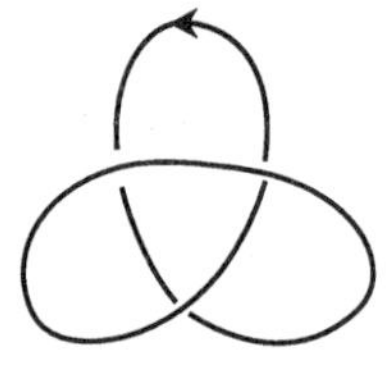
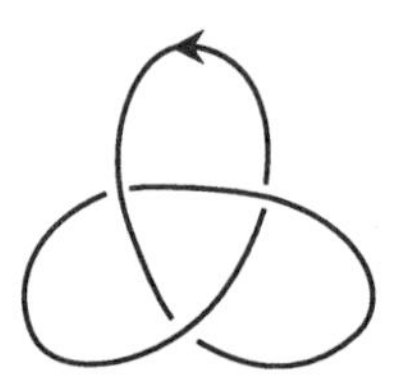
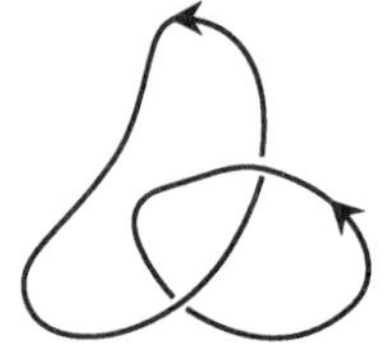

Fig. 2.25

3. Guess the Homfly polynomial of the right-hand trefoil, and check your guess with a computation.

2.5 The Alexander polynomial

J.W. Alexander introduced the polynomial that now bears his name (Alexander 1928) as a simple and effective invariant for distinguishing knot types. The Alexander polynomial may now be obtained from the Kauffman or Homfly polynomials by a change of variable, so is not as effective as these parvenus, but it retains a great deal of intrinsic interest because of its connection with knot groups, which we shall see in Chapter 8, and with covering spaces in Chapter 12. Here we discuss Alexander's original treatment of his polynomial, which is based on an ingenious calculation involving determinants, and uses a labelling of knot diagrams that is a forerunner of the idea of *state* used by Kauffman.

Let D be an oriented link diagram with c crossings. Then by Lemma 2.6 D divides the plane into $c + 2$ regions. We construct a matrix $M(D)$ whose rows are indexed by the crossings of D and whose columns are indexed by the regions, so that $M(D)$ has size $c \times (c + 2)$.

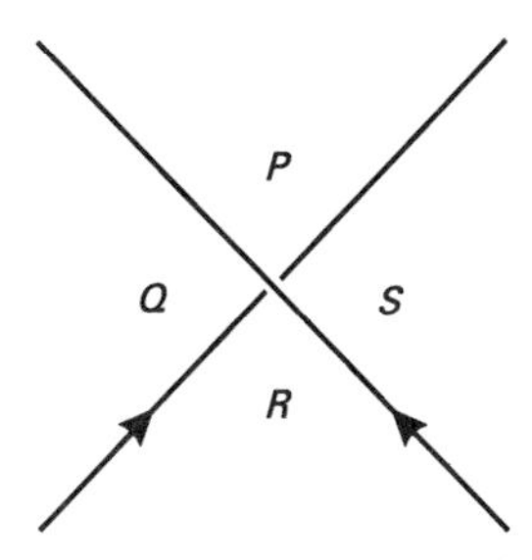

Fig. 2.26

Suppose that regions P, Q, R, and S meet at the ith crossing of D (Fig. 2.26). Then in row i of $M(D)$ we enter t, $-t$, $1, -1$ in columns P, Q, R, and S and zero in all other columns. Note that according to our sign convention for regions, Q and S are positive. At an isthmus, with only three regions adjacent to a crossing, the column indexing the repeated region receives the sum of the symbols allocated to that region (Fig. 2.27).

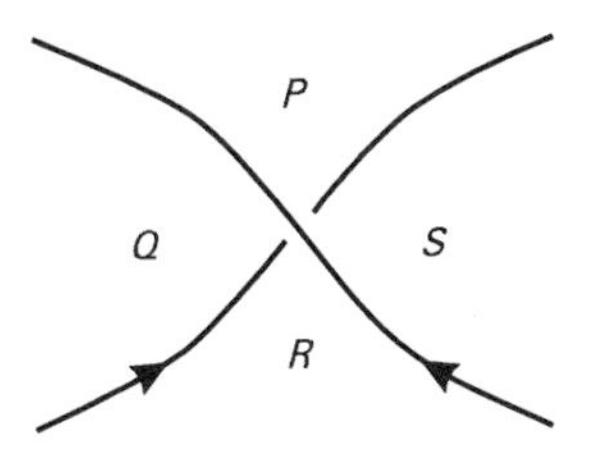

Fig. 2.27

For example, the figure-eight knot with crossings and regions labelled as shown in Fig. 2.28 has the associated matrix

$$M(D) = \begin{array}{c} \\ 1 \\ 2 \\ 3 \\ 4 \end{array} \begin{array}{c} \begin{array}{cccccc} P & Q & R & S & T & U \end{array} \\ \begin{pmatrix} 0 & 0 & -1 & t & 1 & -t \\ 0 & -t & -1 & 1 & t & 0 \\ t & -t & 0 & 0 & 1 & -1 \\ t & -1 & 0 & 1 & 0 & -t \end{pmatrix} \end{array}.$$

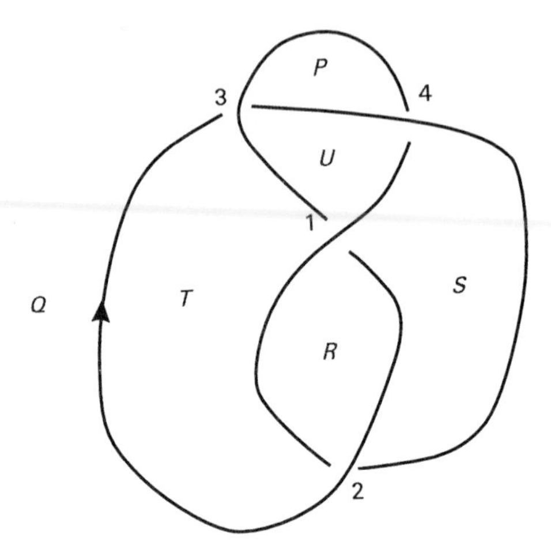

Fig. 2.28

Now delete from $M(D)$ any two columns indexed by a region which have a common boundary arc in D; call the resulting $(c \times c)$ matrix $M_0(D)$. The *Alexander polynomial* Δ_D of D is defined to be the determinant of $M_0(D)$: $\Delta_D = \det(M_0(D))$. It is clear that Δ_D is a polynomial in t with integer coefficients. If, in the above example, we delete columns T and U to obtain

$$M_0(D) = \begin{array}{c} \\ 1 \\ 2 \\ 3 \\ 4 \end{array} \begin{array}{c} \begin{array}{cccc} P & Q & R & S \end{array} \\ \begin{vmatrix} 0 & 0 & -1 & t \\ 0 & -t & -1 & 1 \\ t & -t & 0 & 0 \\ t & -1 & 0 & 1 \end{vmatrix} \end{array}$$

and expand the determinant by the first row, we find

$$\Delta_D(t) = -1(-t + 2t^2) - t(t - t^2) = t^3 - 3t^2 + t.$$

If instead we delete columns R and S to obtain

$$M_0(D) = \begin{array}{c} \\ 1 \\ 2 \\ 3 \\ 4 \end{array} \begin{array}{c} \begin{array}{cccc} P & Q & T & U \end{array} \\ \begin{vmatrix} 0 & 0 & 1 & -t \\ 0 & -t & t & 0 \\ t & -t & 1 & -1 \\ t & -1 & 0 & -t \end{vmatrix} \end{array}$$

and again expand by the first row, we find

$$\Delta_D(t) = t^2 - t^3 + t(-t^2 - t^2 + t^3) = t^4 - 3t^3 + t^2.$$

You will observe that this is a different answer. However, it differs from the first answer by a factor of t. This is a typical outcome, as explained by a theorem proved by Alexander.

2.12 Theorem. *Given an oriented link diagram D, the determinant* $\det(M_0(D))$ *is independent of the choice of columns deleted from the matrix $M(D)$, up to a factor of $\pm t^m$ for some integer m. Furthermore,* $\det(M_0(D))$ *is then an isotopy invariant.* ■

We shall not prove the theorem here, but recommend the reader to enjoy the discursive style of Alexander (1928). Allowing for adjustments of sign and powers of t, we now have an isotopy invariant for a link L computable from any oriented diagram, namely the *Alexander polynomial of L*, denoted $\Delta_L(t)$. The convention is to adjust $\det(M_0(D))$ by a term $\pm t^m$ to give a positive constant term and a positive highest power of t.

Exercise 2.5

1. Choose your favourite knot or link from the book so far, and compute its Alexander polynomial.

3
Topological spaces

The purpose of this chapter is to provide the theoretical background for a rigorous discussion of knots and surfaces, against which the vaguely expressed ideas lying behind our previous discussion can be made precise. The central notion is that of a *topological space*. The precision that topological spaces bring to the ideas of continuity and nearness is, of course, essential to any proper treatment of knots and surfaces. Moreover, topological ideas will prove crucial in the development of more sophisticated invariants in later chapters. Since our interest in topological spaces stems from specific applications that we already have in mind, we make no attempt to give a comprehensive account of even the basic theory of topological spaces here. There are several textbooks that do the job well, and we mention Armstrong (1983) and Brown (1988) as providing suitable further reading.

3.1 Topological spaces

Topological spaces provide an abstract framework for the notion of *continuity*, and as is usually the case with abstraction, there is a great gain in the clarity of the concept, even if some of the examples do not accord with our intuitive expectations. However, the definition of continuity in terms of ε and δ familiar from introductory analysis is now seen to arise from a particular *topology* chosen for the real numbers. This topology is the one appropriate for the analytic viewpoint, and we call it the *usual topology* on $\mathbb{R}^n$ to emphasize its significance.

What are the concepts that we need to make precise in order to put our study of knots and surfaces on a firm base? A glance back at our informal definition of a knot in Chapter 1 shows immediately that we need to define continuous maps from the circle S^1 to $\mathbb{R}^3$ in order to describe a knot as a continuous copy of S^1, and for links we shall need to define a continuous map from the union of several copies of S^1 into $\mathbb{R}^3$. In the treatment of surfaces in Chapter 4, we shall have need of a notion of a subset of $\mathbb{R}^n$ 'looking like' $\mathbb{R}^2$ 'near' any point. What do we mean by 'looking like' and 'near'? In $\mathbb{R}$, 'near' often means 'within ε of'; that is, we fit a small, open interval around the given point. What might 'open' mean in more generality? With these questions in mind, let us proceed to the definition of a topological space without more ado.

Let X be a non-empty set. A *topology on* X is a collection of subsets of X, called *open sets*, satisfying

OS1: the empty set and X itself are open sets;

OS2: an arbitrary union of open sets is an open set;

OS3: the intersection of finitely many open sets is an open set.

The set X, together with the collection of its special subsets that we have called open, is a *topological space*. We shall usually denote a topology by a script symbol such as $\mathcal{T}$, so that a space is a pair $(X, \mathcal{T})$. However, it is customary to suppress explicit mention of $\mathcal{T}$ and refer to X as a topological space, keeping the topology in the back of our minds. Note that a given set can have many different topologies; there is nothing *intrinsically* special about the open sets, other than that we have selected them to play a certain distinguished role. However, in many circumstances, there will be good reasons to select one topology above all others.

Examples

1. Given any set X, we may define every subset of X to be open. This is the *discrete topology* on X.
2. Again on any set X, define a subset $A \subseteq X$ to be open if either $A = \emptyset$ or A has finite complement in X, that is $|X \backslash A| < \infty$. This is the *finite complement topology on* X. What do we get if X is a finite set?
3. *The usual topology on*, $\mathbb{R}^n$, $(n \geqslant 1)$.

For $\boldsymbol{x}, \boldsymbol{y} \in \mathbb{R}^n$, define

$$\|\boldsymbol{x} - \boldsymbol{y}\| = \left(\sum_{i=1}^{n} (x_i - y_i)^2 \right)^{1/2}.$$

This is the customary notion of the distance between two points in $\mathbb{R}^n$. The *open ball with centre* $\boldsymbol{x}$ *and radius* r, denoted by $B(\boldsymbol{x}; r)$, is the subset $\{\boldsymbol{y} \in \mathbb{R}^k \mid \|\boldsymbol{x} - \boldsymbol{y}\| < r\}$; here $r \in \mathbb{R}$ and $r > 0$. If $k = 1$, we see that an open ball in $\mathbb{R}$ is an open interval:

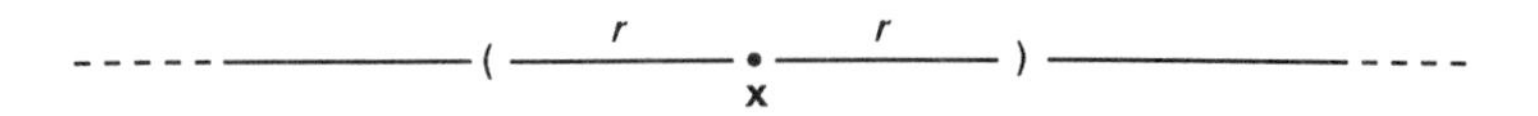

whilst if $k = 2$, an open ball in $\mathbb{R}^2$ is an open disc. The usual topology on $\mathbb{R}^n$ is now defined in terms of open balls in the following fashion. We say that a subset $U \subseteq \mathbb{R}^n$ is open if, for every $\boldsymbol{x} \in U$, there exists $r \in \mathbb{R}$ with $r > 0$ such that $B(\boldsymbol{x}; r) \subseteq U$. Note that r depends on $\boldsymbol{x}$.

Let X be a topological space and let $A \subseteq X$ be a subset. We say that A is *closed* if its complement $X \backslash A$ is open. If X is a topological space, any subset $A \subseteq X$ can be made into a topological space by defining the open sets in A to be all sets $A \cap U$ where U is an open set in X. This is the *subspace topology on A* and we say that A is a *subspace* of X.

The following result will be of use to prove the important *gluing lemma* later in this chapter.

3.1 Lemma. *If A is a closed subspace of a topological space X, and if $C \subseteq A$ is closed in A, then C is closed in X.*

Proof. By definition, $A \backslash C$ is open in A, whence $A \backslash C = A \cap U$ for some open set $U \subseteq X$. Therefore,

$$X \backslash C = (X \backslash A) \cup (A \backslash C) = (X \backslash A) \cup (A \cap U) = (X \backslash A) \cup U.$$

Now $X \backslash A$ and U are open in X and so by **OS2** $X \backslash C$ is open in X. Therefore C is closed in X. ∎

Exercises 3.1

1. Let X have the discrete topology. Which subsets of X are closed? What if X has the finite complement topology?
2. In $\mathbb{R}^n$ with its usual topology, show that the *closed ball with centre $\boldsymbol{x}$ and radius r*, defined as $\bar{B}(\boldsymbol{x}; r) = \{\boldsymbol{y} \in \mathbb{R}^n \mid \|\boldsymbol{x} - \boldsymbol{y}\| \leqslant r\}$ is indeed a closed subset of $\mathbb{R}^n$.
3. Show that the subspace topology *is* a topology on a subset A of a topological space X.
4. Show that any subspace of a topological space with the discrete topology also has the discrete topology. Prove that, as a subspace of $\mathbb{R}$ with its usual topology, $\mathbb{Z}$ has the discrete topology.
5. Show that the union of finitely many closed subsets of a topological space is again a closed subset.
6. (a) Let X be a topological space. We call a subset $N \subseteq X$ a *neighbourhood* of $x \in X$ if there exists an open set $U \subseteq X$ such that $x \in U \subseteq N$. Let $\mathbb{N}(x)$ denote the collection of all neighbourhoods of $x \in X$. Show that $\mathbb{N}(x)$ is non-empty and satisfies the following four properties:

NHD1: if $N \in \mathbb{N}(x)$ then $x \in N$;

NHD2: if $N \in \mathbb{N}(x)$ and $M \supseteq N$ then $M \in \mathbb{N}(x)$;

NHD3: if $N, N' \in \mathbb{N}(x)$ then $N \cap N' \in \mathbb{N}(x)$;

NHD4: if $N \in \mathbb{N}(x)$ and $N^0 = \{y \in N \mid N \in \mathbb{N}(y)\}$ then $N^0 \in \mathbb{N}(x)$.

(b) Suppose now that we are given a set Y and for each $y \in Y$ a non-empty collection $\mathbb{N}(y)$ of subsets of Y satisfying **NHD1**, . . . , **NHD4**. Define $U \subseteq Y$ to be open if, for every $u \in U$, $U \in \mathbb{N}(u)$. Show that this is a topology on Y.

(c) Part (b) opens up the embarrassing possibility of going round in ever-decreasing circles, repeatedly defining 'open sets' and 'neighbourhoods' in terms of one another. Suppose we start with a topological space X, with neighbourhoods as in (a), and define open sets as in (b). Are these new open sets the same as the open sets of X that give the topology? What if we start with a set Y as in (b), define a topology on it as in (b), and then define neighbourhoods as in (a)? Are these neighbourhoods the same as those in the given collections $\mathbb{N}(y)$ for $y \in Y$?

7. Find all the possible topologies on a set with precisely three elements. How many different topologies are there for a set with precisely four elements?

3.2 Continuity

Let X and Y be topological spaces. (Note that we suppress any mention of the topology on the underlying sets, but you should keep in mind that the topologies may be very different.) A function $f{:}X \to Y$ is *continuous* if, for every open set V in Y, its preimage $f^{-1}(V)$ is open in X. (Recall that the preimage $f^{-1}(V)$ is the subset $\{x \in X \mid f(x) \in V\}$; we do *not* assume that an inverse f^{-1} to f exists.) Since continuity is defined in terms of open sets, it depends on the choice of topology for each of the sets X and Y. For example, if $\mathcal{T}$ and $\mathcal{T}'$ are different topologies on X then the identity map $x \mapsto x$ need not be a continuous function $(X, \mathcal{T}) \to (X, \mathcal{T}')$. However, the familiar notion of '$\varepsilon - \delta$' continuity is identical to the notion of continuity with respect to the usual topologies on $\mathbb{R}^m$ and $\mathbb{R}^n$.

3.2 Proposition. *Let $\mathbb{R}^m$ and $\mathbb{R}^n$ have their usual topologies. Then $f\colon \mathbb{R}^m \to \mathbb{R}^n$ is continuous if and only if, given any $\varepsilon > 0$, there exists $\delta > 0$ such that $\|\mathbf{x} - \mathbf{y}\| < \delta$ implies that $\|f(\mathbf{x}) - f(\mathbf{y})\| < \varepsilon$.*

Proof. Suppose that f is continuous and let $\varepsilon > 0$ be given. The open ball $B(f(\boldsymbol{x}); \varepsilon)$ is an open set in $\mathbb{R}^n$ and therefore $f^{-1}(B(f(\boldsymbol{x}); \varepsilon))$ is an open set in $\mathbb{R}^m$: since $\boldsymbol{x} \in f^{-1}(B(f(\boldsymbol{x});\varepsilon))$ there exists $\delta > 0$ such that $B(\boldsymbol{x}; \delta) \subseteq f^{-1}(B(f(\boldsymbol{x}); \varepsilon))$, that is $\|\boldsymbol{x} - \boldsymbol{y}\| < \delta$ implies that $\|f(\boldsymbol{x}) - f(\boldsymbol{y})\| < \varepsilon$. The proof of the converse we leave as an exercise for the reader. ■

3.3 Lemma. *Let X, Y, and Z be topological spaces. If $f: X \to Y$ and $g: Y \to Z$ are continuous functions, then $gf: X \to Z$ is continuous.*

Proof. Let W be an open set in Z. By the continuity of g, we know that $V = g^{-1}(W)$ is open in Y, and by the continuity of f we know that $f^{-1}(V)$ is open in X. Thus $(gf)^{-1}(W) = f^{-1}(g^{-1}(W))$ is open in X and gf is continuous. ■

3.4 The gluing lemma. *Given a function $f: X \to Y$, suppose that there are closed sets A and B such that $X = A \cup B$ and such that the restrictions $f|_A: A \to Y$ and $f|_B: B \to Y$ are continuous. Then f is continuous.*

Proof. Let V be an open set in Y. We want to show that $f^{-1}(V)$ is open in X. Continuity of the restricted functions implies that $f^{-1}(V) \cap A$ is open in A and that $f^{-1}(V) \cap B$ is open in B; thus $A \backslash f^{-1}(V)$ is closed in A and $B \backslash f^{-1}(V)$ is closed in B. By Lemma 3.1, each of $A \backslash f^{-1}(V)$ and $B \backslash f^{-1}(V)$ is closed in X. Using Exercise 3.1.5 we see that $X \backslash f^{-1}(V) = (A \backslash f^{-1}(V)) \cup (B \backslash f^{-1}(V))$ is closed in X and so $f^{-1}(V)$ is open in X. ■

Let X and Y be topological spaces. A function $f: X \to Y$ which is bijective, continuous, and has a continuous inverse $f^{-1}: Y \to X$ is called a *homeomorphism*. A function $f: X \to Y$ which is injective, continuous, and such that the bijection $f: X \to f(X)$ has a continuous inverse is called an *embedding*. Simple examples of embeddings include $\mathbb{R}^2 \to \mathbb{R}^3$ given by $(x, y) \mapsto (x, y, 0)$ and $S^1 \to \mathbb{R}^3$ given by $(x, y) \mapsto (x, y, 1)$ where $x^2 + y^2 = 1$.

If X and Y are topological spaces, is there a natural topology on the set $X \times Y = \{(x, y) \mid x \in X, y \in Y\}$? The first question to answer is this: what do we mean by 'natural'? Clearly the topology will have to be defined in terms of the given topologies for X and Y. Furthermore, there are the canonical projection functions $p: X \times Y \to X$ given by $(x, y) \mapsto x$ and $q: X \times Y \to Y$ given by $(x, y) \mapsto y$ and these should surely be continuous. So let us try to construct a topology meeting these requirements. Let $U \subseteq X$ be open: then $p^{-1}(U)$ is $U \times Y$ so that $U \times Y$ needs to be one of our open sets in $X \times Y$; similarly we need $X \times V$ for any open set V in Y. However, $(U \times Y) \cap (X \times V) =$

$U \times V$ and so if we are to satisfy **OS3** we require that all subsets $U \times V$, where U is open in X and Y is open in Y, are open in $X \times Y$. Then to satisfy **OS2** we require arbitrary unions of such subsets to be open, and this is enough. If we define $A \subseteq X \times Y$ to be open if A is a union of subsets $U \times V$, where U is open in X and V is open in Y, then we do obtain a topology on $X \times Y$: it is called the *product topology on* $X \times Y$.

Exercises 3.2

1. Let X be a topological space. Prove that every function $f: X \to Y$, where Y is any topological space, is continuous if and only if X has the discrete topology.
2. Let $(X, \mathcal{T})$ and $(X, \mathcal{T}')$ be topological spaces. Find a condition on the topologies $\mathcal{T}$ and $\mathcal{T}'$ that ensures that the identity map $x \mapsto x$ is a continuous function $(X, \mathcal{T}) \to (X, \mathcal{T}')$.
3. Given sets X and Y, their disjoint union $X \amalg Y$ is the set obtained by taking the union of disjoint copies of X and Y, so that in $X \amalg Y$ we have two copies of any element in $X \cap Y$. If $X \cap Y = \varnothing$ then $X \amalg Y$ is the usual union $X \cup Y$. How might we define a topology on the disjoint union $X \amalg Y$ of topological spaces X and Y? The simplest idea is to take a subset $U \subseteq X \amalg Y$ to be open if $U = V \amalg W$ where V is open in X and W is open in Y. Show that this does define a topology, and makes the inclusions of X and Y into their disjoint union continuous functions.
4. Find an example of a continuous bijection whose inverse is not continuous. [**Hint.** You have already seen examples if you have done the exercises so far.]
5. Let X be a topological space and let $U \subseteq X$ be a subspace. Show that the inclusion $U \to X$ is an embedding. If X is a topological space and U is a *subset* of X with a topology such that the inclusion *is* an embedding, need the topology on U be the subspace topology?
6. Verify that the product topology is indeed a topology on $X \times Y$.
7. Describe the neighbourhoods of a point $(x, y) \in X \times Y$ in the product topology.
8. A topological space X is called *Hausdorff* if, given any distinct elements $x, y \in X$, there exist open sets U containing x and V containing y with U and V disjoint; that is, we have $x \in U$ and $y \in V$ with $U \cap V = \varnothing$. (The concept is named in honour of F. Hausdorff (1868–1942) and a punning *aide-mémoire* is to say that the elements x and y are *housed off* by the opens sets U and V). The *diagonal* subspace of $X \times X$ is $\Delta(X) = \{(x, x) \mid x \in X\}$. Show that X is Hausdorff if and only if the diagonal subspace is closed

in $X \times X$. Suppose that the finite complement topology on a set X makes X a Hausdorff space. What must be true of X? Find some non-Hausdorff topologies.

9. Let $U \to X$ be a continuous injection between topological spaces. Show that if X is Hausdorff then so is U.

3.3 Connectedness

In addition to providing a framework for continuity, topological notions also allow us to consider connectedness. Intuitively we want to make precise the idea of a topological space being in one piece, and we do this by considering what we might be able to do with a space that was not in one piece. We surely expect to be able to divide such a space into at least two subspaces that do not intersect, and this leads to the following definition.

A *disconnection* of a topological space X is a pair (A,B) of non-empty closed subspaces of X such that $X = A \cup B$ and $A \cap B = \varnothing$. A space X is *connected* if no disconnection of it exists.

A key property of connected spaces is that the only subsets which are both open and closed are the empty set and the whole space. We shall make use of this fact in Chapter 12. We shall show here that this scarcity of subsets that are both open and closed is *equivalent* to connectedness, so that we could have chosen it as the definition. Would this have been a sensible thing to do?

3.5 Proposition. *A topological space X is connected if and only if the only subsets of X that are both open and closed are the empty set and X itself.*

Proof. We first suppose that X is connected, and let A be a subset of X that is both open and closed. Let B be the complement of A in X, so that B is also both open and closed. If both A and B are non-empty, we see that (A, B) is a disconnection of X: since X is connected, no disconnection exists. Therefore one of A and B is empty, and the other is then X itself.

Conversely, suppose that the only subsets of X that are open and closed are $\varnothing$ and X but that X has a disconnection (A, B). Then A is closed, so that its complement B is open; but B is also closed and since B is non-empty we must have $B = X$. But then $A = \varnothing$, which is a contradiction. ■

Given a continuous function f: $X \to Y$ with X connected, might we expect Y to be connected? A little thought should convince you that there is no reason to expect such a result for Y itself, but that it does seem reasonable to suppose that the *image* of f will be connected. This reasonable supposition turns out to be true.

3.6 Preservation of connectedness

Let X be a connected space and let $f: X \to Y$ be a continuous function. Then the image $f(X)$ of f is connected.

Proof. Suppose that A is a subset of $f(X)$ which is both closed and open. Since f is continuous, $f^{-1}(A)$ is both closed and open. By connectedness of X, $f^{-1}(A)$ is either X or the empty set, and it follows that A is either $f(X)$ or the empty set. Therefore $f(X)$ is connected, by Proposition 3.5. ∎

Exercises 3.3

1. Show that $\mathbb{R}$ (with the usual topology) is connected, but that the subspace $\mathbb{Q}$ is not connected.
2. Given any topological space X, define a relation $\sim$ on X as follows. For any $x, y \in X$ we say $x \sim y$ if and only if x and y belong to some connected subspace of X. Show that $\sim$ is an equivalence relation on X. The equivalence classes are called the *connected components* of X. A space in which every connected component contains just one element is said to be *totally disconnected.* Show that any space with the discrete topology is totally disconnected. Is the converse true?

3.4 Identification spaces

Let X be a topological space, and let $f: X \to Y$ be a function to a set Y. Can we equip Y with a topology that makes f continuous? Of course we can! We can give Y the *indiscrete* topology for which the only open sets are Y itself and the empty set. Evidently this is something of a swindle, because the topology has nothing to do with the function f and has made Y into a dull sort of space that accepts every function into it as continuous. Perhaps we should reformulate the problem. It is better to have as rich a structure of open sets as we can on Y subject to the condition that f be continuous, and this idea motivates the following definition.

Given a topological space X, a set Y, and a function $f: X \to Y$, the *identification topology* on Y is obtained by defining $U \subseteq Y$ to be open if and only if its inverse image $f^{-1}(U)$ is open in X. It is immediately obvious that this definition makes f continuous.

An important application of the identification topology is to the quotient set of an equivalence relation. Let X be a topological space, and suppose that $\sim$ is an equivalence relation on X. Is there a natural topology on the set of equivalence classes $X/\sim$? As in our discussion of the product topology, we require the topology on $X/\sim$ to be defined in terms of the given topology on X. Further, there is a function that we shall require to be continuous: namely, the quotient function $X \to X/\sim$ that takes an element $x \in X$ to its class $[x] \in X/\sim$. We achieve all that we require by giving $X/\sim$ the identification topology defined by the quotient function.

We can now take *any* equivalence relation on a topological space X and construct a new space, namely the quotient set of the equivalence relation equipped with the identification topology. In particular, we can identify certain elements of a space with one another by an equivalence relation; in picturesque terminology, we think of gluing one part of a space to another part. Here are some examples of this process in action; it is tempting to recommend the acquisition of paper and a pot of glue before reading further!

Examples

(a) Take a rectangle, and identify two opposite ends after performing a half-twist. The result is a *Möbius band* (or *strip*). Formally we might begin with a rectangle in $\mathbb{R}^2$, say $R = \{(x, y) \in \mathbb{R}^2 \mid 0 \leqslant x \leqslant 2, 0 \leqslant y \leqslant 1\}$, where $\mathbb{R}^2$ has its usual topology and R has the subspace topology, and take the equivalence relation $\sim$ on R generated by equality and the additional relations $(0, y) \sim (2, 1 - y)$ for all y satisfying $0 \leqslant y \leqslant 1$. Then $R/\sim$ with the identification topology is a mathematical model for a Möbius band.

(b) Begin again with a rectangle, and glue together corresponding points of two opposite sides. The result is an open-ended cylinder. Now bend the cylinder so as to bring together the boundary circles of the two open ends, and glue these circles together. The result is a hollow, doughnut-shaped surface called a *torus*.

More formally, we take R as in (a) and impose the equivalence relation generated by equality and the further relations $(0, y) \sim (2, y)$ and $(x, 0) \sim (x, 1)$.

(c) What happens if we try to combine (a) and (b) by introducing a twist into our construction of the torus? We impose on R the equivalence relation generated by equality and the further relations $(x, 0) \sim (x, 1)$ and $(0, y) \sim (2, 1 - y)$. This appears to be harmless enough, but if you try and envisage what happens to a rectangle

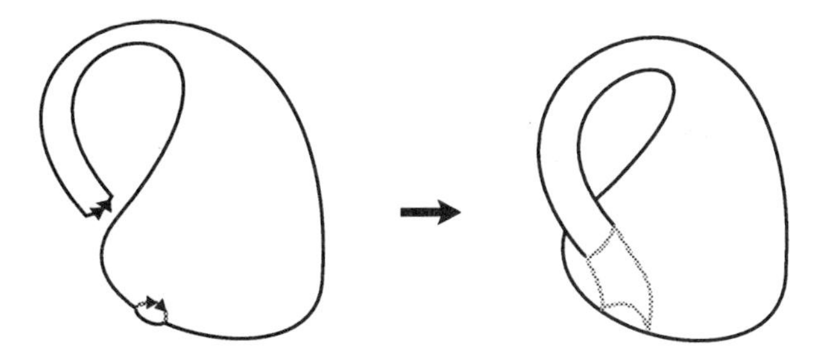

Fig. 3.1

under the corresponding bending and gluing, a difficulty at once emerges. The nicest picture is obtained by imagining the cylinder formed by one gluing to pass through itself before the boundary circles of the cylinder are identified (Fig. 3.1).

The resulting surface is known as the *Klein bottle*. Figure 3.1 does not truly represent the Klein bottle because of the self-intersection, but no model of it can be made in three dimensions (nor any picture drawn in two dimensions!) without such self-intersections. The Klein bottle is an example of a *non-orientable surface*, and as such will reappear in Chapter 4.

(d) Let $f\colon X \to Y$ be a continuous function between topological spaces, and let I denote the unit interval $[0, 1]$ with the subspace topology in $\mathbb{R}$. Form the space $X \times I$ with the product topology, and then the disjoint union $(X \times I) \amalg Y$ with the topology given in Exercise 3 of Section 3.2. Now impose the equivalence relation generated by equality and the further relations $(x, 1) \sim f(x)$. The resulting identification space is called the *mapping cylinder of* f and it is a useful construction in the further theory of topological spaces; see (Brown 1988, chapter 7).

Exercises 3.4

1. Given topological spaces X and Y and a continuous function $f\colon X \to Y$, let (Y, f) denote the set Y given the identification topology. Show that the identity map $(Y, f) \to Y$ is continuous.
2. Find an embedding of the torus into $\mathbb{R}^3$.

3.5 Knots and isotopy

Hereafter, $\mathbb{R}^n$ will always have the usual topology, and we denote by S^1 the unit circle $\{(x, y) \mid x^2 + y^2 = 1\} \subseteq \mathbb{R}^2$ with the subspace topology.

We now have the necessary theoretical background to give our precise definition of a knot. Here it is: a *knot* is an embedding $k\colon S^1 \to \mathbb{R}^3$.

Our definition thus defines a knot as a function k from S^1 that has certain nice properties. Firstly, k is injective; this corresponds to our informal insistence in Chapter 1 that the string of a knot does not intersect itself. Secondly, k is continuous and there is a continuous function back from its image to S^1. This takes care of our requirement that the string forms an unbroken, albeit entangled, copy of the unit circle.

A *link of c components* is an embedding

$$S^1 \amalg S^1 \amalg \ldots \amalg S^1 \to \mathbb{R}^3$$

of c disjoint copies of S^1 into $\mathbb{R}^3$. The topology on $S^1 \amalg S^1 \amalg \ldots \amalg S^1$ is of course that for a disjoint union given in Exercise 3 of Section 3.2.

When should we consider two knots k and l to be the same? We want to capture the notion of their knottedness or interlacing being the same, and in $\mathbb{R}^3$ we might think of moving their images around in a continuous fashion until one image coincides with the other, just as we might deform a physical piece of string from one interlaced form into an equivalent one. The idea that we move the images in $\mathbb{R}^3$ continuously leads to the concept of *ambient-isotopy* in which deformations of the images are effected by applying homeomorphisms of the ambient space $\mathbb{R}^3$, and in the following definition it may be helpful to think of the parameter t as representing time, so that the definition provides a continuous sequence of homeomorphisms of $\mathbb{R}^3$ from time $t = 0$ to time $t = 1$. Now to the definition: two knots k and l are *ambient-isotopic* if there exists a continuous function $H\colon \mathbb{R}^3 \times [0, 1] \to \mathbb{R}^3$ such that

(a) $h_0 = H(-, 0)$ is the identity $\mathbb{R}^3 \to \mathbb{R}^3$

(b) for all $t \in [0, 1]$, $h_t = H(-, t)$ is a homeomorphism $\mathbb{R}^3 \to \mathbb{R}^3$

(c) if $h_1 = H(-, 1)$, then $h_1 k = l$.

If k and l are ambient-isotopic, then the knots $h_t k$ ($t \in [0, 1]$) form a continuously varying family deforming k into l. There is a corresponding definition of ambient-isotopy for links, which we leave the reader to formulate.

Ambient-isotopy of knots and links is precisely the topological analogue of the isotopy of knots that we considered in Chapter 1. We state this crucial fact as a theorem below, but we shall not give a proof. The theorem is due to Reidemeister, who has given a proof in his book (Reidemeister 1932). A more conceptual account is to be found in Burde and Zieschang (1985).

3.7 Theorem. *Two knots are ambient-isotopic if and only if they possess isotopic diagrams.* ■

There are other equivalence relations between functions $S^1 \to \mathbb{R}^3$ that might be used to define a notion of two knots being essentially the same. We can consider a homeomorphism $\mathbb{R}^3 \to \mathbb{R}^3$ as analogous to a change of coordinates, and so define two knots $k: S^1 \to \mathbb{R}^3$ and $l: S^1 \to \mathbb{R}^3$ to be *equivalent* if there exists a homeomorphism $h: \mathbb{R}^3 \to \mathbb{R}^3$ such that $l = hk$. An equivalence class of knots under the equivalence relation of equivalence is commonly called a *knot type*. It is immediate that ambient-isotopic knots are then equivalent, and it should come as no surprise that equivalence is a much weaker notion than ambient-isotopy. For example, it is obvious that the right-hand and left-hand trefoils are equivalent.

A knot k is *piecewise linear* if its image in $\mathbb{R}^3$ is a union of a finite number of line segments. A piecewise linear knot can be represented by a diagram whose arcs are straight lines. A knot is said to be *tame* if it is equivalent to a piecewise linear knot. Tameness may be considered as a technical restriction on knots to ensure that each has only finitely many crossings, and the theory developed in this book will treat exclusively tame knots. All other knots are *wild*; the theory of wild knots encounters different problems to that of tame knots and we propose to say no more about it.

Exercises 3.5

1. Why is it obvious that the right-hand and left-hand trefoils are equivalent?

The final three exercises in this chapter propose, and demolish, a third possible equivalence relation for knots.

2. Call two knots $k: S^1 \to \mathbb{R}^3$ and $l: S^1 \to \mathbb{R}^3$ *comparable* if there exists a continuous function $H: S^1 \times [0, 1] \to \mathbb{R}^3$ such that

   (a) $h_0 = H(-, 0) = k$

   (b) $h_1 = H(-, 1) = l$.

   Show that comparability is an equivalence relation on knots.
3. Prove that equivalent knots are comparable.
4. Prove that any two knots are comparable [**Hint**. Show that any given knot is comparable with the unknot, by the well-known process of pulling the knot tight.]

4
Surfaces

Knots are one-dimensional objects living in three-dimensional space. Surfaces are two-dimensional objects. Some of the ideas that help in understanding knots can also be adapted to help study surfaces. The flow of ideas is two way and we shall see later on how surfaces help us to study knots. The study of surfaces, however, has no need of knot theory for motivation. There are important applications of the classification theorem for surfaces in other areas of mathematics, perhaps most notably in complex analysis, but we will not be going in that direction here.

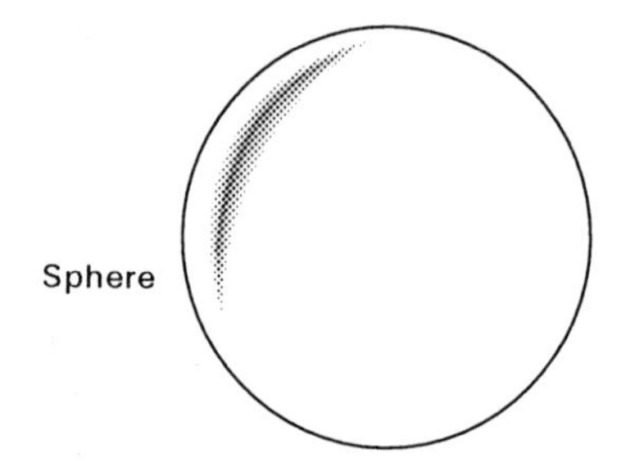

Fig. 4.1

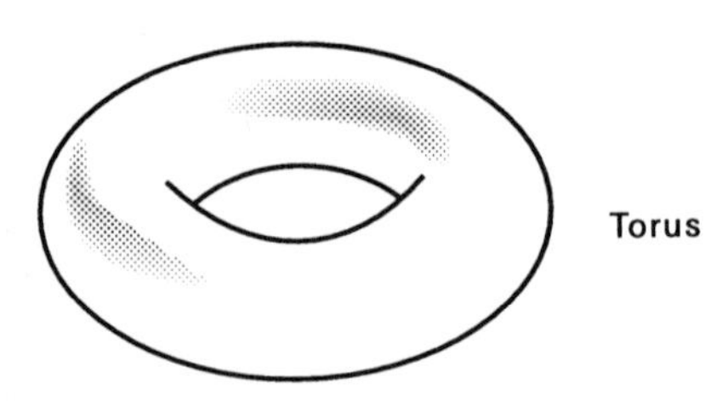

Fig. 4.2

What is a surface? In the last chapter, we saw some surfaces constructed by gluing. We have examples such as those shown in Figs 4.1–4.3.

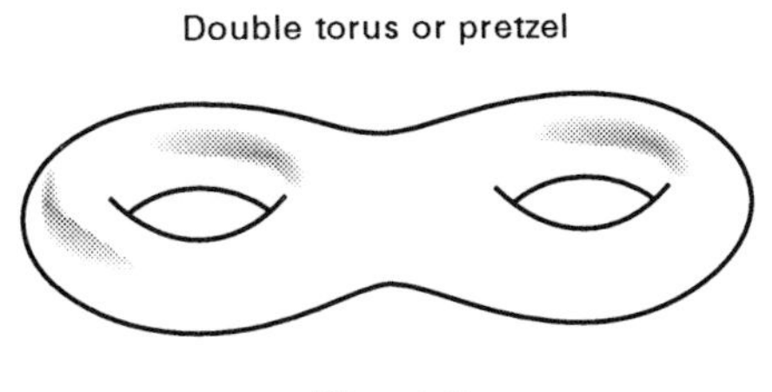

Fig. 4.3

Here we also want to allow for holes or edges in surfaces, as in Figs 4.4–4.7.

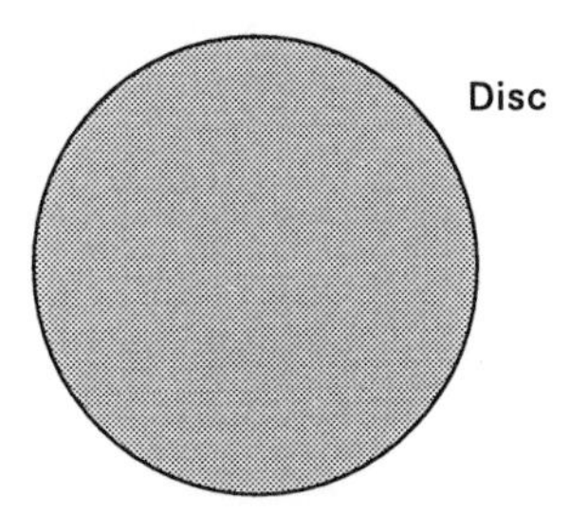

Fig. 4.4

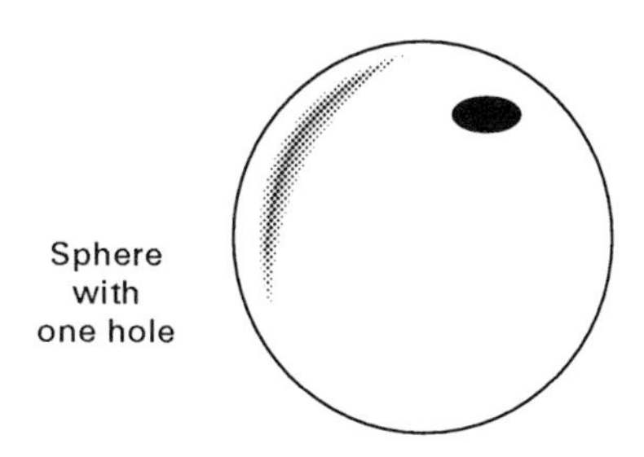

Fig. 4.5

Recall from Chapter 3 that the Möbius band (Fig. 4.8) is obtained by twisting and gluing a rectangle (Fig. 4.9).

Here we have seen some of the simpler examples of surfaces. Can we list all of them? This question needs tightening up before it really

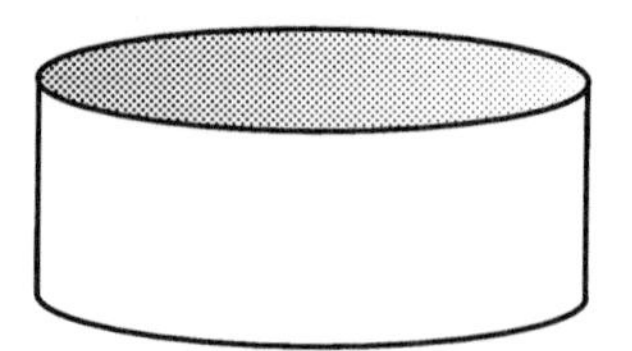

Fig. 4.6

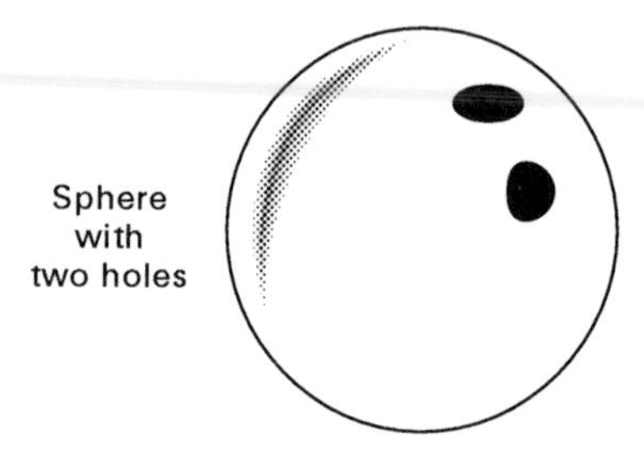

Fig. 4.7

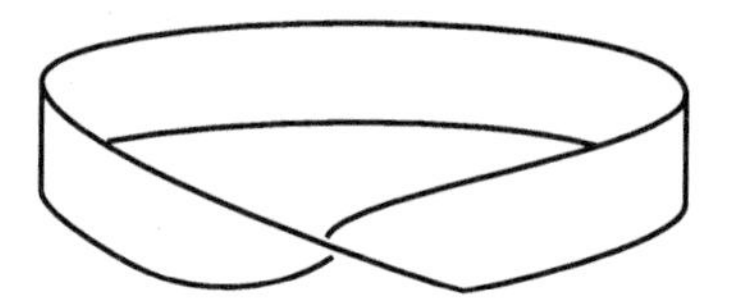

Fig. 4.8

will make sense. There are infinitely many surfaces; for instance, consider a multiple torus with n 'handles', as in Fig. 4.10.

For different values of n, it seems likely that the corresponding multiple tori will be different. Likewise a sphere with n holes is not likely to be the same as a sphere with m holes (unless $m = n$, of

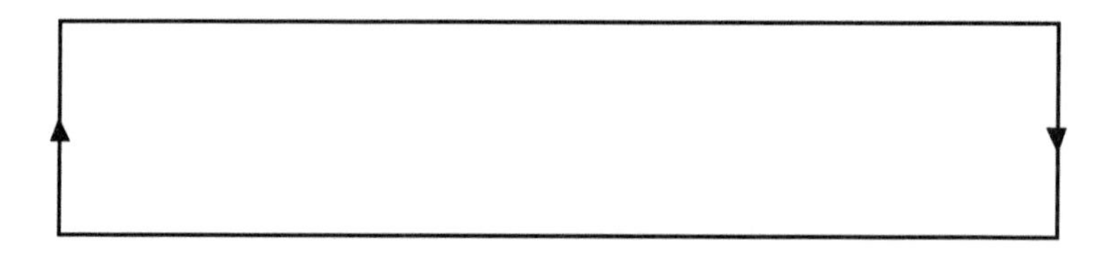

Fig. 4.9

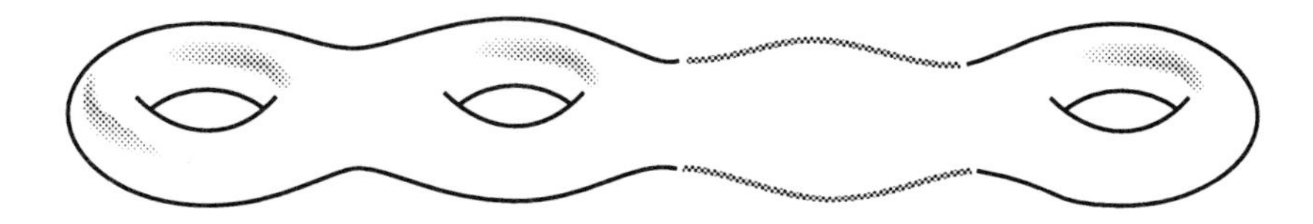

Fig. 4.10

course!). Thus thinking of the problem a bit more we might try to classify surfaces according to certain invariants such as the number of holes or the number of tori in the above examples. Then, given any surface, we would hope to be able to calculate its invariants and assign it some standard name such as an n-torus with m holes in it. This basic idea in fact works quite well but of course our view of it here is a bit primitive. For instance, we have used 'same as' several times without worrying about what that means. To get that point out of the way, we will be more precise and say that we want to classify surfaces 'up to homeomorphism' and so we may tacitly use phrases such as 'S_1 is the same as S_2' as a more relaxed way of saying 'There is a homeomorphism between S_1 and S_2'.

4.1 Combinatorial surfaces

We want to use geometric and combinatorial arguments wherever possible. To do this we are first of all going to consider combinatorial surfaces.

A *combinatorial surface*, S, is a finite union of (solid) triangles satisfying the following rules:

S.1: Any two triangles meet in one edge, one vertex, or not at all.

S.2: Any edge is an edge of one or two triangles.

S.3: For any vertex v, we write St v for the union of all triangles having v as a vertex; then St v is homeomorphic to a disc (St v is called the star of v).

S.4: Given any two vertices in S, then there is a path along edges of triangles joining them.

We now give some comments and further explanation of the rules.

S.1: Two of the triangles in S can meet as in Fig. 4.11, or as in Fig. 4.12, or they may not meet at all, as in Fig. 4.13.

Two triangles cannot meet as in Fig. 4.14, or as in Fig. 4.15, or as in Fig. 4.16.

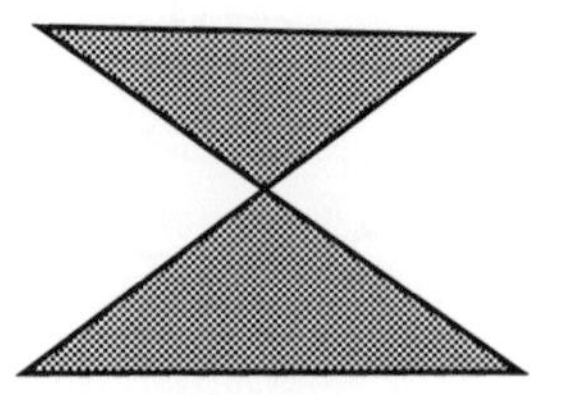

Fig. 4.11

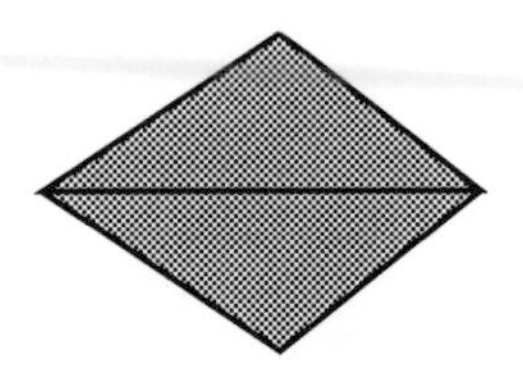

Fig. 4.12

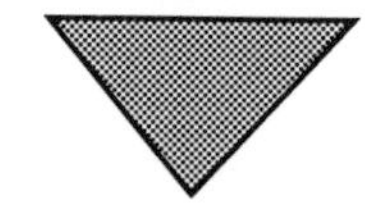

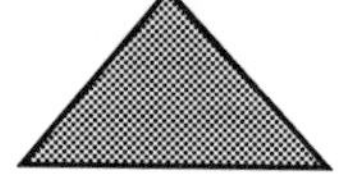

Fig. 4.13

S.2: An edge, e, might only be in one triangle. This happens if it is on the boundary of some hole (Fig. 4.17).

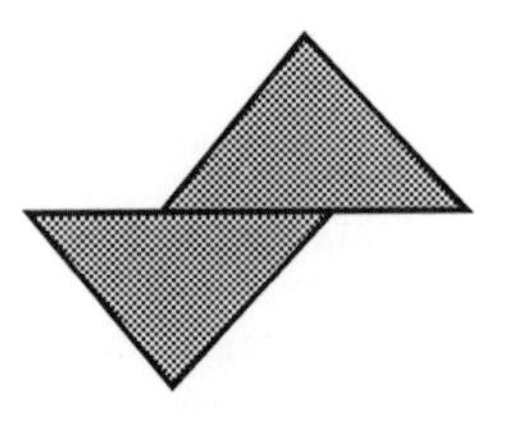

Fig. 4.14

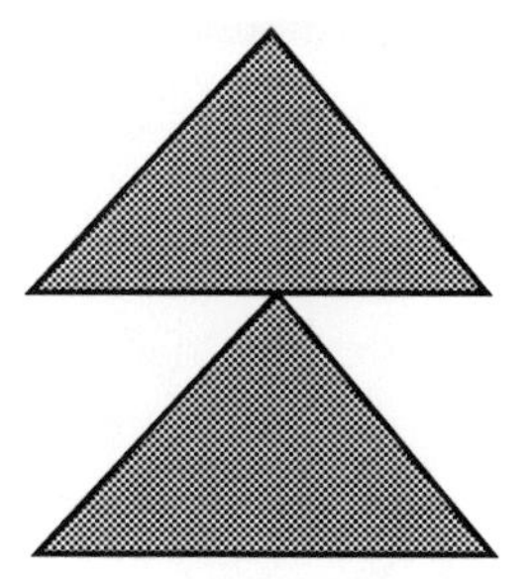

Fig. 4.15

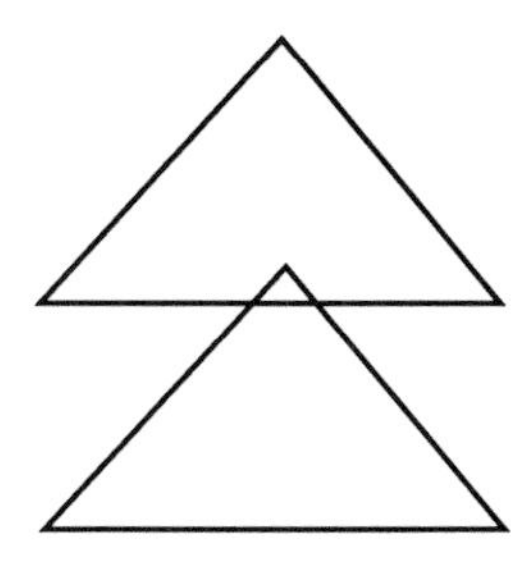

Fig. 4.16

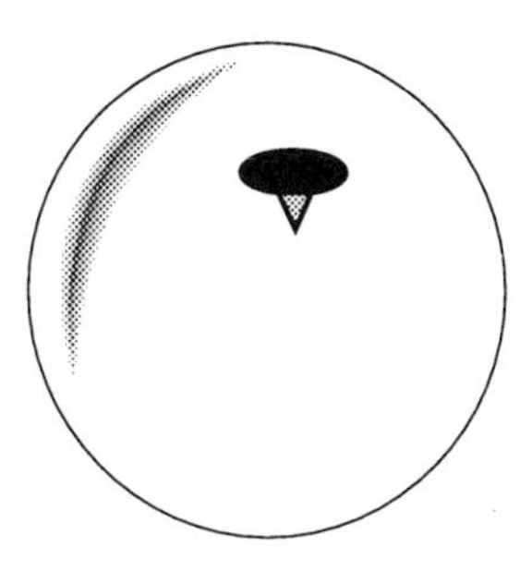

Fig. 4.17

Many edges will be in two triangles, but none may be in three or more.

S.3: The star of a vertex, v, will normally look as shown in Fig. 4.18. If the vertex is in a boundary edge, its star may be different, as in Fig. 4.19, but it cannot be as shown in Fig. 4.20.

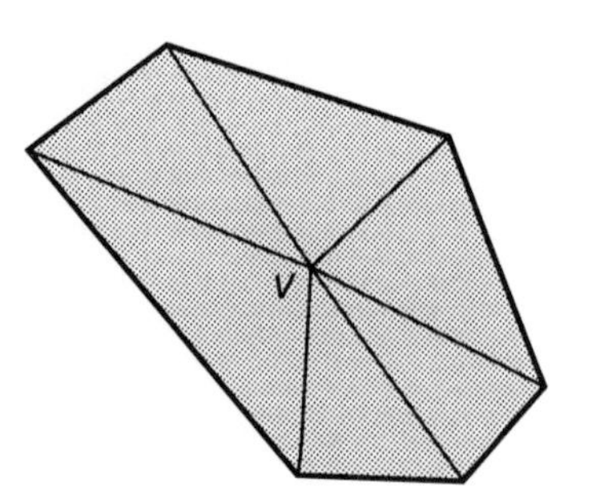

Fig. 4.18

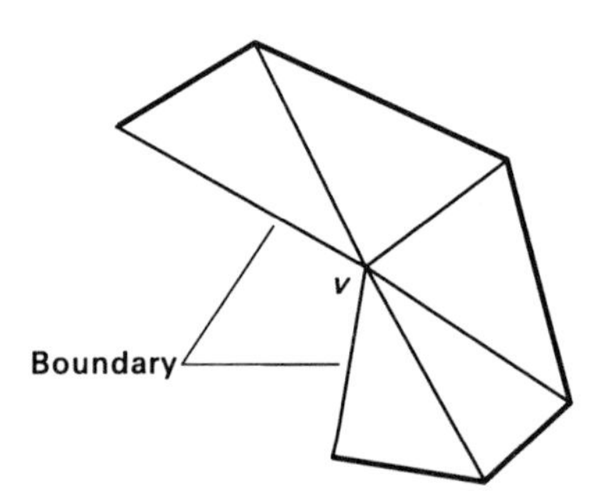

Fig. 4.19

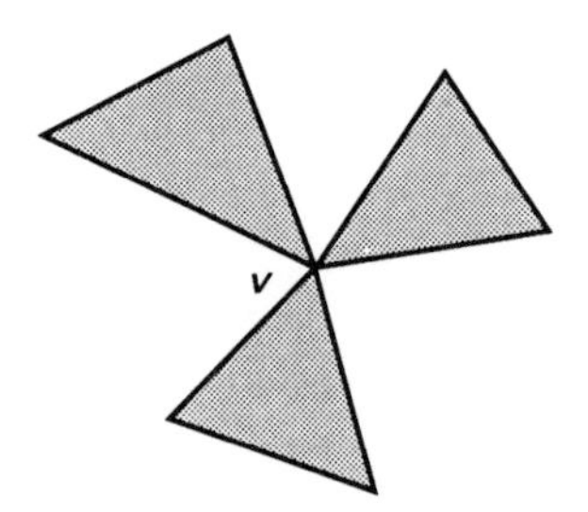

Fig. 4.20

A *path* in a topological space X is a continuous map a: $[0, 1] \to X$. We say a joins the points $a(0)$ and $a(1)$ and that X is path connected if any two points can be joined by a path; that is, given $x, y \in X$, there is a path a in X such that $a(0) = x$, $a(1) = y$. We will be looking

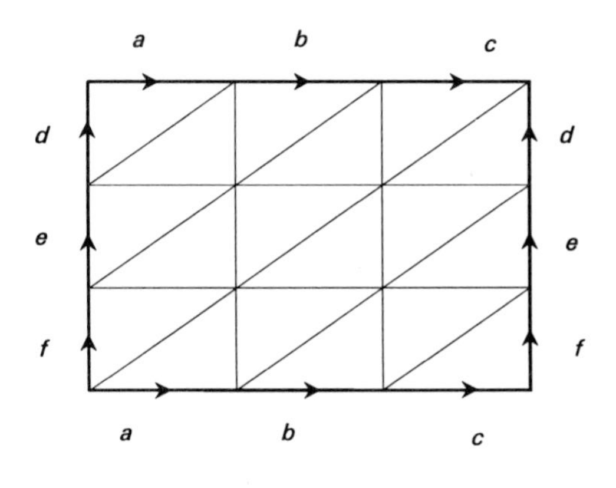

Fig. 4.21

at paths much more closely in Chapter 9. For the moment, we merely note that **S.4** implies that S is path connected.

As we saw in Chapter 3, we can indicate the construction of surfaces by diagrams giving gluing instructions, for example as in Fig. 4.21.

In such a diagram the edges with the same label are to be identified with the arrows matching. Forgetting for the moment the edges inside the rectangle, we can imagine performing these identifications within $\mathbb{R}^3$ (in this case) and the result is a torus (Fig. 4.22).

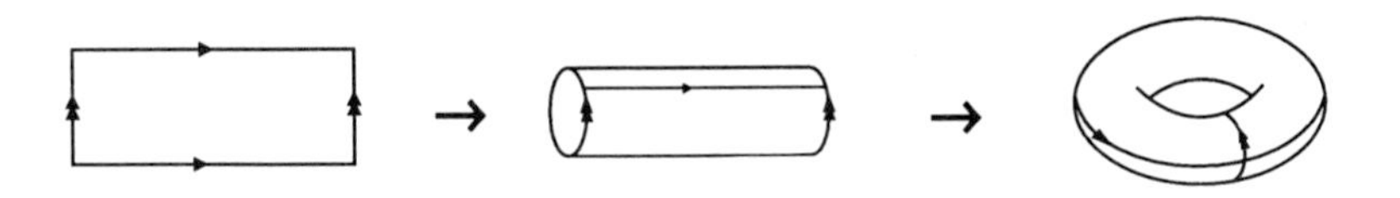

Fig. 4.22

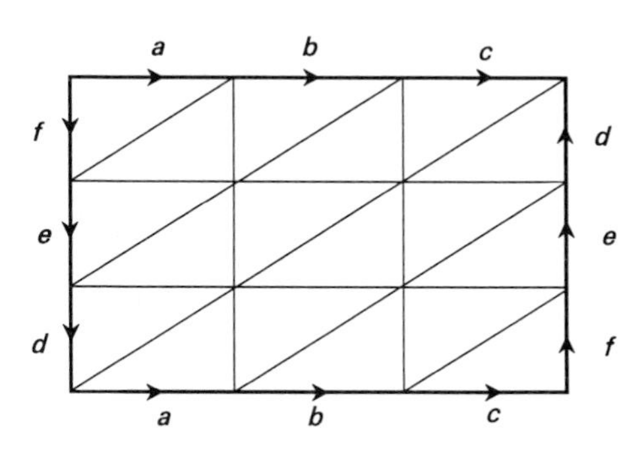

Fig. 4.23

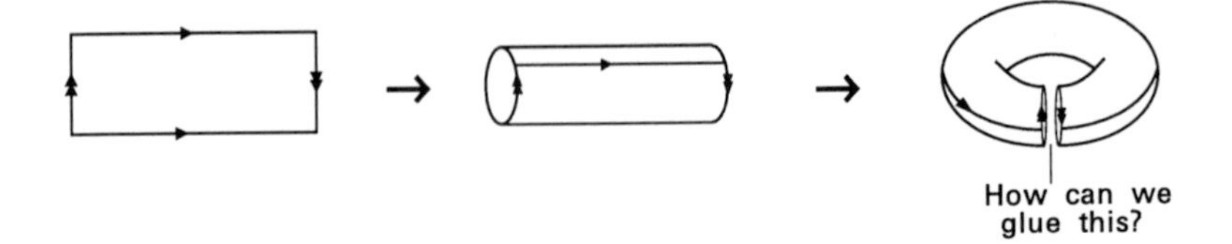

Fig. 4.24

The reason why we had to put 'in this case' is that a different diagram such as in Fig. 4.23 still yields a surface, the Klein bottle (Fig. 4.24), but the identification cannot be done in $\mathbb{R}^3$.

To obtain some idea of what this space looks like it is usual to draw it with a self-intersection as in Fig. 4.25.

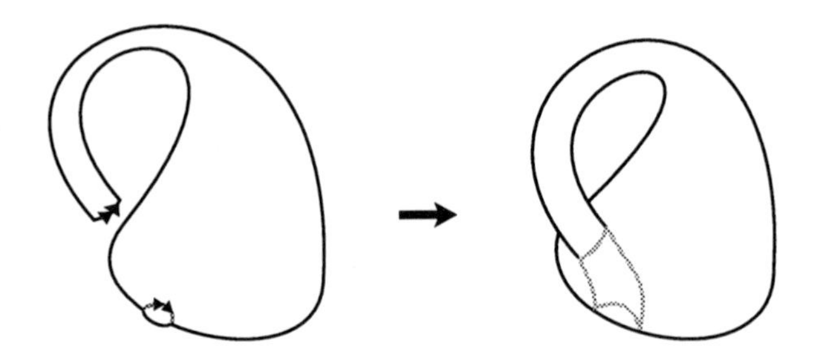

Fig. 4.25

We will see some of the strange properties of the Klein bottle later. To see the restrictions that the rules place on the way the triangles fit together, let us try to triangulate the cylinder (represented as a rectangle with identification in Fig. 4.26).

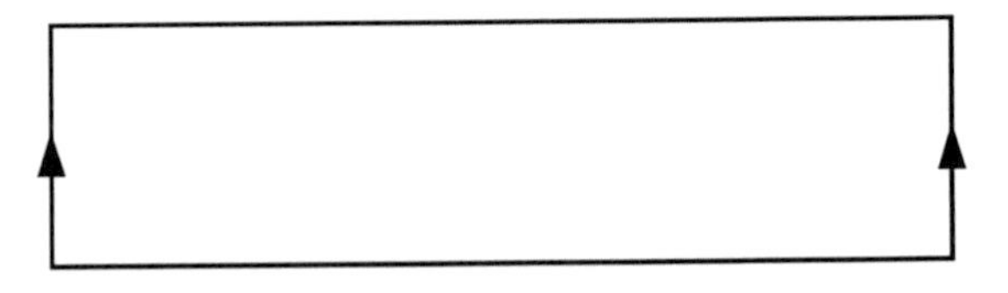

Fig. 4.26

Attempt 1 (Fig. 4.27)

This simple decomposition contravenes **S.1** as the two triangles meet in two edges, so it does not give a triangulation.

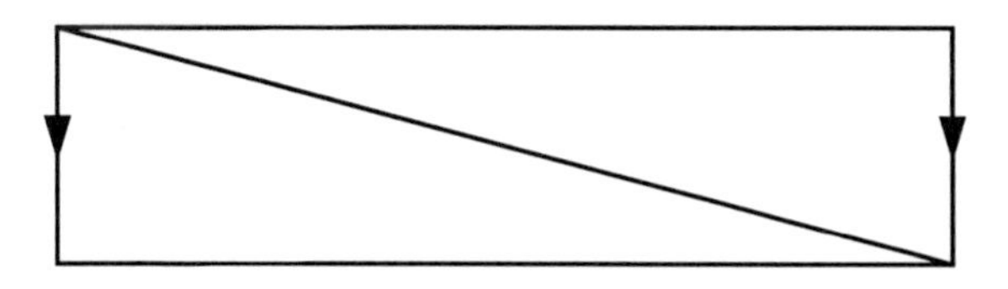

Fig. 4.27

Attempt 2 (Fig. 4.28)

Triangles 1 and 2 meet in three vertices, so this also contravenes **S.1**.

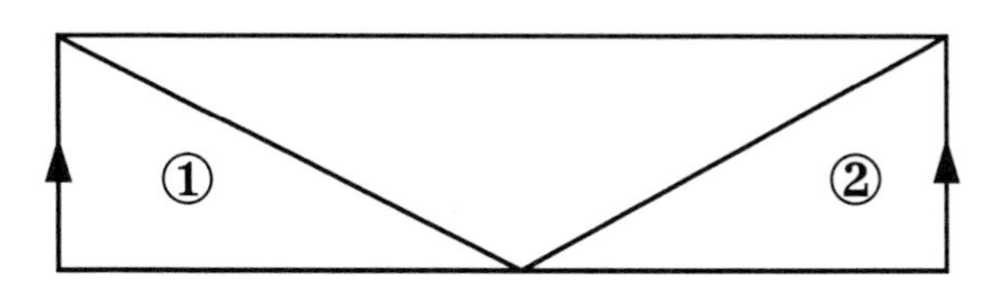

Fig. 4.28

Attempt 3 (Fig. 4.29)

Triangles 1 and 3 meet in two vertices, so again this is disallowed.

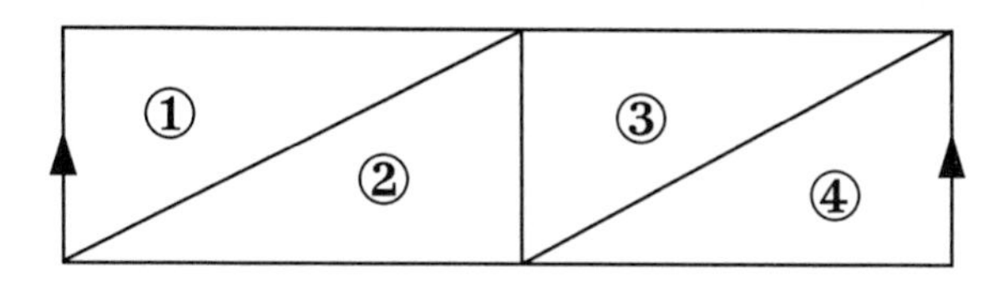

Fig. 4.29

Attempt 4 (Fig. 4.30)

Again the intersection of two of the triangles (2 and 4 this time) consists of just two vertices.

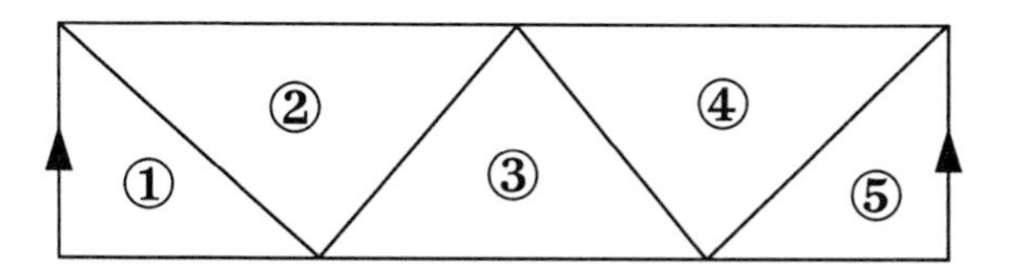

Fig. 4.30

Attempt 5 (Fig. 4.31)

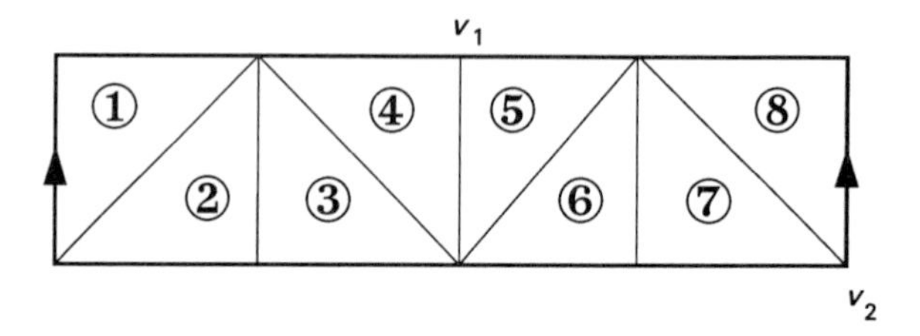

Fig. 4.31

S.1 is now satisfied as is **S.2**. Path connectedness is clear, so we are left to check **S.3**, the condition on stars of vertices. We leave you to check this in detail but it is clear that St v_1 is the union of triangles 4 and 5, so is a triangle and thus homeomorphic to a disc, whilst St v_2 consists of triangles 1, 2, 7, and 8 and so similarly is a rectangle and hence homeomorphic to a disc. This leaves six other vertices for you to check.

So we managed to find a triangulation of the cylinder. It was not as obvious as it may first have seemed. We do not claim to have found the smallest number of triangles. Can you improve on our number of eight? What is the smallest number of triangles needed to get a triangulation of the cylinder? How might you attack this question?

Our main aim will be to classify surfaces up to homeomorphism. To do this we will assume that we have triangulated our surface, but we are interested in the surface, not primarily in the triangulation, and so we can and will change the triangulation as needed, perhaps chopping up some of the triangles into smaller triangles (subdivision). Our search will be for invariants of the surface, that is invariants that do not depend on which triangulation we are using. We will also see certain moves on combinatorial surfaces which change the triangulation. This is much like the situation we saw earlier where moves on knot diagrams allowed us to check if certain quantities were invariants.

The first invariant property we will look at is orientability. There are two basic classes of surfaces: those which are orientable and those which are non-orientable. For instance, the sphere, torus, and multiple tori are orientable whilst the Möbius band and the Klein bottle are non-orientable. We met the Möbius band in Chapter 3. It has a strange single-sidedness property. You can draw a line from any point on the surface to any other without going over an edge. The same is true of the Klein bottle. These strange properties are due to the non-orientability of these surfaces.

Another important non-orientable surface is the projective plane. This first arose in projective geometry. The set $\mathbb{P}^2$ is the set of lines through 0 in $\mathbb{R}^3$. This set can be thought of in several other ways. We can consider the set of non-zero points in $\mathbb{R}^3$ and put on them an equivalence relation defined by

$$(x, y, z) \sim (x', y', z')$$

if there is some non-zero $\lambda \in \mathbb{R}$ such that $\lambda x = x'$, $\lambda y = y'$, $\lambda z = z'$, that is (x, y, z) and (x', y', z') are on the line through 0. The set of equivalence classes is $\mathbb{P}^2$. Alternatively since each line through the origin in $\mathbb{R}^3$ cuts the unit sphere, S^2, in two points, (x, y, z) and $(-x, -y, -z)$ say (where $x^2 + y^2 + z^2 = 1$), we can consider an equivalence relation on S^2 generated by

$$(x, y, z) \sim (-x, -y, -z).$$

Again the set of equivalence classes form $\mathbb{P}^2$. This gives us a clue as to how to give $\mathbb{P}^2$ a topology. If (x, y, z) is in S^2, write $[x, y, z]$ for the corresponding point of $\mathbb{P}^2$. This gives a function $f: S^2 \to \mathbb{P}^2$ defined by $f(x, y, z) = [x, y, z]$. If U is a set in $\mathbb{P}^2$ and $[x, y, z] \in U$, then $f^{-1}(U)$ contains both (x, y, z) and $(-x, -y, -z)$. If $f^{-1}(U)$ is open in S^2, then we will say it is open in $\mathbb{P}^2$. It is a good exercise to check that this does give a topology on $\mathbb{P}^2$.

A final way of looking at $\mathbb{P}^2$ is to note that the set

$$D^2_+ = \{(x, y, z) \mid (x, y, z) \in S^2, z \geqslant 0\}$$

is a disc. Each equivalence class $[x, y, z]$ with $z \neq 0$ meets D^2_+ in one point, but all the $[x, y, 0]$ meet D^2_+ in two diametrically opposite points. Thus $\mathbb{P}^2$ can be obtained from a disc by identifying according to the rule shown in Fig. 4.32. This surface cannot be embedded in $\mathbb{R}^3$.

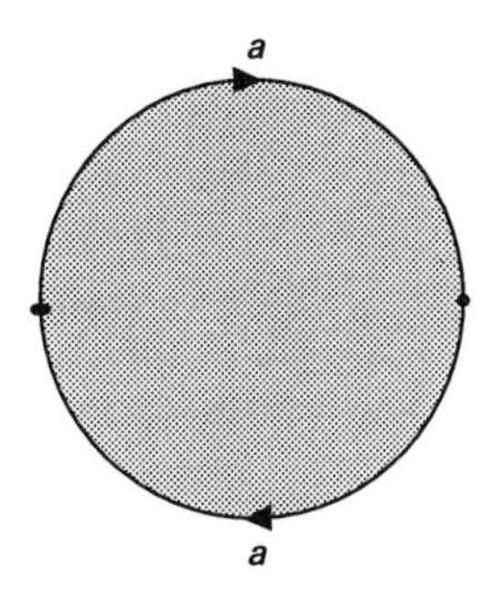

Fig. 4.32

We must now examine what orientability means. As our surfaces are assumed to be triangulated we will first consider what orientation of a triangle means.

The intuitive answer must be that there are two choices of orientation, as shown in Figs. 4.33 and 4.34.

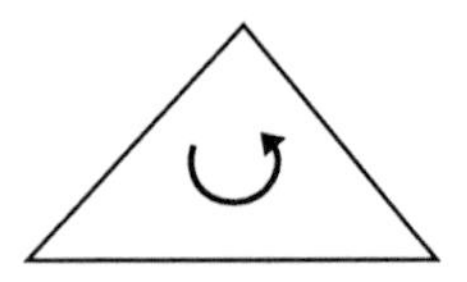

Fig. 4.33

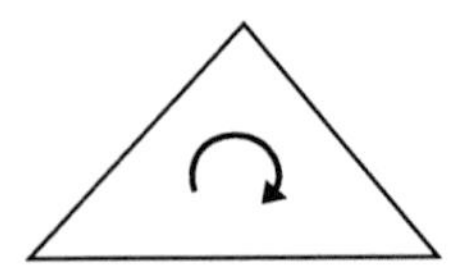

Fig. 4.34

An orientation on a triangle induces one on its edges, as in Fig. 4.35. (Thinking of the triangle as being full of a swirling liquid may help!)

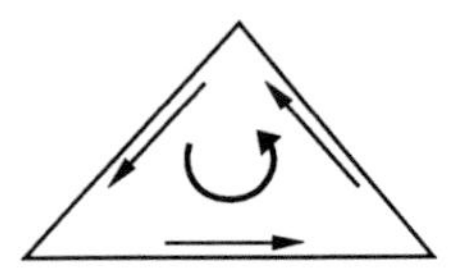

Fig. 4.35

Two triangles are adjacent if they have a common edge. Looking at the diagrams in Figs 4.36 and 4.37 we see that adjacent triangles with similar orientations induce different orientations on their common edge, whilst if the induced directions are the same then the triangles are oppositely oriented.

A surface S is *orientable* if one can choose orientations for all triangles in S such that adjacent triangles are similarly oriented. Otherwise the surface is non-orientable.

Suppose someone presents us with a combinatorial surface; how are we to find out if it is orientable or not? We clearly cannot try all the possible combinations of orientations on the different triangles.

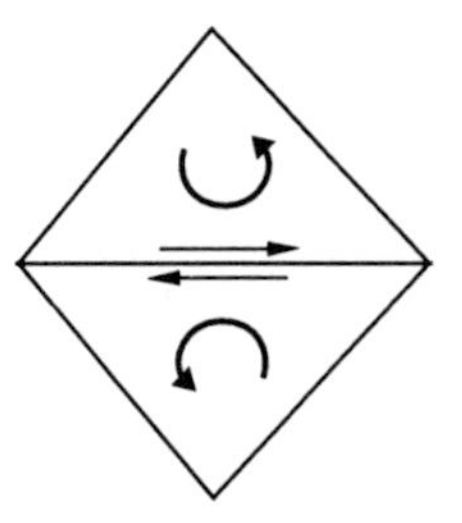

Fig. 4.36

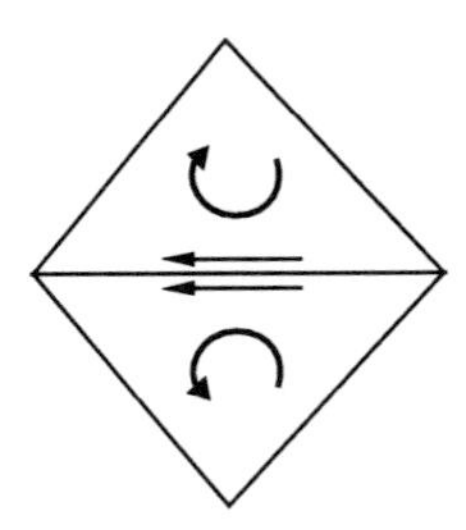

Fig. 4.37

The way to do it is to propagate an orientation starting at some point.

Pick any triangle in S and orient it in either sense. Now go to each of the adjacent triangles and orient them similarly (if you can). Go on to any new triangles adjacent to those already oriented and orient them similarly if possible and so on. Either eventually you will have given an orientation to all the triangles and end up with an orientation of S or at some stage you end up with a situation like that in Fig. 4.38.

There cannot exist an orientation of the new triangle similar to

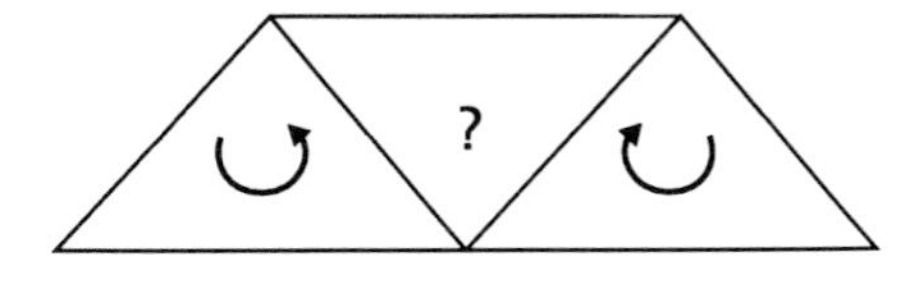

Fig. 4.38

both those already chosen, so the surface must be non-orientable. These cases are mutually exclusive.

As a simple example consider the triangulation of the Möbius band in Fig. 4.39.

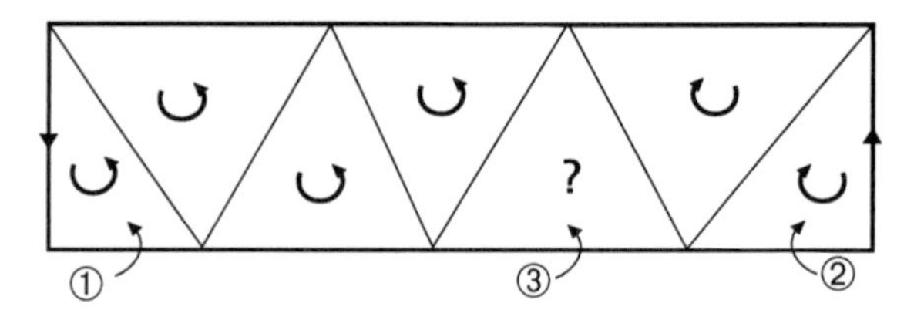

Fig. 4.39

Starting with an anticlockwise orientation of 1, we propagate. (Why is 2 oriented in a clockwise way?) It is impossible to choose an orientation for 3 which is similar to that on adjacent triangles.

We will return later on to look in more detail at orientations.

4.2 Cutting and pasting

The method we will use for analysing surfaces further is called *cutting and pasting*. The idea is simple. Label both sides of each edge and then cut along the edge so as to get a collection of triangles. The fact that the edges are labelled means that we can reconstruct the surface by gluing along edges having the same label. The usefulness of this is that by choosing a neat way of regluing step by step we can obtain a convenient representation of the surface as a polygon with identified edges. To be slightly more formal about this process we will give a step-by-step description of it.

Step 1

Label and orient each edge of all triangles in S with a letter and an arrow. Distinct edges must be labelled with distinct letters. We let

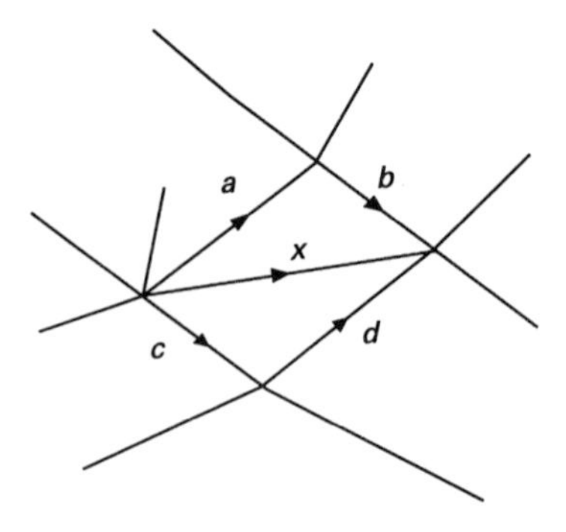

Fig. 4.40

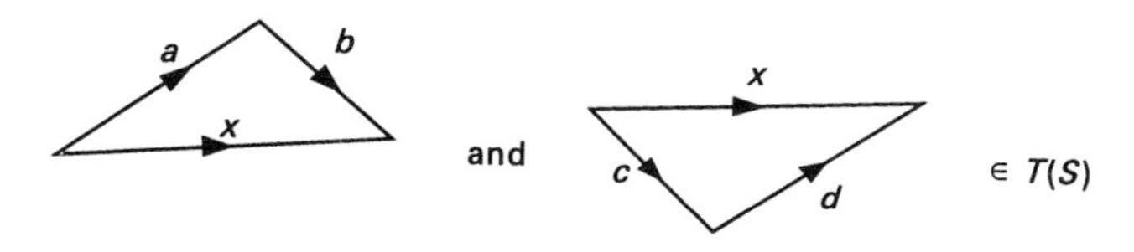

Fig. 4.41

$T(S)$ be the set of all labelled triangles in S. For example, Fig. 4.40 gives Fig. 4.41.

Clearly we can reconstruct S from $T(S)$.

Step 2

Pick any triangle $\tau_0 \in T(S)$. Find a $\tau_1 \in T(S)$ which has an edge label in common with τ_0. Stick τ_0 and τ_1 together, matching the common edge by letter and arrow to form a disc, ρ_1. Find $\tau_2 \in T(S) \backslash \{\tau_0, \tau_1\}$ with one or two edge labels in common with adjacent edge labels in the boundary of ρ_1. Stick τ_2 to ρ_1 to get a disc ρ_2. Continue in this way matching $\tau_i \in T(S)$ to one or two adjacent labelled edges of ρ_i to form a disc ρ_{i+1} until $T(S)$ is exhausted. The final result is a polygon with labelled oriented edges. The process can be illustrated as shown in Figs 4.42–4.48.

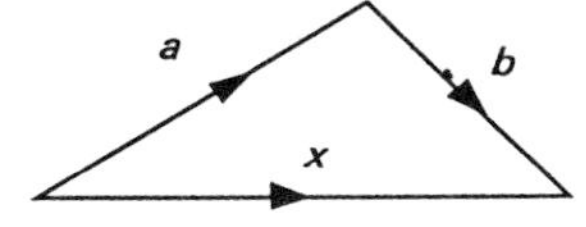

Fig. 4.42

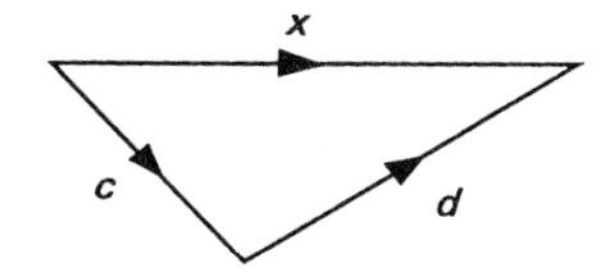

Fig. 4.43

As we always glue along a proper connected segment of the boundary we can never get a hole. (The proof that this always gives a disc is too technical to be included here.)

A typical polygon one might obtain would be something like that in Fig. 4.49.

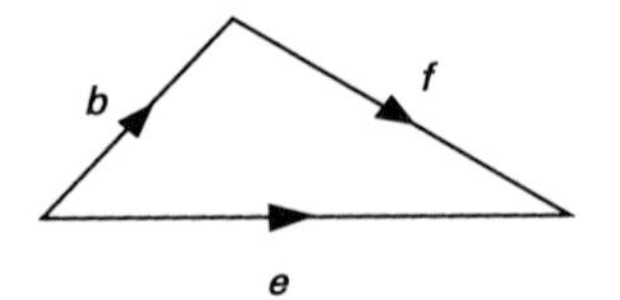

Fig. 4.44

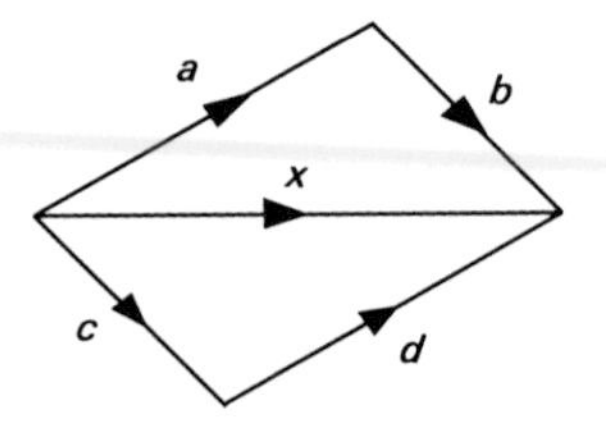

Fig. 4.45

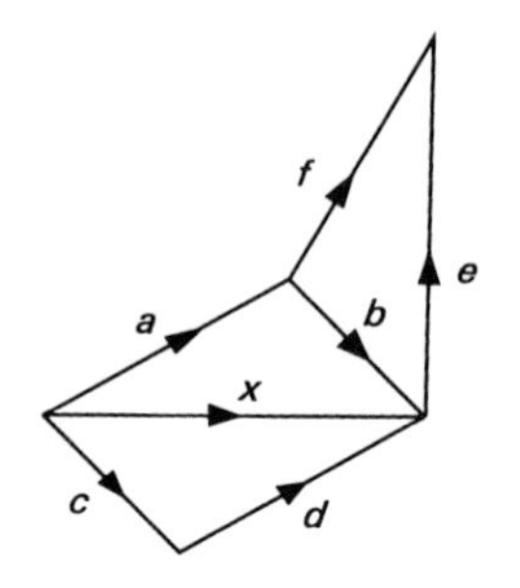

Fig. 4.46

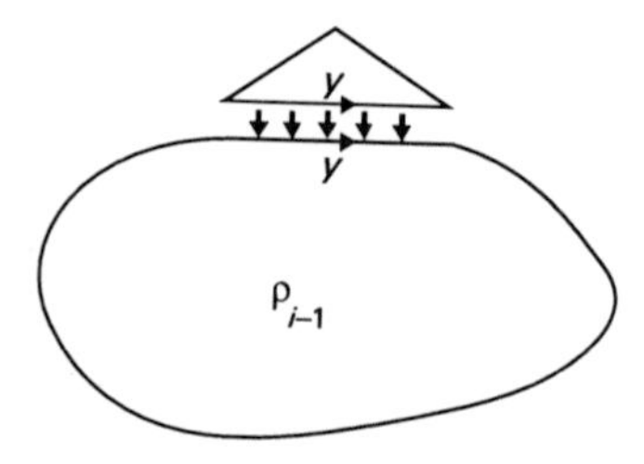

Fig. 4.47

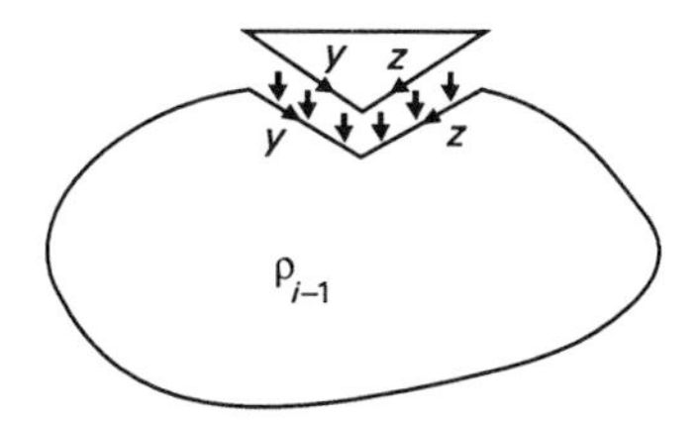

Fig. 4.48

Labels can occur once or twice but never more since each label corresponds to an edge of one or two triangles, never more.

The information stored in such a diagram tells us how to identify edges and hence to reconstruct S. However, the key information is contained in the letters a, b, etc., of the boundary edges of the polygon and the order and direction in which they go; this is encoded in the surface symbol of S.

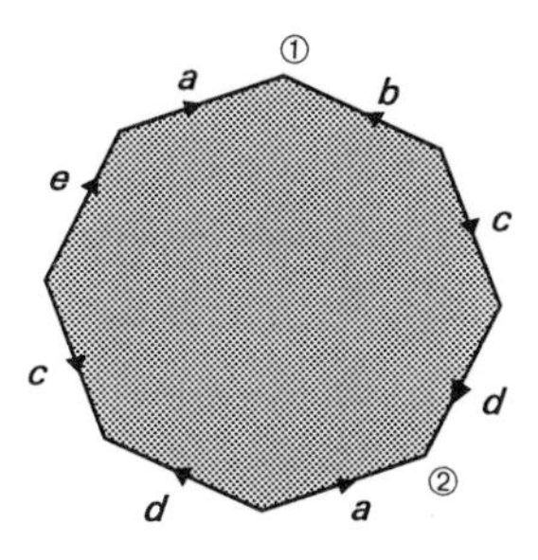

Fig. 4.49

To work out the surface symbol of a polygon, pick any vertex (we will use 1 in the diagram in Fig. 4.49); then, reading off clockwise, write down the edge label if the arrow is pointing clockwise, or write down that label to the power minus one if the arrow is pointing anticlockwise. Thus in our diagram, we get

$$b^{-1}cda^{-1}dc^{-1}ea.$$

The choice of start vertex is ours to make so we might have got (by starting at 2)

$$a^{-1}dc^{-1}eab^{-1}cd$$

and we must therefore consider these two symbols to be 'the same', that is equivalent. We will play with these symbols much as we would play with relations in a group presentation (see Chapter 6) or with a knot diagram using Reidemeister moves. We must, of course, specify the rules of the game we shall play.

Step 3: **operations**

(We will write capital letters for words in small letters and their inverses.)

Operation 1 (Fig 4.50)

$$ABC \underset{\text{paste}}{\overset{\text{cut}}{\rightleftarrows}} AxC \text{ and } Bx^{-1}$$

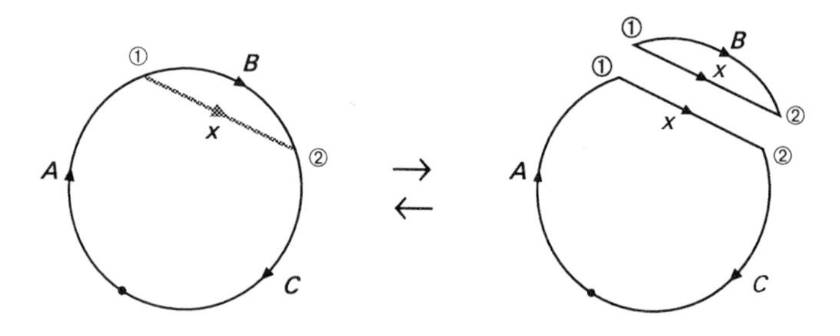

Fig. 4.50

The usefulness of this operation is that one can cut along one edge and paste back along another one, provided of course that labels and arrows permit this. We will see this in action shortly.

Operation 2 (Fig. 4.51)

$$Axx^{-1}B \underset{\text{cut}}{\overset{\text{paste}}{\rightleftarrows}} AB$$

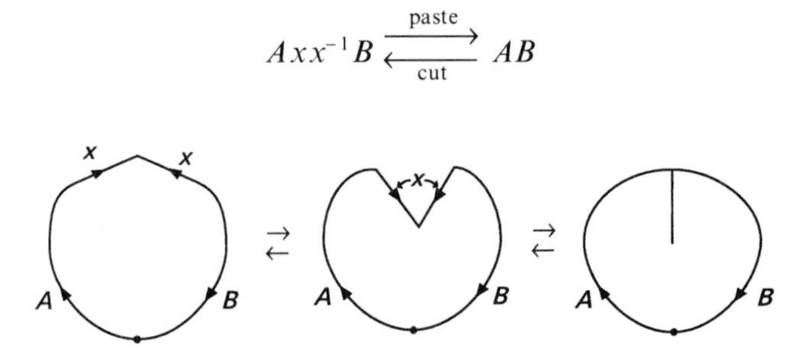

Fig. 4.51

Operation 3

$$x \longrightarrow uv$$

where u, v occur nowhere else in the symbol and the replacement is at every occurrence of x.

The inverse of this is

$$uv \longrightarrow x$$

and can be done if u and v always occur together in the symbol. (Here the order is very important: uv or $v^{-1}u^{-1}$ is all right but vu or $u^{-1}v^{-1}$ does not count.)

This corresponds to splitting an edge (Fig. 4.52).

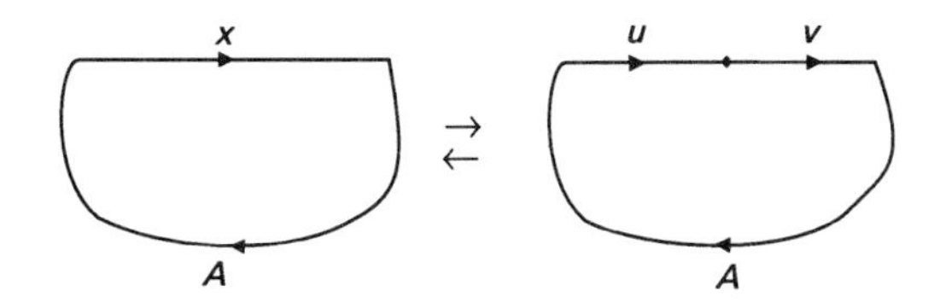

Fig. 4.52

Remark on the operations

You may be wondering why, in operation 1, we can assume that only a single edge is needed to join the two ends of B. We mentioned earlier the idea of chopping or subdividing the triangulation further. By doing this if necessary we can find a path along edges inside the polygon, joining the vertices 1 and 2. Cutting along this path we have a series of labels which always occur in the same order and so we can apply operation 3 to them to replace them by a single edge label. This may alter our triangulation slightly but this does not affect the arguments or ideas we will be using later on.

Composite operations

By combining the above elementary operations we can build up some very useful composite operations. The validity of these moves is indicated by diagrams.

Operation 4

Replace $ABxCxD$ by $AyCB^{-1}yD$, where y is a new symbol not in A, B, C, or D (Fig. 4.53).

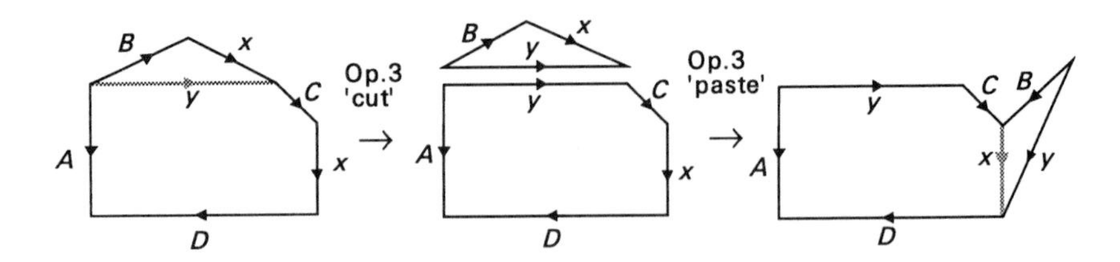

Fig. 4.53

Operation 5

Replace $ABxCx^{-1}D$ by $AyCy^{-1}BD$, where y is a new symbol not in A, B, C, or D.

The proof is much as above, so try and draw the sequence of diagrams for yourselves.

Operation 6

Replace $AxBCxD$ by $AyCyB^{-1}D$.

Operation 7

Replace $AxBCx^{-1}D$ by $AyCBy^{-1}D$.

Again for these two composite operations, we leave it up to you to draw the diagrams which check the validity of the operation. This is an extremely useful exercise to do, especially if you do not see immediately why one of the operations is valid.

Step 4

We call this step assembling the crosscaps. A *crosscap* is a projective plane contained in S. The projective plane has symbol aa and in any surface symbol the occurrence of a repeated letter $\ldots a \ldots a \ldots$ or $a^{-1} \ldots a^{-1} \ldots$ indicates the presence of a projective plane and hence of non-orientability (more on this later). Our first task now that we have these composite operations at hand is to get all the crosscaps to the start of the symbol.

We note that if we have $ABxCxD$ in a symbol, we can use operation 4 to replace it by $AyCB^{-1}yD$ and then operation 6 to obtain $AzzBC^{-1}D$. (In operation 6 we take C to be empty and B to be CB^{-1}.)

Now suppose we can find some recurring letter, say s, in our symbol. Taking A to be empty, B the block of letters before the first s, and so on, we can use the above pair of operations to replace the symbol with one of the form ttB', that is one which starts with a crosscap.

Next examine B' to see if some other letter recurs. If it does (e.g. we might find $B' = BuCuD$, for new meanings of B, C, and D), we can apply the pair of operations, this time with $A = tt$ to replace the symbol with $ttvvBC^{-1}D$. Repeating this process as many times as proves necessary, we will eventually get the symbol in the form

$$a_1a_1a_2a_2 \ldots a_na_nB$$

where B has only edges that occur singly or as $\ldots x \ldots x^{-1} \ldots$.

Step 5

A torus has, as we have seen, a description as a rectangle with opposite sides identified. Thus a surface symbol for the torus will be $aba^{-1}b^{-1}$. If a surface symbol contains the pattern

$$\ldots a \ldots b \ldots a^{-1} \ldots b^{-1}$$

it means that there is an attached torus hidden in the surface. As an attached torus looks like a handle, we say that the surface contains a *handle* if such a pattern is contained within the symbol (Fig. 4.54)

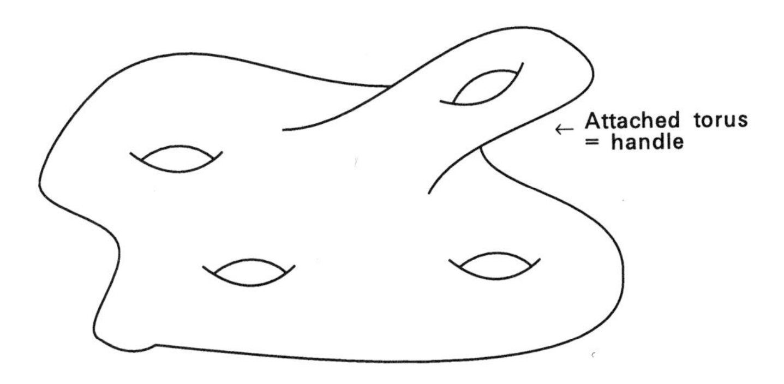

Fig. 4.54

The fifth step is to bring any handles together and then collect them up just behind any crosscaps. To do this we use the following series of operations

$$\begin{aligned} ABaCbDa^{-1}Eb^{-1}F &\xrightarrow{\text{Op. 5}} AxCbDx^{-1}BEb^{-1}F.\\ &\xrightarrow{\text{Op. 7}} AxCyBEDx^{-1}y^{-1}F\\ &\xrightarrow{\text{Op. 7}} AuBEDCyu^{-1}y^{-1}F\\ &\xrightarrow{\text{Op. 5}} Auvu^{-1}v^{-1}BEDCF. \end{aligned}$$

So taking A to be the string of crosscaps, B to be the block before the first letter of the chosen handle, and so on, we can manipulate the symbol so that it has the form

$$a_1 a_1 \ldots a_n a_n u v u^{-1} v^{-1} B'.$$

Of course B' may still contain handles, so we repeat as many times as is necessary to get the symbol in the form

[crosscaps] [handles] [the remainder].

Step 6

You might expect us to attack 'the remainder' next; however, there is a curious phenomenon that occurs if both crosscaps and handles are together in a symbol. The crosscaps influence the handles, changing each handle to a pair of crosscaps, that is

crosscap + handle = 3 crosscaps.

The sequence of moves necessary to check this is

$$aabcb^{-1}c^{-1} \xrightarrow[\text{of Op. 8}]{\text{reverse}} bdbc^{-1}dc^{-1}$$

followed by the collection process of the crosscaps as described previously. It is a valuable exercise to write down all the details to check that $aabcb^{-1}c^{-1}$ can be replaced by $xxyyzz$.

Using this as many times as need be, we can reduce any symbol either to

$$a_1 a_1 \ldots a_n a_n H \quad \text{as soon as any } \ldots a \ldots a \ldots \text{ occurs}$$

or otherwise

$$b_1 c_1 b_1^{-1} c_1^{-1} \ldots b_m c_m b_m^{-1} c_m^{-1} H \quad \text{if no } \ldots a \ldots a \ldots \text{ occurs.}$$

In both cases H contains no crosscaps and no handles. (It may be empty in which case we stop.)

Step 7: assembling the cuffs

We now need to look at H which we shall assume is non-empty. One obvious form we might expect to find in H is a *cuff* corresponding to cdc^{-1} (Fig. 4.55).

Is this the only form within H? To find out we suppose x occurs twice in H; thus we have either $\ldots x \ldots x^{-1}$ or $\ldots x^{-1} \ldots x \ldots$. Choose such an x with as few letters between x and x^{-1} as possible. Then the letters occurring between x and x^{-1} can occur nowhere else.

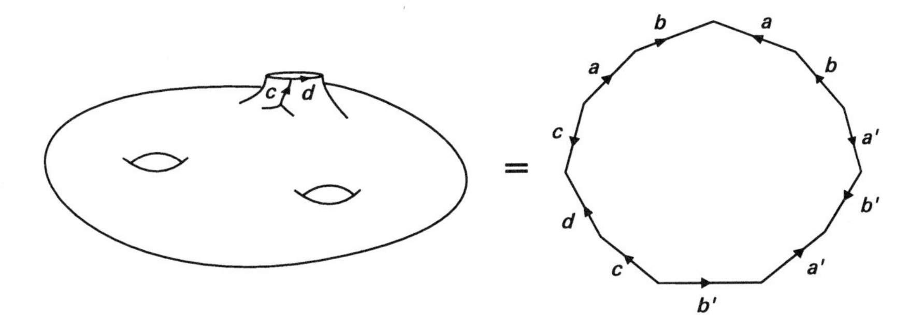

Fig. 4.55

(Why? Remember there are no crosscaps or handles in H.) We can therefore replace (using operation 3) this string of letters from x to x^{-1} by a single letter

$$\dots x \dots x^{-1} \dots \longrightarrow \dots xbx^{-1} \dots .$$

Then H becomes $Axbx^{-1}B$ which in turn can be changed to $yzy^{-1}AB$ (by which operations?).

Continuing in this way we can replace H by

$$y_1z_1y_1^{-1}y_2z_2y_2^{-1} \dots . \, y_sz_sy_s^{-1}W.$$

Now W may be empty in which case we stop and breathe a sigh of relief. If, however, W is not empty, it must consist of single occurrences of letters. These can be amalgamated using operation 3, being replaced by a single letter, w say, and will eventually give us another cuff. At the moment, however, it is more a hole than a cuff. To make another cuff we use operation 2 to add $u^{-1}u$ at the end of the symbol; thus we now have $AHwu^{-1}u$ which, since cyclic permutations of the letters make no difference (corresponding to a change of starting vertex), we can replace by $uAHwu^{-1}$. Now A consists of crosscaps or handles, whilst H consists of cuffs. If a_1a_1 is the first crosscap in A, we get

$$ua_1a_1 \longrightarrow a_1'u^{-1}a_1' \longrightarrow a_1''a_1''u$$

(using which operations?). Repeating as many times as is necessary we can pass u through the crosscaps. If A consists of handles, we note that $ubcb^{-1}c^{-1}$ can be replaced by $fgf^{-1}g^{-1}u$ (again we leave it up to you to work out why), so that we may pass u through all the handles.

Finally we use that $uyzy^{-1}$ can be replaced by $vzv^{-1}u$ to show that u can be passed through all the cuffs. Thus from $uAHwu^{-1}$ we obtain $AHuwu^{-1}$, that is another cuff as promised. We have proved the following proposition.

4.1 Proposition. *Any surface symbol can be transformed by the operations* 1, 2, *and* 3 *to one of the form AH where A is a string of crosscaps or handles (not both) and H is a string of cuffs. (This form is called the normal form of the surface symbol.)*

If any two surfaces S_1 and S_2 yield surface symbols with the same normal form then as their polygons are homeomorphic by a homeomorphism which recognizes the labels and orientations of each edge (i.e. the instructions for gluing the polygon to get the surface), the two surfaces must themselves be homeomorphic. As the converse is clearly true—homeomorphic surfaces have symbols with the same normal form—we get the following first form of the classification theorem.

4.2 The classification theorem for surfaces. *Two surfaces are homeomorphic if and only if the normal forms of any symbols representing them have the same number of crosscaps/or handles and also the same number of cuffs.*

Examples of normal form

(a) $\varnothing$ the empty normal form. In this form it is hard to visualize, but using operation 2 we can replace it by aa^{-1}. This is a sphere.

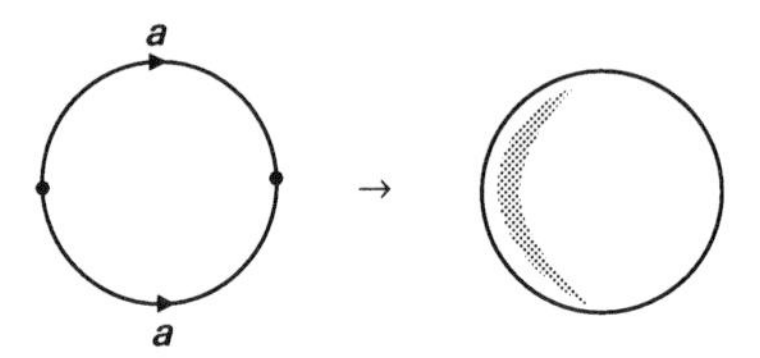

Fig. 4.56

(b) yzy^{-1}.

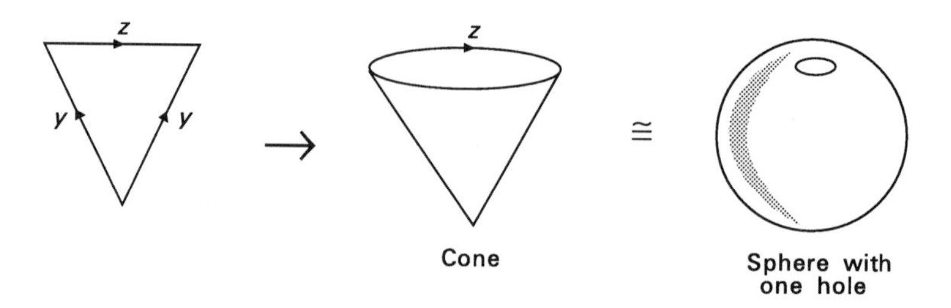

Fig. 4.57

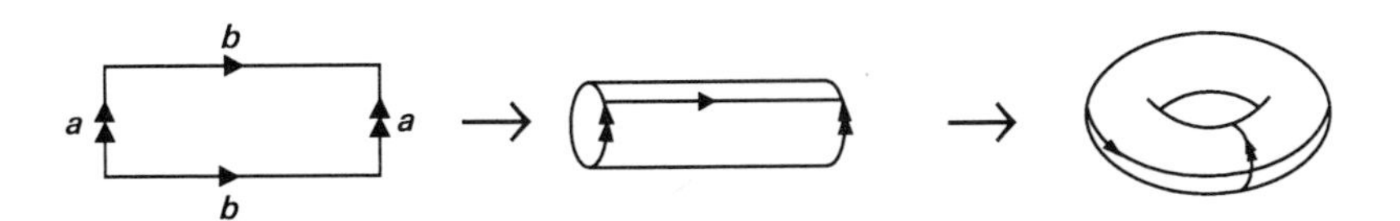

Fig. 4.58

(c) $aba^{-1}b^{-1}$. As we have already seen, this is a torus (Fig. 4.58).
(d) aa is the projective plane, $\mathbb{P}^2$
(e) $a_1a_1a_2a_2$ two crosscaps = Klein bottle.
This is not obvious, but look at Fig. 4.59.

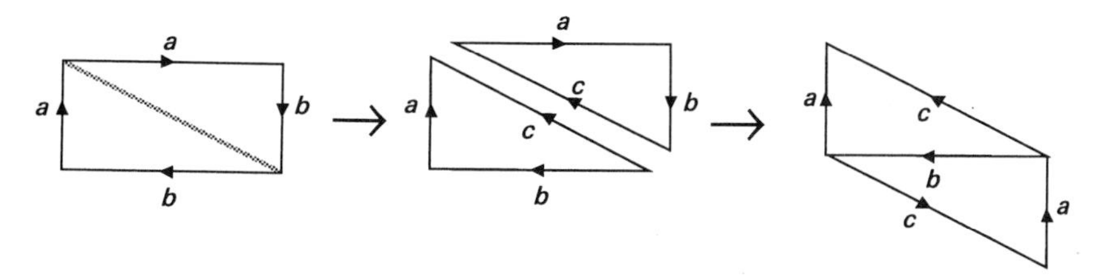

Fig. 4.59

Exercises 4.2

1. Classify the symbol $b^{-1}cda^{-1}dc^{-1}ea$.
2. The Möbius band can be represented as shown in Fig. 4.60. Find the normal form of its surface symbol $abac$.

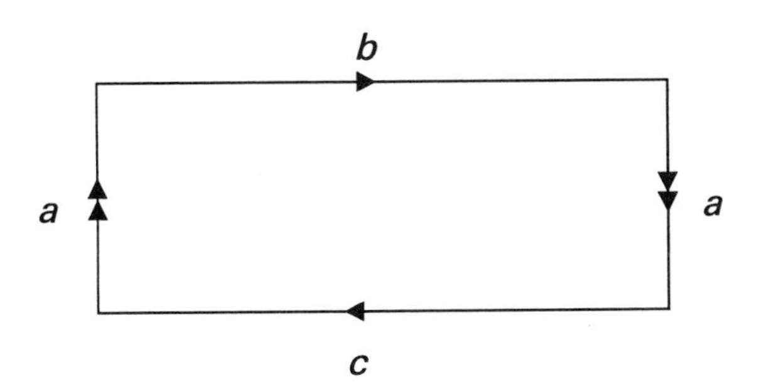

Fig. 4.60

3. Find the normal form of the following symbols:

(a) $a_1a_2a_3a_1^{-1}a_2^{-1}a_3^{-1}$

(b) $a_1a_2a_3a_4a_1^{-1}a_2^{-1}a_3^{-1}a_4^{-1}$

(c) $a_1a_2a_3a_1^{-1}a_2^{-1}a_3$

(d) $a_1a_2a_3a_4a_1^{-1}a_2^{-1}a_3^{-1}a_4$

When you have done a few of these calculations and you can imagine the tedium of performing all the operations step by step on some long surface symbol, you will probably find yourself asking: is there a simpler way of doing this? Happily the answer is yes. The use of invariants of surface symbols allows the normal form to be found very quickly and easily, so let us get on with this straight away.

4.3 Invariants of surfaces: orientability and boundary curves

If we choose a triangulation, we can get a surface symbol and thus eventually a normal form. We can use the triangulation or the surface symbol to obtain certain information on the surface, but any information we thus obtain should not depend on the choices we have made. For instance, the number of letters that occur in a surface symbol is clearly no good as a measure of the geometric properties of the surface since if we have a disc, we can subdivide it in many different ways to get different numbers of edges, but topologically it is still a disc.

Orientability revisited

The first invariant property is one we have already considered briefly, namely orientability.

4.3 Proposition. *A surface S is non-orientable if and only if the normal form of a surface symbol for S contains a crosscap. This occurs if and only if any symbol for S contains a crosscap.*

Proof Suppose in a symbol for S we have a crosscap, so somewhere in S we have the pattern $\ldots x \ldots x \ldots$. Choose a path within the polygon between these two occurrences of x, passing through the edges of triangles and through each triangle at most once. Orient successive triangles along this path similarly; one finds that the two triangles that share the edge x are oppositely oriented (Fig. 4.61), so the surface must be non-orientable.

If, on the other hand, no such repeated x occurs, then as the operations we have used can neither create nor destroy orientability we find no crosscaps in the normal form. However, we can easily construct an orientation for any surface whose normal form has only

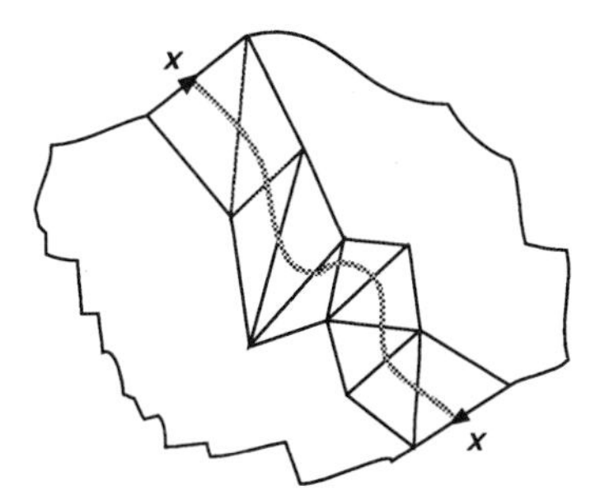

Fig. 4.61

handles or cuffs. (Try some simple examples such as a torus and a torus with one cuff etc., to see how you might find an orientation in general.)

The number of boundary curves

In the normal form it is clear that the number of cuffs is an important source of information. This number is precisely the number of edges in the normal form that occur exactly once. Unfortunately because operation 3 allows us to subdivide an edge, the number of edges that occur exactly once in a symbol may not be the same as the number of cuffs in the final normal form. For instance, if the symbol is $aba^{-1}b^{-1}cdec^{-1}$, the number of edges that occur exactly once is two, but we clearly manufactured this by using operation 3 on the normal form $aba^{-1}b^{-1}cfc^{-1}$. With this example it was easy to guess what the normal form was, but if the symbol had been all mixed up using others of the operations this might not have been so easy. How can we attack this in general?

We start by finding out how many distinct vertices there are around the boundary of the polygon. For instance, in the torus symbol $aba^{-1}b^{-1}$, all the edges start and end at the same vertex, but in $aba^{-1}b^{-1}cfc^{-1}$ one needs two vertices as we shall see. Looking at the polygon of the symbol, we start by choosing one edge and labelling one end, say

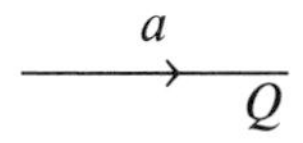

If the label a occurs twice on the boundary of the polygon, we insert Q as a label at the appropriate end of the second edge. The next edge to a, say b, now starts or finishes at Q, so we must insert Q again, that is

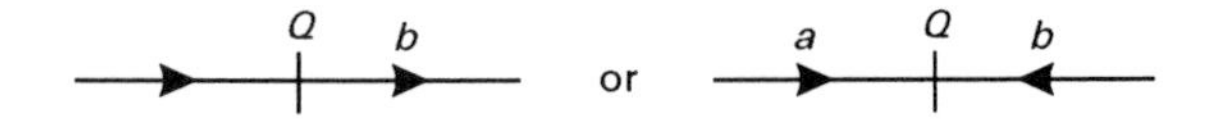

wherever b reoccurs (if indeed it does). Often this will insert Q at at least one new position allowing us to repeat the process until no further insertions of Q are called for. Now label any unlabelled end of an edge with a new label R, say, and insert R wherever required. Continue until all vertices on the boundary of the polygon are labelled.

As a simple example look at a surface with two crosscaps and a cuff. We will not draw the polygon, merely work with the symbol. We will show intermediate stages:

$$\begin{array}{ccccccccc} a & & a & b & b & c & d & c^{-1} \\ & Q & & & & & & & . \end{array}$$

Put Q as the label of the end of a; this occurs twice

$$\begin{array}{ccccccccc} a & & a & & b & b & c & d & c^{-1} \\ & Q & & Q & & & & & \end{array}$$

so a starts as well as finishes at Q, b starts at Q, and c^{-1} ends at Q (since the symbol as a whole must start and finish at the same vertex). This means c starts at Q and at this stage we get

$$\begin{array}{cccccccccccc} & a & & a & & b & & b & & c & d & c^{-1} & \\ Q & & Q & & Q & & Q & & Q & & & & Q. \end{array}$$

No more Q can be fitted in, so we insert a label R as the end of c (and thus the start of c^{-1}). This gives

$$\begin{array}{ccccccccccccccc} & a & & a & & b & & b & & c & & d & & c^{-1} & \\ Q & & Q & & Q & & Q & & Q & & R & & R & & Q \end{array}$$

so there are two distinct vertices on the boundary of the polygon.

An edge whose label occurs just once in a symbol is called a *boundary edge*. Earlier we looked at the example of a polygon with symbol $b^{-1}cda^{-1}dc^{-1}ea$; this has two boundary edges, b and e. If we work out the vertex labels on this example we find

$$\begin{array}{ccccccccccccccccc} & b^{-1} & & c & & d & & a^{-1} & & d & & c^{-1} & & e & & a & \\ Q & & R & & Q & & Q & & Q & & Q & & R & & Q & & Q \end{array}$$

so we find that b and e share vertices. They fit together to give a boundary curve:

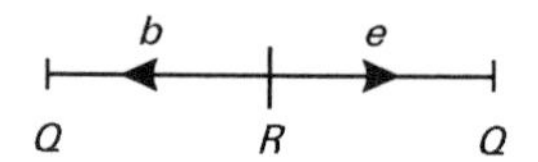

In general, we join up each string of boundary edges to make boundary curves, but there may be more than one of these. As we hinted at earlier, the number of the boundary curves does not depend on the triangulation we chose.

4.4 Proposition. *The number of boundary curves is the same as the number of cuffs in the normal form for the surface symbol. It is therefore an invariant of the surface.*

Proof. We will show that the number of boundary curves is unchanged by the processes that the symbol goes through as it is transformed into its normal form. As the number of boundary curves in the normal form is easily checked to be the number of cuffs (do check this), this will prove the result.

Firstly we note that the choice of starting vertex and direction in the reading of the surface symbol does not affect the boundary edges. Now consider what happens under each of the elementary operations in turn.

Operation 1

$$ABC \longleftrightarrow AxC \text{ and } Bx^{-1}.$$

The new label x occurs twice, so this does not alter the number of boundary edges or curves.

Operation 2

$$Axx^{-1}B \longleftrightarrow AB.$$

Again it is clear that as x occurs twice, no change occurs in the number of boundary curves.

Operation 3

$$x \longleftrightarrow uv.$$

If x is not a boundary edge then no change occurs. If, however, x is a boundary edge, the number of boundary edges will increase by one, but the number of boundary curves will stay the same.

Exercises 4.3

1. Find the number of distinct vertices for a polygon with symbol having g crosscaps or handles and r cuffs.
2. How many vertices are needed for the symbol

$$abc^{-1}adbc^{-1}efd^{-1}?$$

4.4 Invariants of surfaces: Euler characteristic and genus

Suppose S is a (combinatorial) surface with α_0 distinct vertices, α_1 distinct edges, and α_2 triangles. The *Euler characteristic* $\chi(S)$ is given by

$$\chi(S) = \alpha_0 - \alpha_1 + \alpha_2.$$

The usefulness of this number in various contexts and guises will be one of the themes of the coming chapters. Its usefulness here is that it does not depend on the triangulation used and is an invariant of the surface. This is perhaps a bit of a surprise, until you see why it happens.

4.5 Proposition. *Let P be a polygon representing a surface S. Suppose that the surface symbol of S gives v distinct vertices and e distinct edges on the boundary of P; then $\chi(S) = v - e + 1$.*

Proof. We will prove this by induction on the number, k, of triangles making up P. If $k = 1$, then either there are no identifications, so $S = P$, $\alpha_0 = v$, $\alpha_1 = e$, and $\alpha_2 = 1$, or we have a cuff and again $\alpha_0 = v$, $\alpha_1 = e$, and $\alpha_2 = 1$. In this case it is clearly true.

Now suppose $k > 1$ and that we know the result as long as P has fewer than k triangles in it. Recall that P was constructed by adding triangles in one of two ways to the disc already constructed. The two ways were as shown in Figs 4.62 and 4.63.

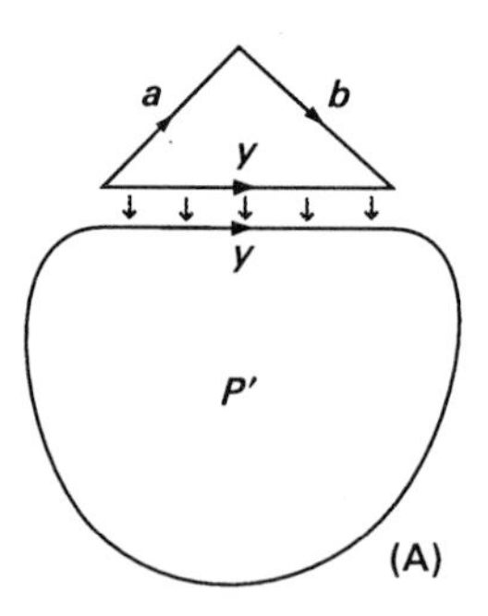

Fig. 4.62

We handle the two cases separately and each will lead to some 'subcases'.

(A) We first look at what happens if a and b occur nowhere else in the symbol; then the vertex between a and b is a new vertex. On recording the change in v and e from P' to P, we find $v = v' + 1$, one

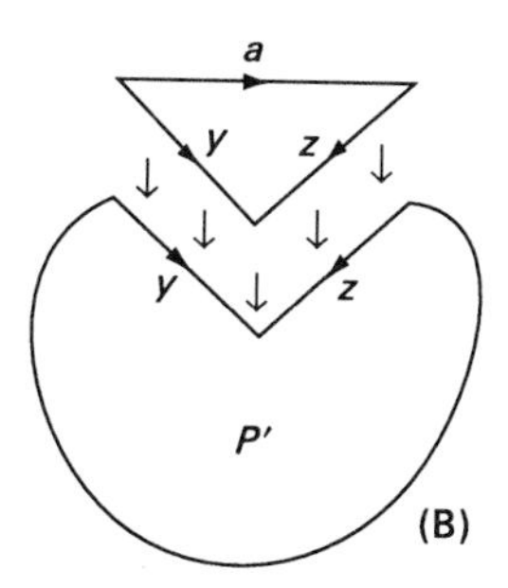

Fig. 4.63

new vertex, and $e = e' + 1$ since although we have gained a and b we have lost y from the boundary. Thus $v - e + 1 = v' - e' + 1$ and so stays unchanged. Now P' represents some surface, say S', and we have, by hypothesis, that since P' has only $k - 1$ triangles,

$$\chi(S') = \alpha_0' - \alpha_1' + \alpha_2' = v' - e' + 1.$$

We can, however, work out $\alpha_0, \alpha_1, \alpha_2$ from $\alpha_0', \alpha_1', \alpha_2'$, the corresponding numbers for S'.

We had one new vertex, so $\alpha_0 = \alpha_0' + 1$; two new edges, a and b, so $\alpha_1 = \alpha_1' + 2$; and one new triangle, so $\alpha_2 = \alpha_2' + 1$. Thus

$$\chi(S) = (\alpha_0' + 1) - (\alpha_1' + 2) + (\alpha_2' + 1)$$

that is, $\chi(S) = \chi(S') = v' - e' + 1 = v - e + 1$, as required.

The other subcases are similar. We will use the same notation but will explain less along the way, leaving the details for you to check.

Suppose now that a is not a label on the boundary of P' but b is; then $v = v'$, $e = e'$ (a is in, y is out!), so $v - e + 1 = v' - e' + 1$ as before. Also we have $\alpha_0 = \alpha_0'$, $\alpha_1 = \alpha_1' + 1$, $\alpha_2 = \alpha_2' + 1$, so $\chi(S) = \chi(S')$ and we complete the argument as before.

If a is a label on the boundary of P' and b is not, then the same argument applies.

Finally if both a and b occur on the boundary of P', then $v = v'$, $e = e' - 1$, $\alpha_0 = \alpha_0'$, $\alpha_1 = \alpha_1'$, $\alpha_2 = \alpha_2' + 1$.

We now look at case (B). Here there are only two subcases: either a is in the boundary of P' or it is not. If it is, then we get $\alpha_1 = \alpha_1'$, $\alpha_2 = \alpha_2' + 1$, whilst $v = v' - 1$, $e = e' - 2$. The equations balance as before: since by hypothesis $v' - e' + 1 = \alpha_0 - \alpha_1' + \alpha_2'$, one gets $v - e + 1 = \alpha_0 - \alpha_1 + \alpha_2$. If a is a new edge, not in the boundary of P', then we get $\alpha_0 = \alpha_0'$, $\alpha_1 = \alpha_1' + 1$, $\alpha_2 = \alpha_2' + 1$, whilst $v = v' - 1$, $e = e' - 1$ which again balances the equations. ■

The next stage in checking that this Euler characteristic is an invariant is to see what happens under the operations.

4.6 Lemma. *The transition to normal form does not change the values of the expression $v - e + 1$ determined by vertex and edge labels.*

Proof. (As the detailed proof is a bit like that of the invariance of the number of boundary curves, we will leave the details up to you (but do write them down!). We will provide just a sketch.)

Both operations 2 and 3 give us one new vertex label and one new edge label, so $v - e + 1$ stays the same.

Any application of operation 1 is followed by one of the type of its inverse. This reglues an edge, reassembling the polygon in a new way, but overall the number of vertex and edge labels stays constant. ■

We have shown that $\chi(S)$ is independent of the triangulation and can be calculated from the polygon in its normal form. We now need only to work out $\chi(S)$ for the various normal forms to have complete information.

Type: Non-orientable, g crosscaps, r cuffs.

Normal form: $a_1a_1 \ldots a_ga_gy_1z_1y_1^{-1} \ldots y_rz_ry_r^{-1}$.

From the exercise earlier in this section, we have:

number of vertex labels $= r + 1$

number of edge labels $= 2r + g$

hence Euler characteristic $= \boxed{2 - g - r.}$

Type: Orientable, g handles, r cuffs.

Normal form: $a_1b_1a_1^{-1}b_1^{-1} \ldots a_gb_ga_g^{-1}b_g^{-1}y_1z_1y_1^{-1} \ldots y_rz_ry_r^{-1}$.

We get (again from the exercise):

number of vertex labels $= r + 1$

number of edge labels $= 2r + 2g$

hence Euler characteristic $= \boxed{2 - 2g - r.}$

We have seen that (i) orientability, (ii) number of boundary curves, and (iii) Euler characteristic are not only invariants of the surface, but also able to be calculated from any symbol for the surface. We can therefore find r and $\chi(S)$ and by choosing either the orientable or non-orientable formula from the above, calculate g, the number of crosscaps or handles. This number, g, is called the *genus* of the surface. We have:

$$\text{if } S \text{ is non-orientable, } g = 2 - r - \chi$$

$$\text{if } S \text{ is orientable, } \quad g = \tfrac{1}{2}(2 - r - \chi).$$

We summarize the theoretical aspect of this as follows.

4.7 The classification theorem for surfaces. *Two surfaces are homeomorphic if and only if they are both orientable or both non-orientable, have the same number of boundary curves and the same genus.*

The practical aspect of the result is clear: it saves us work! The method of cutting and pasting does lead eventually to the normal form but the above ideas allow us to classify surfaces quickly and painlessly. As an example, we will look again at our symbol

$$b^{-1}cda^{-1}dc^{-1}ea.$$

Orientability: Non-orientable because of $\ldots d \ldots d \ldots$.
Vertex labels:

$$\begin{array}{ccccccccccccccccc} & b^{-1} & & c & & d & & a^{-1} & & d & & c^{-1} & & e & & a & \\ Q & & P & & Q & & Q & & Q & & Q & & P & & Q & & Q \end{array}$$

Boundary edges: b and e giving one boundary curve, so $r = 1$.
Genus: First calculate $\chi(S) = v - e + 1 = 2 - 5 + 1 = -2$, so $g(S) = 3$. Thus S has normal form $a_1a_1a_2a_2a_3a_3yzy^{-1}$.

Exercises 4.4

1. Classify the following surfaces giving genus, Euler characteristic, orientability, etc.
   (a) $ab^{-1}cdad^{-1}ceb$
   (b) $eabc^{-1}adbc^{-1}ef$
   (c) $ab^{-1}cda^{-1}d^{-1}c^{-1}eb$
2. Given two surfaces S_1 and S_2, their connected sum, denoted $S_1 \# S_2$, is formed by cutting a small circular hole in each surface and then gluing the two surfaces together along the boundary of the holes (Fig. 4.64).

If A_1 is a symbol for S_1 and A_2 one for S_2, one can obtain a symbol for $S = S_1 \# S_2$ by eliminating y from the pair of symbols $Axyx^{-1}$ and $Azyz^{-1}$, (Fig. 4.65) that is by replacing the 'holes' by cuffs.

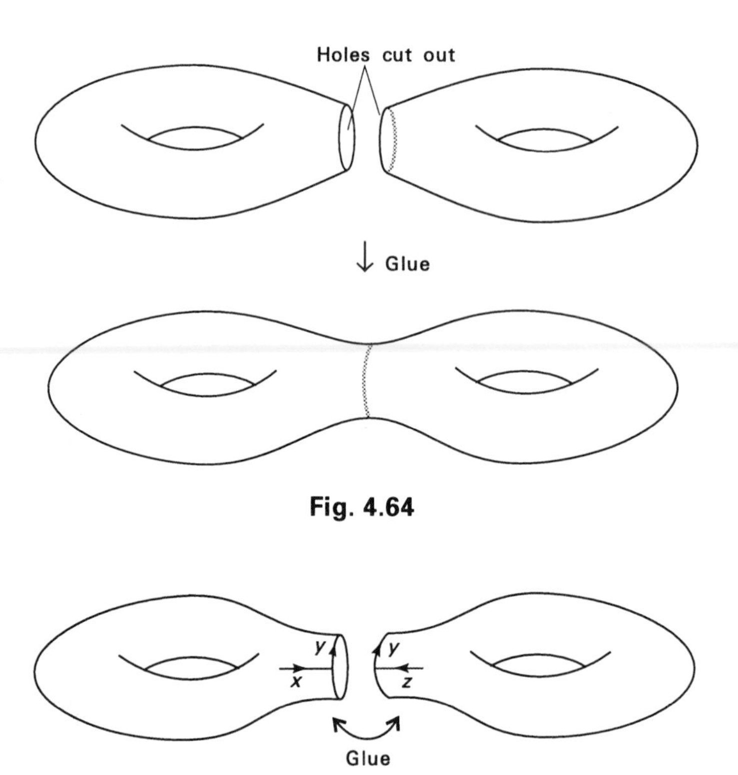

Fig. 4.64

Fig. 4.65

(a) Find $\chi(S)$ in terms of $\chi(S_1)$ and $\chi(S_2)$.

(b) Find $g(S)$ in terms of $g(S_1)$ and $g(S_2)$.

(Note you need to take care if one surface is non-orientable whilst the other is orientable.)

(c) Prove that $\mathbb{P}^2$ # torus is the same as $\mathbb{P}^2 \# \mathbb{P}^2 \# \mathbb{P}^2$.

The classification theorem can be rephrased in terms of connected sums since a genus g surface with r cuffs is either the connected sum of g tori (if orientable) or g projective planes (if non-orientable) together with r cuffs.

3. Let S be a surface without boundary curves (so no cuffs appear in the normal form). Suppose we have a triangulation of S using α_0 vertices, α_1 edges, and α_2 triangular faces. Show that

(a) $3\alpha_2 = 2\alpha_1$

(b) $\alpha_1 = 3(\alpha_0 - \chi(S))$

Now by considering the maximal possible number of edges joining α_0 vertices, show that

(c) $\alpha_0 \geqslant \frac{1}{2}\left(7 + \sqrt{49 - 24\chi(S)}\right)$.

These calculations give a lower bound for the number of vertices needed to triangulate S. Find a lower bound for the number of triangles in a triangulation of (i) the projective plane, (ii) the Klein bottle, (iii) the sphere. Can you find a triangulation with exactly this number of triangles?

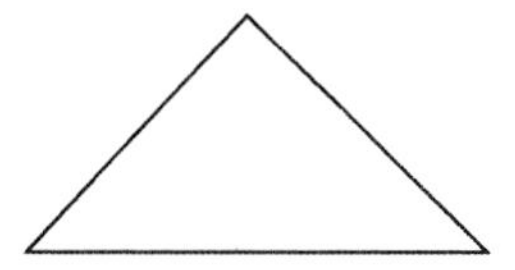

Fig. 4.66

4. Let S_1 and S_2 be two homeomorphic orientable surfaces of genus g and with $r > 0$ boundary curves. Construct a new surface, D, by identifying the boundary edges of S_1 and S_2; this is the *double* of S_1 (or S_2). You might visualize it as being constructed by replacing each triangle (Fig. 4.66) in a surface by two copies near to each other (Fig. 4.67) except if an edge is a boundary edge (Fig. 4.68).

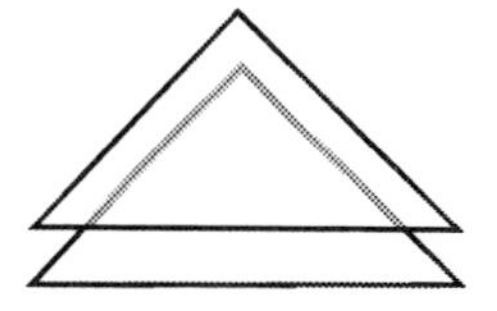

Fig. 4.67

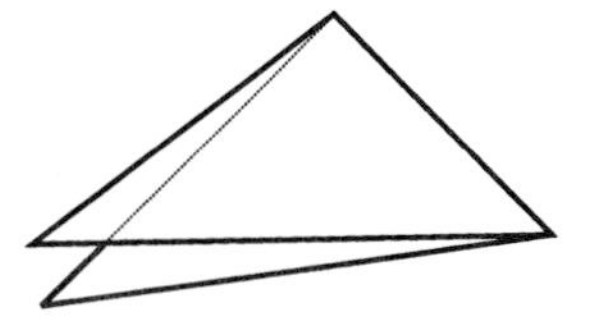

Fig. 4.68

Show that D is orientable, has genus $2g + r - 1$, Euler characteristic $4 - 4g - 2r$, and no boundary curves.

4.5 Knots and surfaces

So far the only connections between knots and surfaces that we have met in this chapter have been ones of methodology: ideas such as the use of invariants, the reduction of classification problems to moves, and so on. In the next few pages we will see how these two classes of object interact at a deeper level and hence that some parts of the classification result for surfaces can be used to produce results in knot theory. In the next chapter these will be used in a study of the 'arithmetic' of knots.

Our main aim here is to prove the following result of Seifert.

4.8 Theorem. *Given any knot K (in $\mathbb{R}^3$) there is an orientable surface S with one boundary curve embedded in $\mathbb{R}^3$ in such a way that the boundary of S is the knot K.*

Proof. Take a projection of the knot, K. Orient the diagram then at each crossing, add two extra directed arcs bypassing the crossing, but compatible with the orientation, as shown in Fig. 4.69.

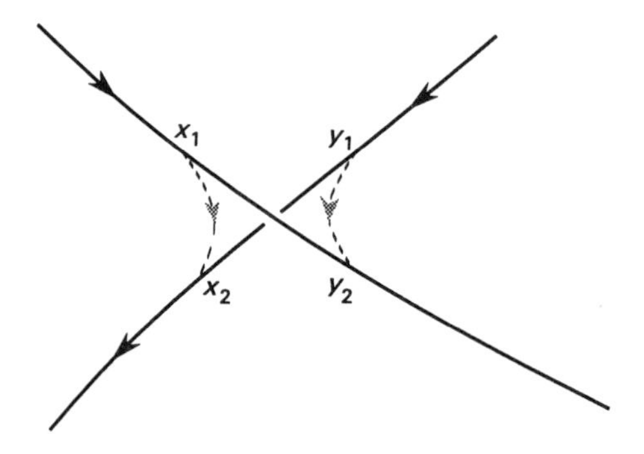

Fig. 4.69

Now delete the crossing leaving a set of oriented circles, called *Seifert circles*. (In our crossing we throw away the parts of the knot between x_1 and y_2 and between y_1 and x_2.)

As an illustration, consider the process outlined above performed on the figure-eight in its usual projection, as in Fig. 4.70.

The Seifert circles may or may not be nested. Starting from any innermost circle and working outwards fill in (the word 'span' is often used) each circle with a disc, the discs to be disjoint.

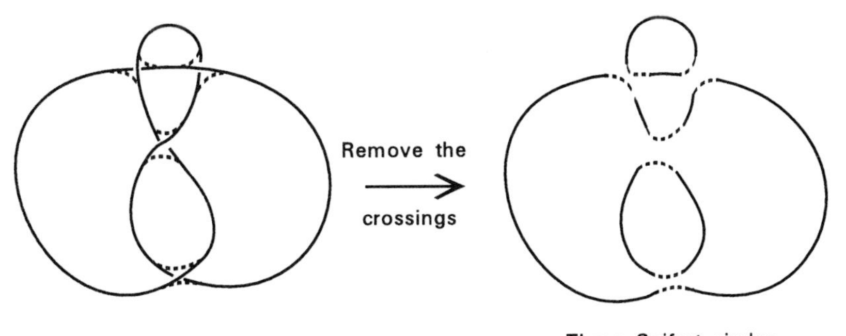

Fig. 4.70

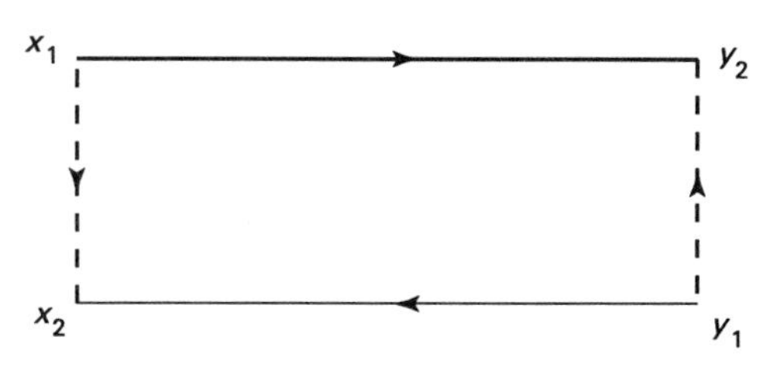

Fig. 4.71

The two arcs and the parts of the crossing removed make up a rectangle as in Fig. 4.71.

Fill this in with a solid rectangle. Do this at each crossing. Thus the crossings now look as in Fig. 4.72, each one filled with a twisted

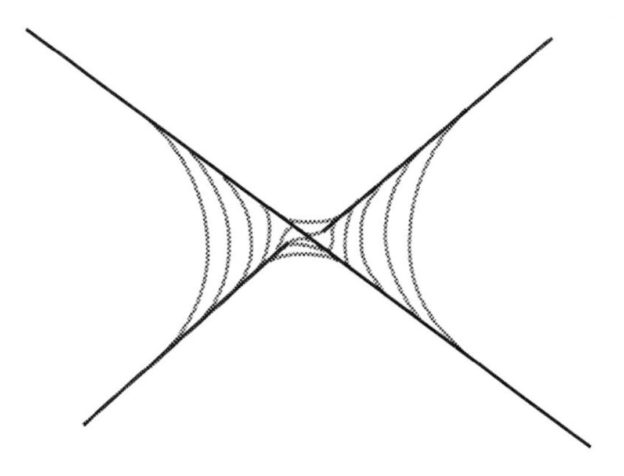

Fig. 4.72

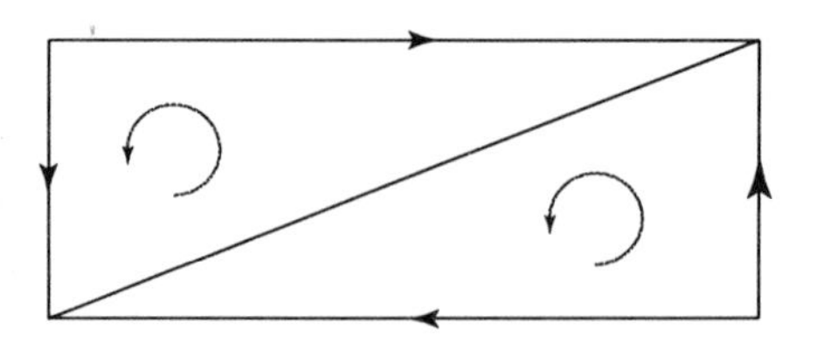

Fig. 4.73

rectangle. Each spanning disc of each Seifert circle has a natural induced orientation coming from the orientation of the knot. As each twisted rectangle (Fig. 4.73) can be oriented coherently with these orientations, the surface obtained by gluing the twisted rectangles onto the Seifert circles will be orientable.

4.9 Proposition. *The surface, S, constructed from K by the above method, has genus*

$$g = \tfrac{1}{2}(d - s + 1)$$

where d is the number of crossings in K and s the number of Seifert circles arising from K.

Proof. A disc has Euler characteristic 1. If we add a rectangle between two discs (Fig. 4.74) then the number of vertices, α_0, does not change since there are no new vertices; the number of edges, α_1, increases by three; and the number of triangles, α_2, increases by two.

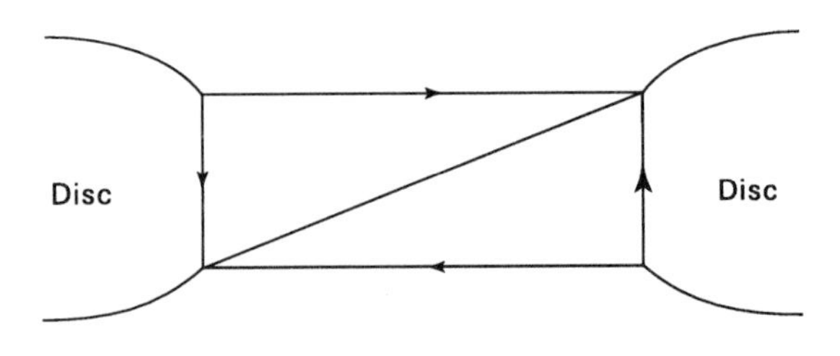

Fig. 4.74

We start with s discs, each contributing 1 to $\chi(S)$, and then add d rectangles, each reducing $\alpha_0 - \alpha_1 + \alpha_2$ by 1; hence $\chi(S) = s - d$ and as S is orientable $g(S) = 1/2\,(d - s + 1)$. ■

You might expect this number to be an invariant of the knot; however, a different projection might give a different surface and hence a different g. To avoid these problems we take the easy way out and make the following definition: the *genus* $g(K)$ of a knot K is the minimal genus of all orientable surfaces which span K.

The genus thus defined gives an invariant of the knot, but the form of definition makes it difficult to calculate. Since the Seifert construction does not necessarily give a minimal genus surface, we can only say that

$$g(K) \leqslant \tfrac{1}{2}(d - s + 1).$$

A theoretical algorithm was developed by W. Haken in 1961 but at the time it was too unwieldy to be useful in practice. With more powerful computing methods and with the renewed interest in knot theory due both to the development of the new polynomials and the perception of potential for applications, the question of developing methods for calculating the genus of knots is again receiving a lot of attention.

It is useful to note the following:

1. It is quite easy to span certain knots with non-orientable surfaces. For example, the trefoil can be spanned by a Möbius band (Fig. 4.75), so BEWARE!

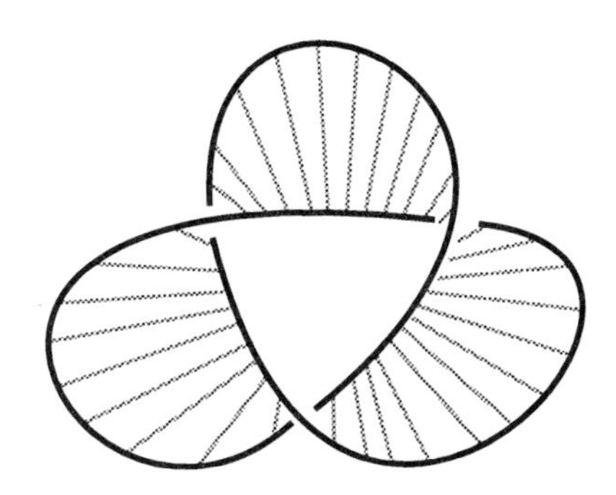

Fig. 4.75

2. If $g(K) = 0$, then K is the trivial knot.
3. If K is known to be non-trivial and one finds some orientable surface of genus 1 spanning the knot then $g(K) = 1$. (In our example, we looked at the figure-eight knot having four crossings and three Seifert circles, so the surface that resulted did have minimum genus 1.)
4. If $g(K) \neq 0, 1$, it can be extremely difficult to calculate the genus. Can one be certain that no smaller genus surface spans the knot?

5. There is a strange link with the Alexander polynomial. If K is an alternating knot (i.e. there is some projection in which, as one goes along the knot, overpasses alternate with underpasses), then $g(K)$ is half the degree of the Alexander polynomial of K. In general, the degree of $\Delta(t)$ is less than or equal to $2g(K)$ (see Burde and Zieschang 1985).

Exercises 4.5

Find bounds for the genus of the following knots:

(a) trefoil (Fig. 4.76)

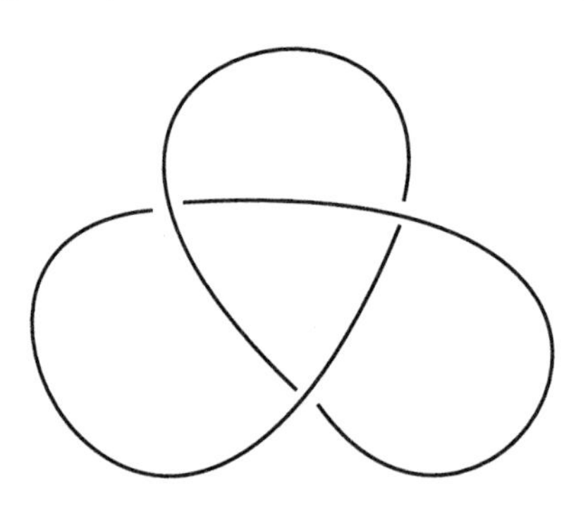

Fig. 4.76

(b) cinquefoil (Fig. 4.77)
(c) any $(2, n)$-torus knot (Fig. 4.78)
(d) any generalized figure-eight knot (Fig. 4.79)

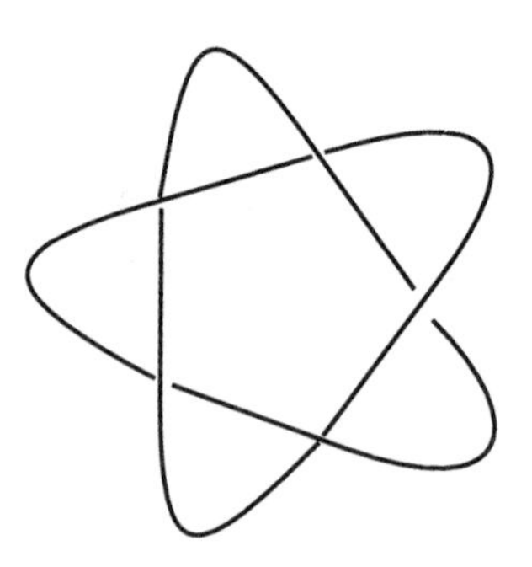

Fig. 4.77

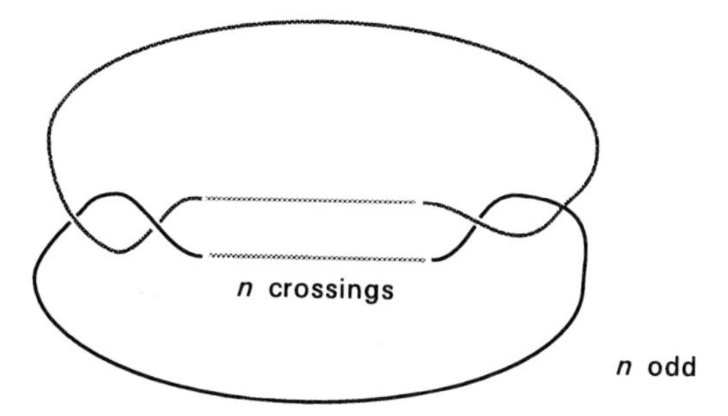

Fig. 4.78

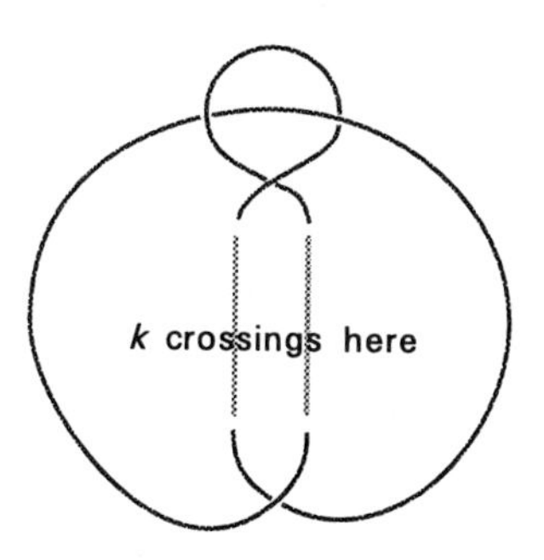

Fig. 4.79

5
The arithmetic of knots

Analogy is a very powerful logical tool. When two different types of situation seem to be running parallel from certain viewpoints, one can try to see if the similar behaviour extends further to give new insight into the different structures. Of course it is as important to notice where the analogy fails as to follow those situations where it works. A basic analogy used in knot theory is that the sum of oriented knots behaves very much like the product of natural numbers; hence the title of this chapter.

5.1 The sum of oriented knots

Suppose that we have two oriented knots K and L. Then we can form their 'sum' in an obvious way following the natural intuition of first tying K and then L in a piece of string. The details are as follows.

Cut each knot as shown in Fig. 5.1.

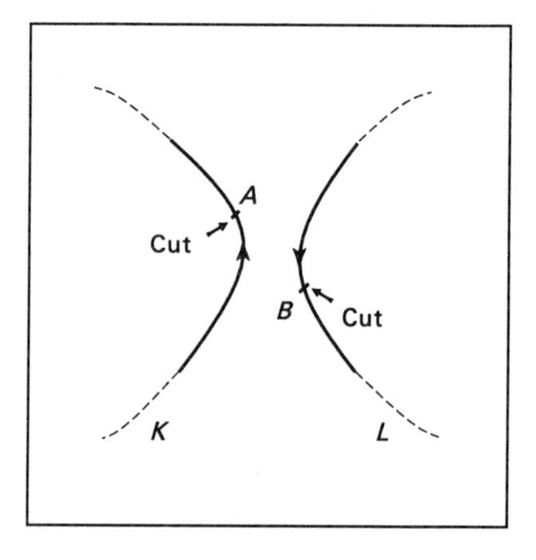

Fig. 5.1

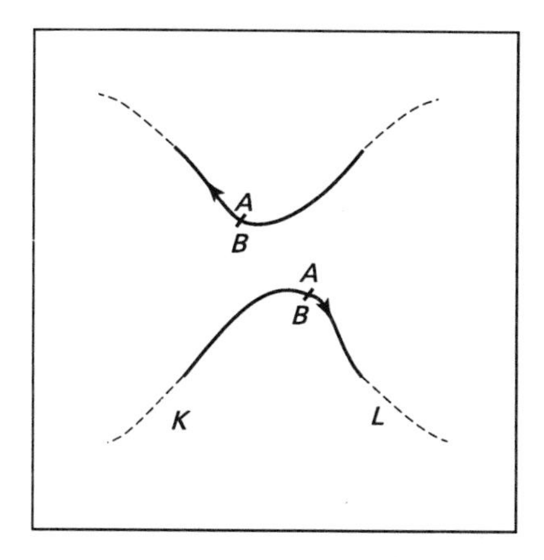

Fig. 5.2

Join them together preserving orientation, as in Fig. 5.2. We will write $K + L$ for the result.

Examples

(Figs 5.3 and 5.4)

granny knot = trefoil + mirror trefoil

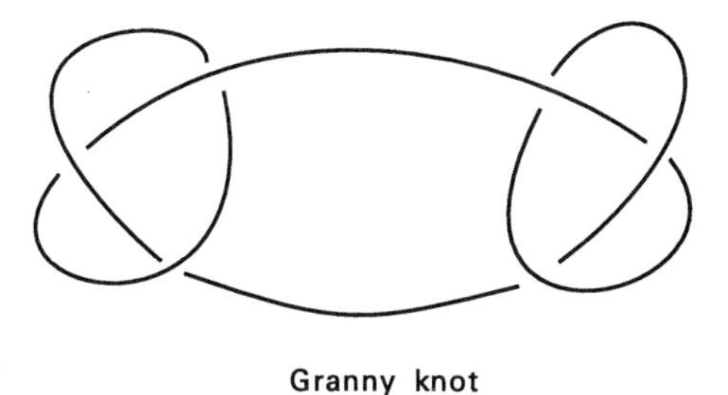

Granny knot

Fig. 5.3

How does this operation + act on the class of all knots? Does it respect equivalence for instance? Does it depend on where the cuts are made? Is + an appropriate choice of symbol? Does + behave like the addition of integers? The following goes some way to answering these questions. We shall denote equivalence of knots by the symbol ~.

reef knot = trefoil + trefoil

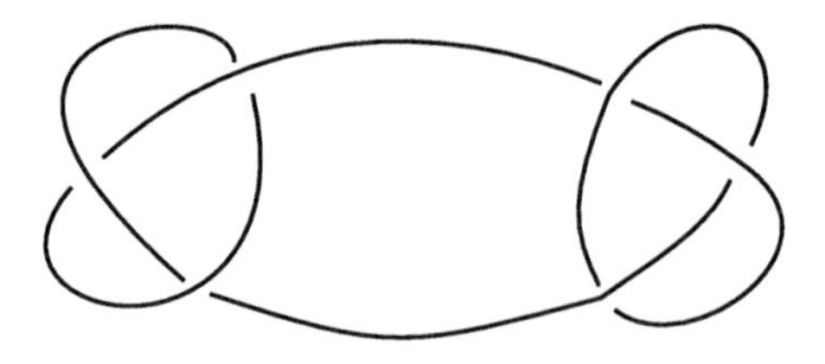

Reef knot

Fig. 5.4

5.1 Proposition

(a) *If* $K \sim K'$ *then for any* L, $K + L \sim K' + L$

(b) $K + L \sim L + K$

(c) $(K + L) + H \sim K + (L + H)$

(d) K + trivial knot $\sim K$.

The full proofs of (a) and (c) are long. For (a), you have to use the homeomorphism sending K onto K' to construct that sending $K + L$ onto $K' + L$. This uses details of the behaviour of homeomorphisms that are deeper than we have yet gone, so we will omit the proof. For (c), the fact itself is unsurprising if not obvious, but again one has to construct things with great care so as to be able explicitly to produce the homeomorphisms. Again we omit the proof.

For (d), a sufficient reason to believe the result is as shown in Fig. 5.5.

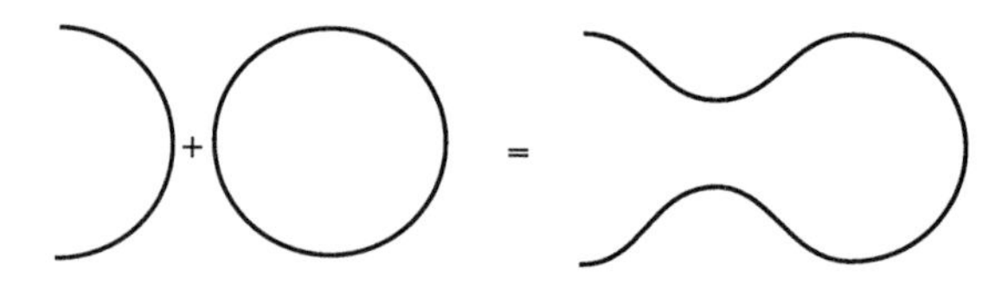

Fig. 5.5

Finally, for (b), the result seems a lot less obvious than any of the others; however, the reason for this property is very pretty and easy to see—once it is pointed out to you. A full proof can be based on this 'sketch' proof, or rather 'proof by sketches'. We illustrate the idea in Fig. 5.6.

Make K smaller and slide it along L as in Figs 5.7 and 5.8.

Now pull K back to the size it was to start with (Fig. 5.9).

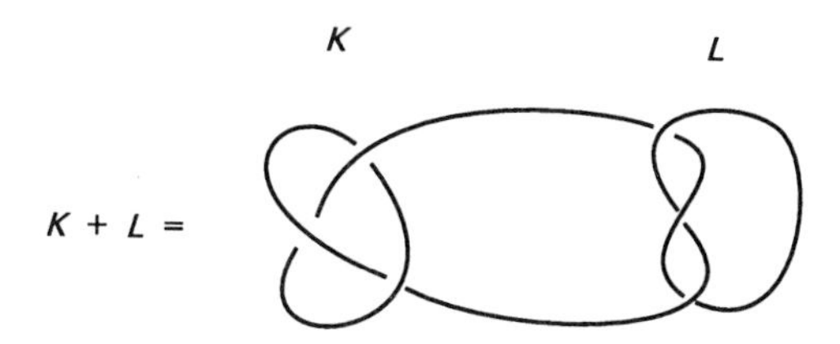

Fig. 5.6

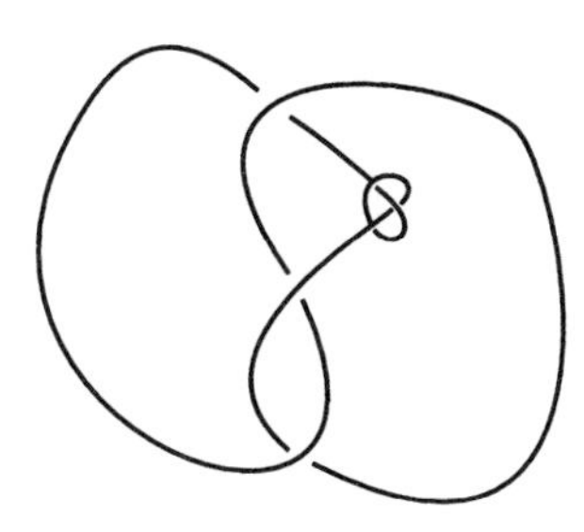

Fig. 5.7

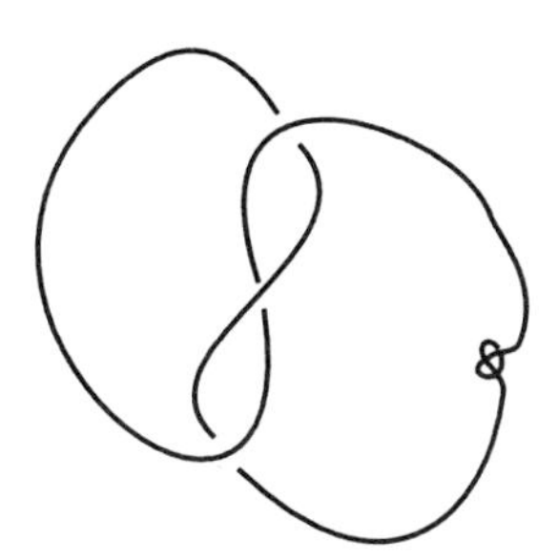

Fig. 5.8

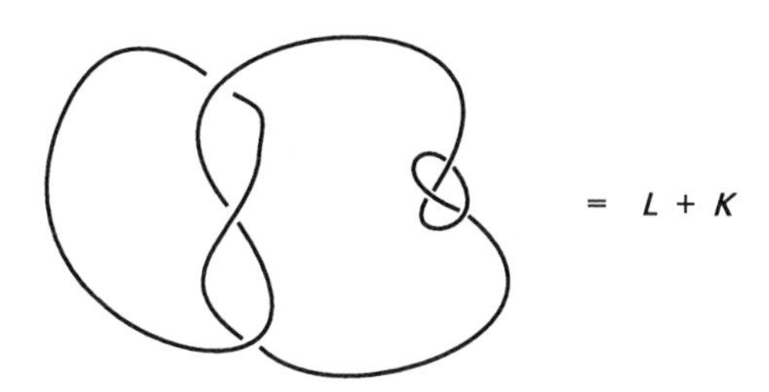

Fig. 5.9

These results then show that the operation + on *knot types* is well defined, commutative, associative, and has an identity which we write **0**, a very suitable symbol when one remembers that the identity is the type of the unknot.

So far we have developed this knot sum showing that it behaves much like *addition* of natural numbers. It is necessary to pause here a moment and consider a couple of points. Firstly that there are many other systems which behave as above. (These structures are called commutative monoids.) As this is the case we must not jump to the conclusion that this 'analogy' is in any way complete. For instance, we do not yet know if it is possible to subtract knots; that is, if we have $K + L$ is there some L' such that $K + L + L' \sim K$? What about cancellation: if $K + L \sim K' + L$, is $K \sim K'$?

Although we have as yet few invariants for knots (bridge number, genus, colouring properties, various polynomials) we can hope that investigation of how the value of an invariant on $K + L$ depends on its values on K and L will at the very least shed some light on these problems and perhaps indicate new properties to investigate further.

5.2 Genus of $\boldsymbol{K + L}$

Suppose we have surfaces S_K and S_L spanning K and L respectively and such that $g(S_K) = g(K)$ and $g(S_L) = g(L)$ that is they are minimal genus spanning surfaces. If we place S_K and S_L in different half-spaces, we can 'pipe' them with a rectangle as shown in Fig. 5.10.

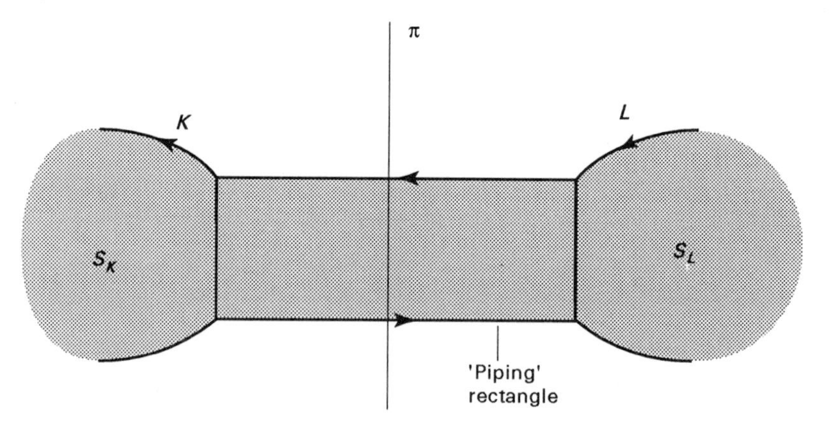

Fig. 5.10

The result will be an orientable surface S with a single knotted boundary component. That boundary component forms a copy of $K+L$, that is S spans $K+L$ and so

$$g(S) \geqslant g(K+L).$$

We can calculate $g(S)$ in terms of $g(K)$ and $g(L)$. We leave the justification to you; the equation to verify is

$$g(S) = g(S_K) + g(S_L)$$

so

$$g(K+L) \leqslant g(K) + g(L).$$

This simple argument thus suggests that maybe $g(K+L)$ is $g(K) + g(L)$. We will investigate this relationship more closely, trying to see if that is the case or if some as yet unknown feature comes in to block equality.

We suppose K and L as before are on different sides of some plane π with the joining arcs crossing cleanly, that is each intersecting π in a single point, as in Fig. 5.11, so $\pi \cap (K+L) = \{A, B\}$, say.

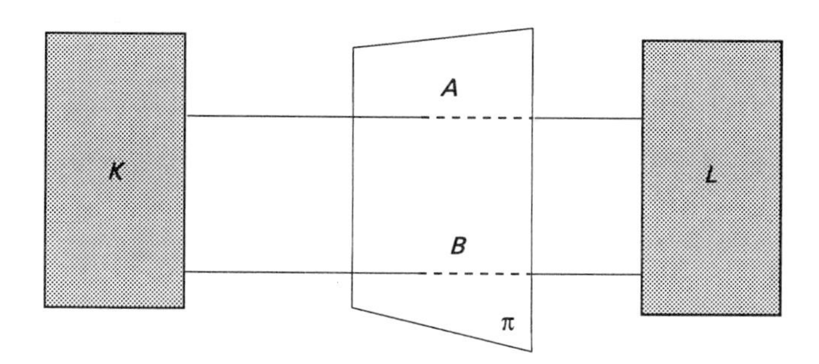

Fig. 5.11

We have some oriented surface S spanning $K+L$ and having genus exactly $g(K+L)$ (Fig. 5.12).

Shifting S slightly if necessary, we can assume that it also cuts π cleanly, that is that $S \cap \pi$ consists of a finite number of disjoint circles and an arc joining A to B (Fig. 5.13). (The proof that this can be done requires a study of *transversality*. If you like this type of argument and want to know more, we suggest you look up Lang (1972).)

Some of these circles may be nested. Choose a circle that does not loop around the arc from A to B and, moreover, does not surround any smaller loop (Fig. 5.14).

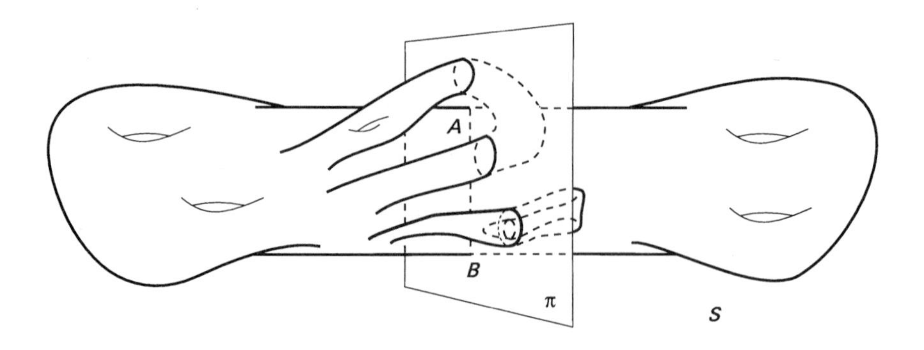

Fig. 5.12

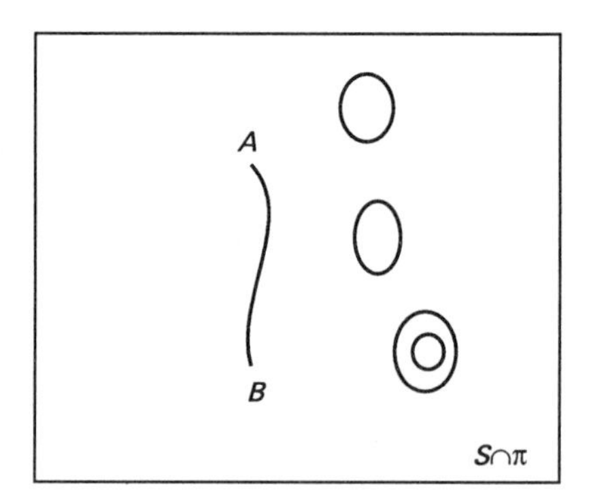

Fig. 5.13

Put a cap on this circle and pull it away from π to get a new surface S_{new} still spanning $K + L$. This process may have disconnected the surface; if so throw away the part not containing $K + L$ in its boundary. The new surface has the same genus as our original S

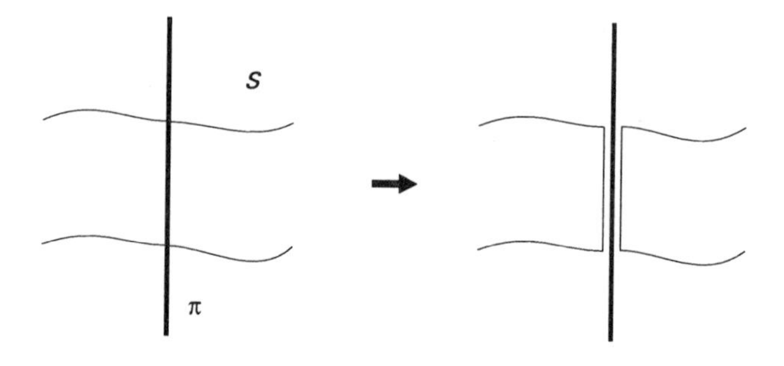

Fig. 5.14

since S_{new} still spans $K + L$, S_{new} has genus at most that of S_{new} (since our 'surgery' above can at best have chopped through a handle or have resulted in some handles in a thrown-away piece), but S has minimal genus among spanning surfaces of $K + L$, that is $g(S_{new}) = g(S) = g(K + L)$.

We repeat this process until we are left only with circles which loop around A and B. Starting with the largest such circle attach discs around the backs of K and L. Continuing in this way we get a minimal genus surface S' spanning $K + L$ such that $S' \cap \pi$ is just the arc joining A to B. Cut along this arc to obtain two surfaces S_1 and S_2 spanning K and L respectively. We have

$$g(K + L) = g(S_1) = g(S_1) + g(S_2) \geqslant g(K) + g(L).$$

Combining this with the result of our earlier work gives a proof of the following theorem.

5.2 Theorem (sum theorem for genus). *If K and L are oriented knots, then*

$$g(K + L) = g(K) + g(L).$$

5.3 Bridge number of $K + L$

The above discussion on $g(K + L)$ leant heavily on the minimality of the genus of the chosen spanning surface. Bridge number was also defined using a minimum over all possible positions. Can a similar type of argument to that used in Section 5.2 give us a similar result for $b(K + L)$ in terms of $b(K)$ and $b(L)$?

Our first step is to use guesswork: is $b(K + L)$ equal to $b(K) + b(L)$? The obvious first reaction is 'probably', but if both K and L are the unknot then $b(K) = b(L) = b(K + L) = 1$, so this first guess is not right. The reason is not hard to see.

Suppose we take a diagram of K and a diagram of L, each having the minimal number of bridges for the respective knot. Now compose the knots as in Fig. 5.15.

The result would seem to have $b(K) + b(L)$ bridges; but is this the smallest possible number?

By choosing the points where K and L are cut we can reduce the number of bridges by one. Pick the cutting point A in K to be between the first underpass and the first overpass and in L to be between the last overpass and the second underpass, as in Fig. 5.16; then the second arc thus created contains no overpass and hence

$$\text{no. of bridges in diagram} = b(K) + b(L) - 1.$$

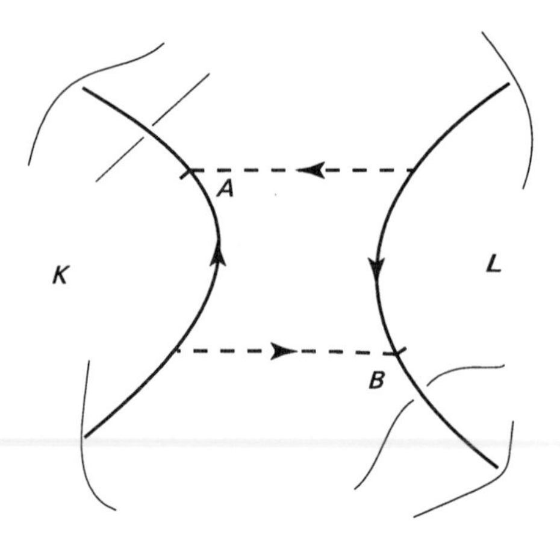

Fig. 5.15

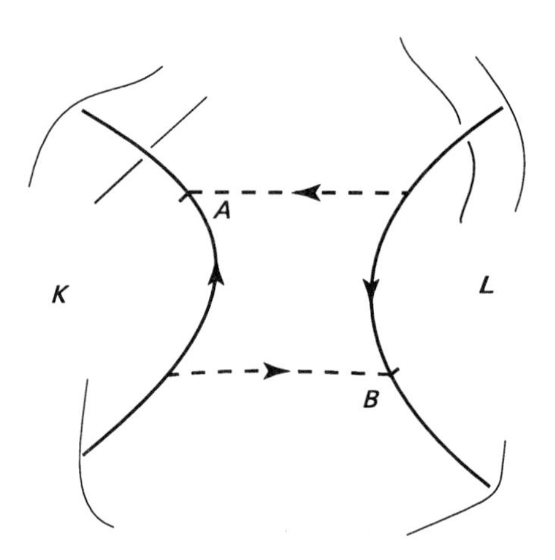

Fig. 5.16

It might be possible to reduce the number of bridges yet more by some 'cunning trick', so we get

$$b(K+L) \leqslant b(K) + b(L) - 1.$$

Our experience with the genus sum formula suggests that perhaps

$$b(K+L) = b(K) + b(L) - 1.$$

We note that if L is trivial, since $b(L) = 1$, then this would be consistent (giving $b(K) = b(K) + 1 - 1$). This formula is correct but we will not attempt to prove the result here.

5.3 Theorem (sum formulae for bridge number). *If K and L are oriented knots, then*

$$b(K + L) = b(K) + b(L) - 1.$$

If you can read German, you could look up the original proof. It is in Schubert (1956).

5.4 Factorizations, prime knots, and the factorization theorem

We saw how the sum of knots was well behaved, much like the sum of numbers. Is there a 'subtraction' of knots? Let us suppose we could find a '$- K$' so that $K + (- K) = 0$; then

$$g(K) + g(- K) = 0$$

and $g(K) = - g(K)$. That is strange: how could a knot have negative genus? So the only knot K for which $- K$ exists in this sense is $K = 0$ itself.

We started out this chapter talking about analogy. Our attempt to make an analogy between the sum of knots and the sum of numbers only goes so far. There is no 'additive inverse', just as there is no multiplicative inverse in the natural numbers. Perhaps the better analogy would be between the sum of knots and multiplication (not addition) on $\mathbb{N}$. Both are commutative, associative, and have a neutral element (1 for $\mathbb{N}$, 0 for knots) and we have just seen that $+$ on knots does not have an inverse operation.

This suggests that an analogy between the factorization of knots and the factorization of numbers would be worth investigation. If we could fracture a complicated knot into simpler factors then presumably knowledge of these simpler factors should tell us everything about the original knot. This would be the case provided certain conditions on this factorization process were satisfied. Ideally we could hope that the fracturing process would terminate just as $18 = 3 \times 6 = 2 \times 3 \times 3$ cannot go further without introducing trivial factors. The reasons that such termination is important are clear. It is not much good if one can go on and on and never know if one more step would give a complete answer. We can also hope that a complete answer is unique. The half-way stage of our fracturing of 18 was given as 3×6 but we could equally well have given 2×9. The uniqueness of the 'prime factorization', $2 \times 3 \times 3$, is what makes it useful. What evidence can we get for 'primes' among knots?

Suppose K is a knot with genus 1; can we fracture it, say as $K_1 + K_2$? We would have

$$1 = g(K) = g(K_1) + g(K_2)$$

so as any genus is non-negative, one of the $g(K_i)$ must be zero, so any genus 1 knot must be 'prime'. We can therefore list quite a few 'prime' knots starting with the trefoil and the figure-eight.

We could equally well use the bridge number formula. We know that $b(K) = 1$ means that $K \sim 0$. What can we say about knots with bridge number equal to 2? Are these prime?

If $b(K) = 2$ and $K = K_1 + K_2$, then

$$2 = b(K) = b(K_1) + b(K_2) - 1$$

so $b(K_1) + b(K_2) = 3$. The only way this can happen is if one of $b(K_1)$ and $b(K_2)$ is 1, that is the corresponding $K_i \sim 0$. We therefore have that all 2-bridge knots are prime. Either argument shows that the trefoil is prime.

It is about time that we gave a formal definition of prime knot as we have been happily (?) discussing them for a while now.

A knot K is said to be *prime* if K is not trivial and whenever there are knots K_1 and K_2 so that $K \sim K_1 + K_2$ one at least of K_1 and K_2 is the unknot.

In studying decompositions of knots we earlier mentioned termination. The first result guarantees this.

5.4 Lemma. *A knot cannot be indefinitely factorized. More precisely, given a knot K, there is a natural number N such that if $K \sim K_1 + \ldots + K_n$ for some $n > N$ then some of the K_i are trivial. (In fact $N = g(K)$ will do, but need not be 'best possible'.)*

Proof. Suppose $K \sim K_1' + K_2'$, both K_1' and K_2' non-trivial; then as $g(K_i') \geqslant 1$ for $i = 1, 2$

$$g(K) > g(K_i'), \quad i = 1, 2.$$

The lemma follows since if $K \sim K_1 + \ldots + K_n$ and all $g(K_i) \geqslant 1$ then, of course, $g(K) \geqslant n$. ■

The second result we note is a 'uniqueness' result.

5.5 Lemma. *If K is prime then $K|L + M$ implies $K|L$ or $K|M$ (where we write $K|L$ etc. to mean K is a factor of L.)*

We only sketch the proof giving an idea of why the result is true. There is a plane π such that $\pi \cup (L + M)$ consists of two points only, L lies on one side of π, and M on the other. The 'knotted part' of K

must also lie totally on one side or the other of π since otherwise it would be simple to decompose K as a sum in a non-trivial way. Thus K is a factor of either L or M as required.

5.6 Theorem (prime factorization of knots). *If K is a knot, then there is a decomposition*

$$K \sim P_1 + \ldots + P_m$$

where each P_i is a prime knot. Such a decomposition is unique up to order.

Proof (instructions only). Really this should be left up to you to do; however, here are instructions for writing out a proof. (Do follow them.)

First look up some proofs of the prime factorization of integers. Most will proceed by saying that if n is not prime then there are n_1, n_2 such that $n = n_1 n_2$ and then will go on splitting n_1 and n_2. Termination and uniqueness use ideas similar to Lemmas 5.4 and 5.5 above. Find a proof which is simple and adapt it to handle knots.

(The moral of the story would seem to be something like: if your analogy is a good one, the underlying logic of the situations will be the same. Certain parts of mathematics study exactly the logic of such situations. If you enjoy a challenge, see if you can formulate a system or structure which contains just enough to be able to prove a prime factorization theorem. The idea of the lattice of divisors of a number may set you on the right lines. Do the 'divisors' or 'factors' of a knot also form a lattice? Are these lattices special in any way? The theory is known, but you should be able to make some headway towards finding a solution. Have fun! If you have not met the notion of a lattice, it is a special type of ordered set. Try to find the definition.)

5.5 *n*-colourability

We have seen that the genus and bridge number are 'well behaved' with regard to the sum of knots. What about other invariants, especially the 'elementary' ones? The easiest to handle is n-colouring.

If K is n-colourable then suppose in forming a sum with L, we cut K on an arc coloured by the colour a. How is one to colour $K + L$? Simply use the colour a for all parts of $K + L$ that were originally in L and check this works.

In some ways, this shows that n-colourability is limited in its usefulness as a property. It can be useful for prime knots, but for composite knots n-colourability is rather a weak property.

What if K is m-colourable and L n-colourable? Is it possible to combine the two indices of colourability in some way? The answer is known but we will not give it here. We suggest that you try some simple examples, put forward a conjecture, and then attempt to prove it. Of course your conjecture may be false!

5.6 Polynomials and the sum of knots

We earlier met three polynomial invariants. How do these react with respect to the sum of knots?

Alexander polynomials

We take K and L and label the diagrams in the usual way. We take the same face label, a, for the unbounded region of the diagrams in Fig. 5.17.

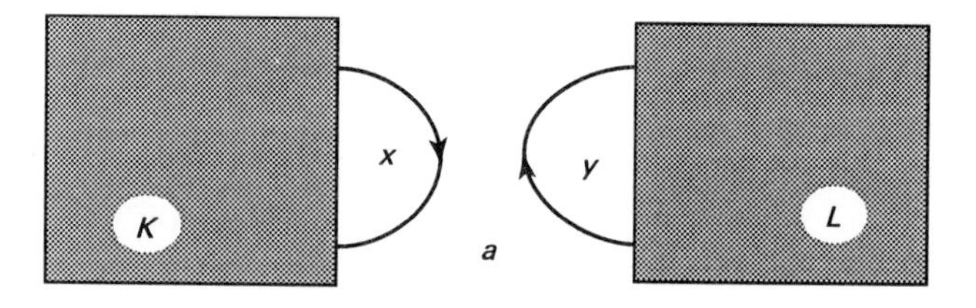

Fig. 5.17

When we have drawn up the $m \times (m+2)$ matrix for K we delete the columns corresponding to the adjacent regions, x and a, thus getting a square $m \times m$ matrix M_K. The Alexander polynomial is $\det(M_K)$ after normalization. We do the similar process for L, deleting the columns corresponding to Y and a, with $\Delta_L(t) = M_L$. The diagram for $K + L$ will look like that in Fig. 5.18.

This time in the $(m+n+2) \times (m+n)$ matrix we delete columns corresponding to z and a. It should be clear that

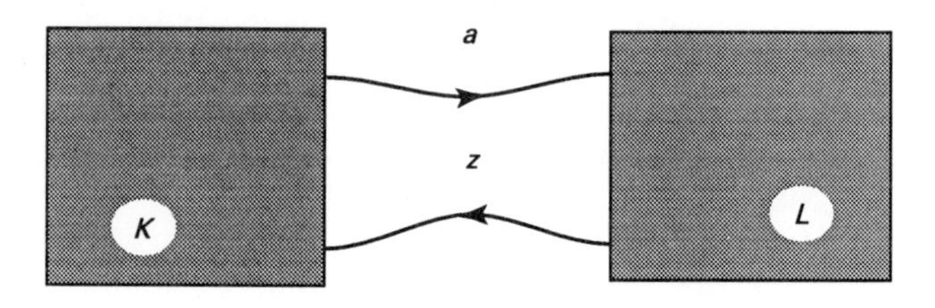

Fig. 5.18

$$M_{K+L} = \begin{pmatrix} M_K & 0 \\ 0 & M_L \end{pmatrix}$$

since no crossing originally in K can contribute non-trivially to a column in the L-part of M_{K+L} since the only possibility of such a contribution would have been via column a or z and these have been deleted.

Of course $\det\left(M_{K+L}\right) = \det\left(M_K\right) \cdot \left(\det M_L\right)$, that is

$$\Delta_{K+L}(t) = \Delta_K(t) \cdot \Delta_L(t).$$

In many ways the Alexander polynomial serves as a testing ground for properties of the more recently defined polynomials, so we may conjecture that these react similarly.

The Homfly polynomial

To prove that the Homfly polynomial of a sum of knots is the product of those of the constituent parts we need to prove one or two preliminary results.

If K and L are two links, we let $K \amalg L$ be their distant (i.e. disjoint, unlinked) union; then

$$P(K \amalg L) = -\frac{\ell + \ell^{-1}}{m} \cdot P(K) \cdot P(L).$$

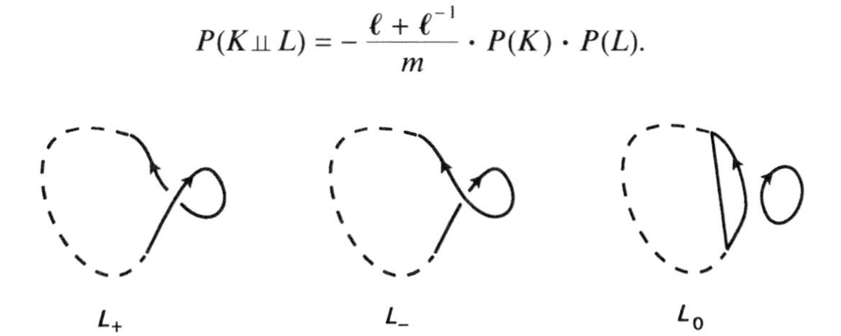

Fig. 5.19

If L is the unknot this comes from the cases shown in Fig. 5.19 since $K + L_+ \sim K \sim K + L_-$, whilst $K \amalg L \sim L_0$. Feeding these into the defining equation for P gives

$$\ell P(K) + \ell^{-1} P(K) + m P(K \amalg L) = 0$$

so

$$P(K \amalg L) = -\frac{\ell + \ell^{-1}}{m} \cdot P(K)$$

If L is more complicated, then use induction on the crossing number of any one of its diagrams; try to set up this induction formally.

To use the Homfly equation we will need to extend the notion of oriented sum of knots to links. One simply chooses one pair of components on which to perform the usual knot sum.

We now can consider the three diagrams (Fig. 5.20), so $\ell P(K+L) + \ell^{-1} P(K+L) + mP(K \amalg L) = 0$.

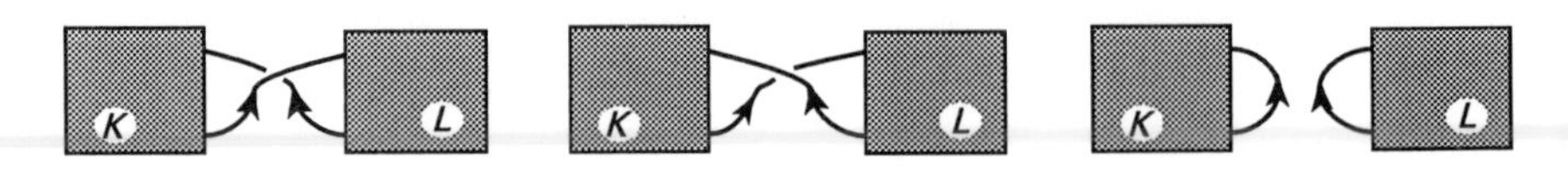

Fig. 5.20

This gives

$$(\ell + \ell^{-1}) P(K+L) = -mP(K \amalg L)$$

$$= (\ell + \ell^{-1}) P(K) \cdot P(L)$$

and so as expected

$$P(K+L) = P(K) \cdot P(L).$$

Given the link between P and Δ, we could retrieve our earlier result on Δ from this one.

We leave you to investigate the Kauffman polynomial.

6
Presentations of groups

We need group theory so as to be able to handle some of the deeper invariants of knots, in particular the group of a knot. We might expect the type of group theory needed would be fairly limited and somewhat specialized. However, the overall process translating geometric problems into algebraic ones is very much two way—algebra pervades the topology and vice versa—so quite a lot of group theory turns out to have implications in topology, whilst the topology can provide new methods and insights in group theory. Attempts to compartmentalize mathematics fail; there are too many important links joining apparently distinct branches. There is an underlying and all-pervading logical structure that is difficult to pin down and to see clearly, but which is none the less there behind all of mathematics.

Remark to the reader

The contents of this section will be needed at different times and how you use them depends on various factors. You may have a course of group theory in which some or all of this material is handled; if so this section merely provides a view of that material that will be useful in this book. If such a course of group theory is not available then it will be necessary to treat some, at least, of the topics to a minimum depth depending on what material from the rest of the book is used in the course. Our intention in this chapter is thus to provide a central source for this material and a delineation of the group theory that will be needed. As explained earlier, we consider that group theory and low-dimensional topology interact so beautifully that it would be a pity not to include some vision of this interaction in any course involving knots, surfaces, etc.

6.1 Examples of presentations

We hope you will learn some useful and appealing group theory from this chapter, but we must remember it is not primarily a group theory textbook, and so we will take a fairly informal approach. We shall assume that you are familiar with the contents of a first basic course in group theory and have met subgroups, normal subgroups, quotient groups, homeomorphisms, and the 'isomorphism theorems'. It will also help if you have some simple examples of groups at your disposal, such as the cyclic groups, the dihedral groups, and the symmetric groups. We will start by examining the dihedral group, D_3, of order 6. This is the symmetry group of an equilateral triangle (Fig. 6.1)

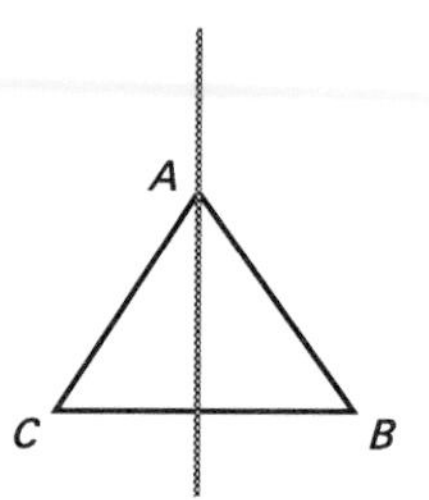

Fig. 6.1

Let R be the clockwise rotation through 120°, so that R permutes the vertices according to the rule $A \to B$, $B \to C$, $C \to A$. Let S be reflection in the axis of symmetry (shown) that goes through A and the midpoint of BC, so that S permutes the vertices according to the rule

$$A \to A, \quad B \to C, \quad C \to B.$$

One easily checks that any symmetry of the triangle can be written (in more than one way) as a composite of R and S. For instance, rotation through 240° (clockwise) is clearly $R^2 = RR$, whilst reflection in the axis through B (which is given by the permutation $A \to C$, $B \to B$, $C \to A$ can be written as R^2SR (= 'do R twice, reflect, finish with R'). The identity symmetry can be written as R^3 or R^6 or R^{-3} etc. or as S^2, S^4, etc.

If you have not yourself 'played' with D_3—and it is not enough to have watched someone else doing so—it would be a good idea to experiment now. For instance, give a description of each of the elements in the form R^iS^j. Write each of them in the form S^kR as well—is it possible? Remember that $R^0 = S^0 = I$, the identity symmetry.

As we mentioned above, every element of D_3 can be written as a 'word' in the elements R and S, that is it can be expressed as a product of these two elements. We say that R and S *generate* the group. This generating set $\{R, S\}$ is certainly not unique. If some set X of elements generates D_3 (so any element of D_3 can be written as a product of elements of X and their inverses in some order and possibly with repeats), then any Y containing X clearly must also generate the group. In particular, we could take $Y = D_3$, so the set of elements of D_3 can act as a set of generators for the group D_3! Usually, of course, we hope to find as small a set of generators as possible for a group, but we also want to try to get as simple an expression as possible for all the elements in terms of the *generators*. Of course the meaning of simple tends to vary somewhat from context to context. For instance, a set of generators for a group arising in a geometric context may be considered to be a good one because it allows the elements of the group to be expressed in a geometrically significant way, but whilst a geometer may prefer those generators, the significance of the same group to an algebraist may be such that a completely different set of generators may be considered much simpler. This suggests that we must develop ways of moving from one set of generators to another, a means of translating words in one 'alphabet' to words in another.

Knowing the generators is not enough. The symmetry group D_4 of a square has order 8; it is also generated by two elements. All symmetric groups are generated by two elements. The cyclic group C_6 or order 6 can be generated by two elements and so on. What distinguishes one group from another are the relationships or equations satisfied by the generators. For instance, in D_3 we saw that R^3 is the identity as is S^2, so R has order 3 whilst S has order 2, but in D_4 the two corresponding generators have orders 4 and 2. As an experiment let us try and generate words from R and S knowing only that $R^3 = 1 = S^2$:

$$1, R, R^2, S, RS, SR, R^2S, SR^2, SRS, SR^2S, \dots.$$

Thinking about this, it is clear that any word in R, R^2, and S can occur and there is no reason why any two of these should coincide. Thus we have not yet found a defining set of relations for D_3.

Using the same notation as before, RS is the composite

$$\begin{array}{ccccccccc} A & \longrightarrow & B & \longrightarrow & B & & A & \longrightarrow & B \\ B & \longrightarrow & C & \longrightarrow & A & = & B & \longrightarrow & A \\ C & \longrightarrow & A & \longrightarrow & C & & C & \longrightarrow & C \end{array}$$

so is the reflection in the axis through C and has order 2, that is $(RS)(RS) = 1$. This gives us

$$RS = (RS)^{-1} = S^{-1}R^{-1} = SR^2$$

(remember $S^2 = 1$ and $R^3 = 1$). We add $(RS)^2 = 1$ into our list of relations and note that we now get at most six words:

$$1, R, R^2, S, RS, R^2S.$$

(For instance, if one tries to form SR we find

$$RSRS = 1 \Rightarrow SR = R^{-1}S^{-1} = R^2S$$

so it is already in our list.)

This suggests that D_3 can be specified by naming two generators R, S satisfying

$$R^3 = S^2 = (RS)^2 = 1.$$

This *is* the case but we have not proved it yet! It might happen that $(RS)^2 = 1$ forces others of our list of six words to be equal. How can we know?

Exercise 6.1

In each of the following cases we have specified two generators and two relations. One extra relation is needed to obtain the desired group—find such a relation

(a) C_6: the cyclic group of order 6

$$\text{generators } x, y \text{ such that } x^3 = y^2 = 1$$

plus another relation.

(b) D_4: the symmetry group of the square = dihedral group of order 8

$$\text{generators } x, y \text{ such that } x^4 = y^2 = 1$$

plus another relation.
(In both cases, the answer will be given later on in this chapter.)

Thus we have arrived at an informal notion of a presentation of a group. We write

$$G = (X: R)$$

where X is a set of generators and R a set of relations.

Examples

(a) $D_3 = (x, y: x^3 = y^2 = (xy)^2 = 1)$
(b) $C_6 = (a: a^6 = 1)$.

6.2 Dihedral groups

The dihedral group D_n of order $2n$ is defined as being the symmetry group of a regular n-gon. This group has a presentation

$$(x, y : x^n = y^2 = (xy)^2 = 1).$$

We will need later on to handle presentations 'effortlessly'. We will also find certain ideas from group theory, such as commutators and conjugacy, useful. We will use this presentation of D_n to illustrate these concepts, thus also indicating how one plays with presentations.

Firstly we will show that any word in x and y can be reduced to one of the form $x^i y^j$ with $0 \leqslant i < n$, $j = 0$ or 1. The idea is simple. As $x^n = y^2 = 1$, any occurrence of x^m for $m > n$ can be reduced to one with x^{m-kn} where $0 \leqslant m - kn < n$. Similarly, any power of y other than 0 or 1 can be reduced to either $y^0 = 1$ or y itself. Now if anywhere in our word we have yx we can use $xyxy = 1$ to get $yx = x^{-1}y^{-1} = x^{n-1}y$ to move each x to the left of y and then to repeat the reduction as before. (For instance, if $n = 5$, $xyx^2 = x(yx)x = xx^4yx = x^5yx = yx = x^4y$.) This really needs to be set out more formally as a proof by induction on the number of letters in the word. If w is a word in x and y, the length of w, denoted $\ell(w)$, is the number of letters in w counted with multiplicities, for example the length of x^2yx^{-3} is 6. (Note: $x^{-3} = x^{-1} \cdot x^{-1} \cdot x^{-1}$ counts as $+3$ not -3.)

If $\ell(w) = 1$ then w is one of x, x^{-1}, y, or y^{-1}. Using the relations these reduce to x, x^{n-1}, y, and y respectively, so if $\ell(w) = 1$, everything is fine.

Assume that provided $\ell(w) < m$ then the result holds. Now look at w with $\ell(w) = m$. The word w can be written in one of four forms $w'x$, $w'x^{-1}$, $w'y$ or $w'y^{-1}$ where w' consists of all the letters except the last one, but $\ell(w') = m - 1$ and so w' reduces to $x^i y^j$ by assumption. The proof now proceeds to check every one of the different cases that can occur. (We will not do them all.)

1. If $w = w'x$ then w reduces to $x^i y^j x$. If $j = 0$ then we have x^{i+1} which either is of the required form or if $i = n - 1$ reduces further to x^0. If $j = 1$, $x^i yx$ reduces to $x^{i+n-1}y$ which in turn reduces to x^{i-1} or $x^{n-1}y$ depending on whether $i > 0$ or $i = 0$.

2. If $w = w'x^{-1}$ then $x^i y^j x^{-1}$ 'reduces' to $x^i y^j x^{n-1}$. This then reduces to $x^{i+n-1}y^j x^{n-2}$ if $j = 1$, and so on. We can safely leave the details to you. Try to find a formula for $x^i y^j x^{-1}$ where all the x to the right of y have been removed.

The last cases, 3 and 4, are easier. Try them yourself.

Given two elements g_1 and g_2 of some group G, we say g_1 and g_2 are *conjugate* if there is some $g \in G$ such that $g_2 = gg_1g^{-1}$. Being conjugate is an equivalence relation on G and the equivalence classes of conjugate elements are called conjugacy classes. Thus g_1 and g_2 are in the same conjugacy class exactly when they conjugate. Note that a subgroup N of G is normal if it is self-conjugate, that is if $N = gNg^{-1}$ for all $g \in G$.

Our next aim is to work out what the conjugacy classes in D_n look like. First we note that $(xy)^2 = 1$ and $y^2 = 1$ together have another consequence:

$$(xy)^2 = 1 \Rightarrow xy = y^{-1}x^{-1} = yx^{-1}.$$

We consider conjugating x^iy^j by x and by y:

(1) $j = 0$: $x(x^i)x^{-1} = x^i$
$j = 1$: $x(x^iy)x^{-1} = x^{i+1}yx^{-1} = x^{i+2}y$
(which may need reducing if $i \geqslant n - 2$)

(2) $j = 0$: $y(x^i)y^{-1} = yx^iy = x^{n-i}$

$j = 1$: $y(x^iy)y = yx^i = x^{n-i}y$.

Now if we try to conjugate x^iy^j by x^ky we have

$$(x^ky)(x^iy^j)(y^{-1}x^{-k})$$

and so we can find the conjugacy class of any element by repeated application of conjugation by the y and the x.

Example

Find the conjugacy class containing x:

x conjugated by any power of itself is x.
x conjugated by y is x^{n-1}.
x^{n-1} conjugated by any power of x is x^{n-1}.
x^{n-1} conjugated by y is x (why does this return on itself?). Thus the conjugacy class is $\{x, x^{n-1}\}$.

If G is a group, and $g_1, g_2 \in G$, then the *commutator* $[g_1, g_2]$ is the element

$$[g_1, g_2] = g_1g_2\,g_1^{-1}g_2^{-1}.$$

Thus $[g_1, g_2]$ measures how much g_1 and g_2 do not commute. The commutators in G generate a subgroup $[G,G]$ of G called the commutator subgroup of G. This subgroup is normal since a conjugate of a commutator is again a commutator. (Check it yourself. Note: $g(ab)g^{-1}$ can be rewritten $(gag^{-1})(gbg^{-1})$.)

The corresponding quotient group G_{ab} is called the *Abelianization* of G. It is an Abelian group and the quotient homeomorphism

$$\varphi : G \longrightarrow G_{ab}$$

has the pleasing and useful property that given any Abelian group H and a homeomorphism

$$\psi : G \longrightarrow H$$

we can factorize ψ uniquely via φ, that is we can find a homeomorphism

$$\psi' : G_{ab} \longrightarrow H$$

such that $\psi = \psi'\varphi$ and ψ' is the only homeomorphism with this property.

What is the Abelianization of D_n?

Remarks

Note that commutators obey some simply checked rules such as

$$[a, b]^{-1} = [b, a]$$

and

$$[ab, c] = a[b, c]a^{-1}[a, c].$$

In fact, if $G = (X : R)$ then $[G, G]$ is the smallest normal subgroup containing the set $\{[x, x'] : x, x' \in X\}$, so the Abelianization G_{ab} has presentation

$$G_{ab} = (X : R \cup \{[x, x'] : x, x' \in X\}).$$

(Again you should try to prove these statements *now* even though you will be given a proof later. In any case you will need the result several times when working with knot groups later.)

Exercises 6.2

1. Prove that a subgroup N of G is normal if and only if it is a union of conjugacy classes.
2. Prove that $[D_n, D_n] = \langle x^2 \rangle$, the subgroup generated by x^2. What is $(D_n)_{ab}$ if n is odd, and if n is even?
3. Find the conjugacy class of y in D_n. Repeat for x^2 and then for xy. Is there a pattern building up? Try to list *all* the conjugacy classes in D_n. (Note that when n is even there is a conjugacy class consisting just of a particular power of x.)
4. We have worked in some detail through the various properties of the dihedral groups, D_n, in order to get you thinking and working

with presentations. To check that you can handle group presentations in this informal way, we suggest you work out the corresponding facts about the dicyclic or generalized quaternion groups. The nth of these groups, denoted Q_{2n}, has presentation

$$(x, y : x^n = y, y^{-1}xy = x^{-1}).$$

To start with, prove that $x^{2n} = 1$, so $y^4 = 1$. (In fact, Q_{2n} has order $4n$.) Next work out the division of the elements of Q_{2n} into conjugacy classes. Using the fact that any normal subgroup must be a union of complete conjugacy classes, try to find *all* normal subgroups of Q_{2n} (and hence all quotient groups) and finally work out $[Q_{2n}, Q_{2n}]$ and $(Q_{2n})_{ab}$.

6.3 Cayley quivers

Given a group G and some set of generators X of G, how is one to find a complete set of relations? 'Complete' clearly means that no more are needed to get G. 'Trial and error' may work but is haphazard. For a start, if one has a set of generators X and some set R of relations, how is one to know even if $(X : R)$ has the same order as G? The group G may, for instance, be finite, yet $(X : R)$ is infinite because we have not yet found all the relations necessary to 'capture' G.

If G is fairly small one can draw a Cayley quiver (also called a Cayley colour graph). This is a neat way of displaying the needed information and is also of great theoretical importance, as we will see later on.

We assume that G is known and X is known to be a set of generators. We construct a graph having the elements of G as vertices. More precisely, we set

$$V = \{v_g : g \in G\}$$

to be a set of vertices in one–one correspondence with G (and thus labelled by the elements of G). If v_g is a vertex and $x \in X$ a generator, we draw an edge (labelled with x) from the vertex v_g to the vertex v_{gx}. We do this for each $g \in G$ and $x \in X$. The resulting graph is both directed (the edge goes 'from v_g to v_{gx}') and hence is a quiver (see Chapter 7) and labelled. The labelling is often thought of as a colouring of the graph, hence the term 'Cayley colour graph'.

1. C_5 generated by x (Fig. 6.2)

$$C_5 = \{1, x, x^2, x^3, x^4\} \quad X = \{x\}$$

There is only one label.

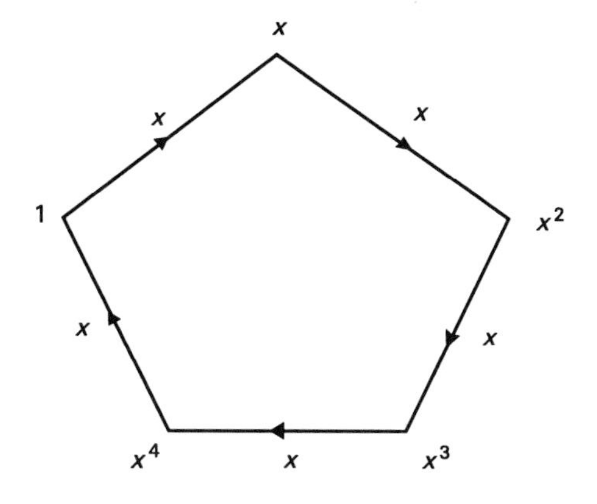

Fig. 6.2

2. Of more interest is D_3 (or S_3) generated by R and S (Fig. 6.3).

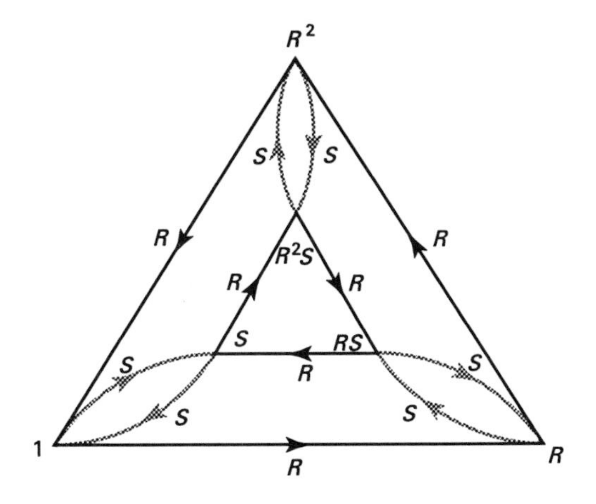

Fig. 6.3

The problem of drawing the Cayley quiver of a group G with respect to generating set X is that it rarely turns out to look as nice as those which have been given to you in a book! Knowing the Cayley quiver does, however, yield the relations. One can 'see' that $R^3 = 1$, $S^2 = 1$ and $RSRS = 1$. To evaluate a word in R and S, we need only follow the path through the quiver specified by that word. For instance, consider

$$SRRS^{-1}RSR.$$

What is it? Reading from left to right and starting at vertex 1, go along the S labelled edge if you meet an S and along the R labelled

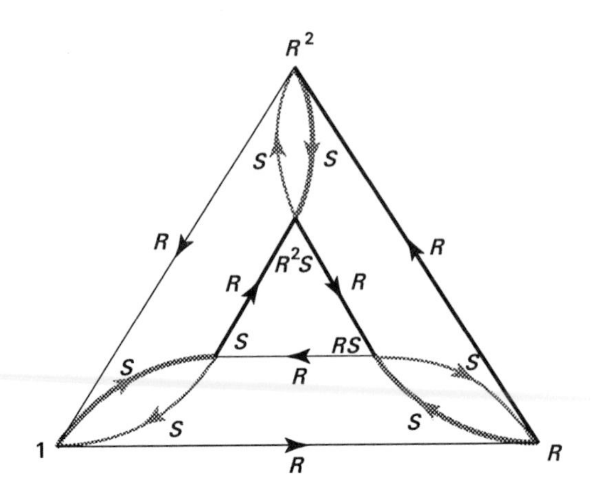

Fig. 6.4

one if you meet an R. The word corresponds to the path as shown in bold in Fig. 6.4. So we end up at the vertex labelled RS.

Using this idea gives the relations listed, $R^3 = 1$, $S^2 = 1$, and $(RS)^2 = 1$, and 'clearly' all other loops at 1 can be considered as consequences of these relations. We only intend this as a vague idea here. We need more machinery before making it precise.

Exercises 6.3

1. Draw up a Cayley quiver for $C_2 \times C_3$ ($\cong C_6$) generated by a of order 2 and b of order 3. (If you prefer, C_6 represented as

$$\{1, c, \underset{\substack{\uparrow \\ b}}{c^2}, \underset{\substack{\uparrow \\ a}}{c^3}, c^4, c^5\};$$

then a and b are as shown.) Compare this with the Cayley quiver of D_3.

2. (a) Draw up a Cayley quiver for

$$D_4 = (x, y : x^4 = y^2 = (xy)^2 = 1)$$

with x and y as the generators.

(b) What will be the structure of the Cayley quiver of

$$D_n = (x, t : x^n = y^2 = (xy)^2 = 1)$$

again with x and y as the generators?

3. Investigate the Cayley quiver of Q_{2n} with the presentation given earlier. In particular draw that of Q_4 which has order 8.

4. An automorphism of a labelled quiver is a bijective mapping φ which assigns vertices to vertices and edges to edges respecting labels (so that if edge e is labelled x then it can only be sent to another edge labelled x) and respecting incidence, that is if e is an edge from a to b, $\varphi(e)$ will be an edge from $\varphi(a)$ to $\varphi(b)$.

For example, taking Γ to be the Cayley quiver of D_3 with generators R and S, a rotation of the graph through 120° is an automorphism of Γ as is a transformation that interchanges the inner and outer triangles reversing directions. List all automorphisms of this quiver. You can compose automorphisms by 'do one and then the other' so that they form a group Aut(Γ). Find the structure of Aut(Γ) when Γ is the Cayley quiver of D_3 as above. Repeat with other Cayley quivers. This should enable you to conjecture a result. Try to prove what you have conjectured.

Remarks

1. We have assumed that the idea of a quiver is fairly intuitive in this section. Before we investigate quivers in greater detail and in particular Cayley quivers, we will take more care with the definitions. Which comes first, examples or definitions? Historically it is usually the examples first, and then definitions followed by more examples. You do not think of a precise definition of a concept until some examples are available, but the precise concept then brings to light other examples.

2. Given a presentation $(X : R)$, it can be impossible to construct the Cayley quiver of the group. The construction is equivalent to solving the 'word problem' for that presentation. The 'word problem' is of deciding when two words in the generators yield the same element of the group. This has been proved in general to be unsolvable!

If the group is known to be finite, then there is a method known as coset enumeration which not only gives the exact size of the group, but also yields a table which can help in drawing a Cayley quiver. Unfortunately the Todd–Coxeter algorithm used in this method will not tell you if a group is not finite, so this method of obtaining a Cayley quiver cannot be extended to infinite groups. This is annoying as most of the groups we will be using later on (knot groups and surface groups) are infinite groups. (For a detailed treatment of the Todd–Coxeter method of coset enumeration, the reader can consult Johnson (1990)).

Notwithstanding this comment about the difficulty of drawing Cayley quivers of infinite groups in general, Cayley quivers will be useful later on both for giving examples of certain topological constructions and for 'spin-off' back in group theory of those same constructions.

6.4 Free groups

There is a problem of an almost philosophical nature that needs to be resolved when dealing with presentations. If we start with a group G and elements $x, y \in G$, we can easily talk about how to generate words from the 'letters' x, y and their inverses, and then look for relations amongst these words. However, we sometimes do not know anything about G except that x and y are in it and possibly that there are some relations. What are these 'relations'? We can think of them as equations satisfied by x and y: the use of the word 'equation' implies that two things are the same. Where are these two 'things' to live? The logical problem involved is important but there is also a practical side. Unless we have a precise meaning for 'word' and 'relation', it will be difficult to avoid circular arguments; we will not know what to believe or what methods are valid. The idea that solves the problem is a very elegant one, namely that of a free group. We start with a set, X, and generate new elements by 'formally' multiplying the elements of X together to get 'words' in the elements of X, continuing until we have constructed a group $F(X)$, the free group on X. For instance, if $X = \{x\}$ then the free group on X must contain all powers of x, both positive and negative, and that is all. It is 'free' since no two of these powers are the same, so there are no constraints imposed on it. We will return to the formalities of the construction later. We will start with a formal definition in terms of the most powerful property that free groups have.

Definition

A group F is said to be free with basis $X \subseteq F$ if for any group H, any mapping ϕ: $X \to H$ can be uniquely extended to a group homeomorphism ϕ': $F \to H$ (i.e. if $x \in X \subseteq F$, $\phi'(x) = \phi(x)$; the diagram

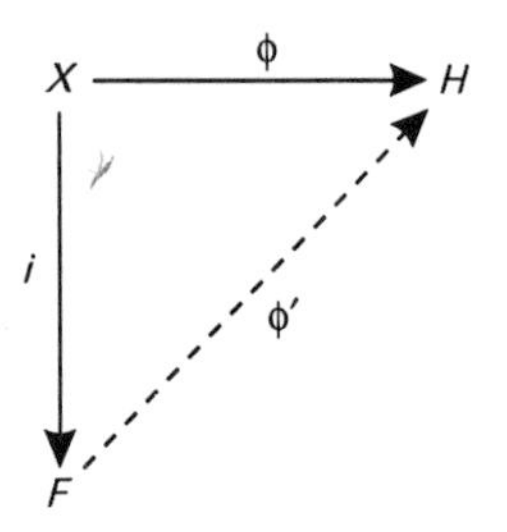

commutes, $\phi' i = \phi$ *as mappings*).

Remarks

1. This situation can be thought of as being a non-linear analogue of the fact in linear algebra that in order to define a linear transformation

$$T: V \longrightarrow V'$$

between two vector spaces, it is sufficient to choose a basis $\{e_i : i \in I\}$ for V and to specify all the vectors, $Te_i \in V'$. Since $\{e_i\}$ is a basis for V we have, if $u \in V$, $u = \Sigma_{i \in I} u_i e_i$ for some unique set of coefficients $\{u_i : i \in I\}$ and as T has to be linear, the only possible value for $T(v)$ is $\Sigma u_i T(e_i)$. Thus, specifying the images of the basis specifies a unique T defined on all of V.

2. If we suppose X consists of a single element x, then $\phi : X \to H$ is determined by the element $\phi(x)$. We suggested that the free group on $\{x\}$ consisted of all positive and negative (formal) powers of x, that is

$$F(\{x\}) = \{x^n : n \in \mathbb{Z}\}.$$

This is an infinite cyclic group. We can see that it is free on X since the fact that ϕ': $F(\{x\}) \to H$ has to be a homeomorphism ensures that $\phi'(x^n) = \phi'(x)^n$ and this is $\phi(x)^n \in H$.

Hopefully you can guess at least some of the construction of $F(X)$ in general. We will call the set X the *alphabet*; the elements of X will be *letters*. We define a symbol a^n to be a *syllable*, so a^0, a^{-1}, s^2, a^{101}, etc., are all syllables. A *word* is a finite ordered sequence of syllables, for example

$$c^3 b^{-3} a^0 a^1 c^2 c^{-3} b^2$$

where $a, b, c \in X$.

There is one special word, namely the empty word, which has no syllables. We denote it by 1.

The set of all possible words on the alphabet X will be denoted $W(X)$, and we can map X into $W(X)$ via the mapping $a \mapsto a^1$. We define the product of words by juxtaposition; for instance, for

$$a^2 b^{-1} c^1 \quad \text{and} \quad b^1 a^3 d^{-2} \in W(X)$$

then we take

$$(a^2 b^{-1} c^1) \cdot (b^1 a^3 d^{-2}) = a^2 b^{-1} c^1 b^1 a^3 d^{-2} \in W(X).$$

We note (i) that multiplication of words is closed: if $w_1, w_2 \in W(X)$ then $W_1 \cdot W_2 \in W(X)$; (ii) that multiplication is associative: if

$w_1, w_2, w_3 \in W(X)$ then $(w_1 w_2)w_3 = w_1(w_2 w_3)$; and (iii) that for any $w \in W(X)$, $1.w = w.1 = w$, so the empty word, 1, acts as an identity.

Since we have treated words as formal objects, $W(X)$ with this multiplication is not a group. Even though a^1 and a^{-1} are in $W(X)$, we do not have that $a^1 \cdot a^{-1}$ is empty. It is clear that if a group is to be constructed from $W(X)$ then $a^1 \cdot a^{-1}$ and 1 must be identified, that is we must form the quotient of $W(X)$ by some equivalence relation so that the set of equivalence classes is a group. The equivalence relation must clearly include the relations $a^1 \cdot a^{-1} \sim 1$, $a^{-1} \cdot a^1 \sim 1$ and must be compatible with multiplication. We construct the equivalence relation in a series of steps.

Definition of the equivalence relation on $W(X)$

1. *If $w = w_1 a^0 w_2$, $w' = w_1 w_2$, where $w_1, w_2 \in W(X)$ and $a \in X$, then we say that the change, w goes to w', is an elementary contraction of type I. Its reverse, w' goes to w, is an elementary expansion of type I.*

2. *If $w = w_1 a^p a^q w_2$, $w'' = w_1 a^{p+q} w_2$ then we say that the change from w to w'' is an elementary contraction of type II etc. (This just means we have introduced the usual laws of indices.)*

3. *If u, v are in $W(X)$, we will write $u \sim v$ if one can be obtained from the other by a finite sequence of the above elementary operations.*

It should be clear that $\sim$ is an equivalence relation; however, it is worth repeating that 'it is clear' should be a signal for the reader to check that it is indeed clear and why. We therefore suggest that you sketch out a proof that $\sim$ is an equivalence relation. We will denote by $[v]$ the equivalence class containing $v \in W(X)$. We then note the following.

6.1 Lemma. *If $u, v, w \in W(X)$ and $u \sim v$ then $uw \sim vw$ and $wu \sim wv$.*

We outline a proof, again leaving it up to you to fill in the details. The first step is to note that if $u = u_1 a^0 u_2$, $v = u_1 u_2$ then uw and vw are linked by an elementary contraction of type I. (Why?) The same holds for wu and wv. The next point is to note a similar result for contractions of type II. Finally we have that there are $u_2, \ldots, u_n$ for some n, $u = u_1$, $v = u_n$, and each u_i linked to u_{i+1} by elementary operations. We can thus use $u_1 w, \ldots, u_n w$ to link uw and vw in a similar way. A repeat of this argument with wu_1, wu_n completes the proof.

As a result of this lemma, we have that the operation

$$[u][v] = [uv]$$

is *well defined*, that is it does not depend on the representatives $u \in [u]$ and $v \in [v]$. If we write $F(X)$ for the set of equivalence classes, we can note the following properties of this operation:

(1) [1] acts as an identity element since $1u = u1 = u$ implies that $[1][u] = [u][1] = [u]$ for all $u \in W(X)$,

(2) the operation is associative as, if $[u], [v], [w] \in F(X)$, we have $u(vw) = (uv)w$ and so $[u]([v][w]) = [u][vw] = [u(vw)] = [(uv)w] = \ldots$, etc.

(3) each element of $F(X)$ has an inverse. To see this, first note that for any syllable a^p, we have $a^{-p} \in W(X)$ and

$$[a^p][a^{-p}] = [a^{p+(-p)}] = [a^0] = [1].$$

Any word $u \in W(X)$ can be written as a finite string of syllables and hence can be expressed in the form $u = a_1^{p_1}a_2^{p_2} \ldots a_n^{p_n}$, where the $a_i \in X$ and the $p_i \in \mathbb{Z}$, the integers. Of course in general the list of letters $a_1, \ldots, a_n$ may contain many repeats, they may, in fact, all be equal, and some of the p_i may be zero. This does not matter as will be clear, we hope. Define a 'reverse' of u to be

$$u^{(r)} = a_n^{-p_n} \ldots a_2^{-p_2} a_1^{-p_1},$$

that is the string in reverse order with the sign of each power changed. An induction on n shows that

$$u^{(r)}u \sim 1 \quad \text{and} \quad uu^{(r)} \sim 1.$$

(As usual we expect you to check that you can do this basic type of argument satisfactorily.) Now we have

$$[u^{(r)}][u] = 1 \quad \text{and} \quad [u][u^{(r)}] = 1$$

so $[u]$ has an inverse. One might ask if $[u^{(r)}]$ depended on the choice of u within the equivalence class, $[u]$. This question can be tackled in two different ways. As they are again usual proof forms we expect you to complete the details. The first involves checking that if u and v are linked by an elementary operation then $u^{(r)}$ and $v^{(r)}$ are linked by the same sort of elementary operation. (The argument is quite simple, do not try to make it difficult!) Then assume u and v are linked by a finite sequence $u_1, \ldots, u_n$ as before and prove $u^{(r)}$ and $v^{(r)}$ are linked by $u_1^{(r)}, \ldots, u_n^{(r)}$.

The other proof is neater but uses the above in a hidden form, that is by using results we have already noted. We assume $u \sim v$ so $[u] = [v]$ and wish to compare $[u^{(r)}]$ and $[v^{(r)}]$. We use that multiplication is associative, that $[u^{(r)}][u] = [1]$ and $[v][v^{(r)}] = [1]$, together with the identity property of [1]:

$$[u^{(r)}] = [u^{(r)}][1] = [u^{(r)}]([v][v^{(r)}]) = ([u^{(r)}][v])[v^{(r)}]$$
$$= ([u^{(r)}][u])[v^{(r)}] = [1][v^{(r)}] = [v^{(r)}]$$

Summing up, we see that $F(X)$ is a group. Is it a free group on X? We can think of X as included in $F(X)$ via the mapping $a \mapsto [a^1]$, so suppose we have a group H and a mapping ϕ as in the definition of freeness. We have to prove that (i) ϕ extends to give some homeomorphism $\phi'\colon F(X) \to H$ and (ii) ϕ' is unique with this property (i.e. if $\psi\colon F(X) \to H$ is any homeomorphism satisfying $\psi(x) = \phi(x)$ for each $x \in X$ then $\psi = \phi'$).

We first look at the existence of ϕ'. We have to decide what $\phi'[u]$ is for an arbitrary $[u] \in F(X)$. There is an obvious way of defining a mapping $\phi_1\colon W(X) \to H$ which extends ϕ, is completely determined by it, and is as near to being a homeomorphism as makes sense (remember $W(X)$ is *not* a group). For instance, if $a \in X$ there is a syllable $a^n \in W(X)$. We clearly would like $\phi_1(a^n) = \phi(a)^n$ since this would have to be true if $W(X)$ *was* a group and ϕ_1 *was* a homeomorphism. An arbitrary word $w \in W(X)$ is a string of syllables

$$w = a_1^{n_1}, \ \ldots, a_r^{n_r}$$

and if ϕ_1 was a homeomorphism we would get

$$\phi_1(w) = \phi(a_1)^{n_1} \ldots \phi(a_r)^{n_r}$$

which makes sense since it is an equation between elements of H. Since this is what we hope will happen, why not explore the consequences of taking it as a definition of ϕ_1? The set $W(X)$ has a multiplication and it is clear that $\phi_1(w_1 w_2) = \phi_1(w_1)\phi_1(w_2)$. At this point we must pause—what is $\phi_1(1)$? We have not dotted our i's and crossed our t's. The empty word, 1, is in $W(X)$ but is an empty string (so is as above with $r = 0$). What, therefore, is $\phi_1(1)$? Clearly it must be the product of an empty family of elements of H, but the exact meaning of that is not clear. However, in H, the usual laws of indices etc. hold so we always have, for instance,

$$h^m \cdot h^n = h^{m+n}.$$

If we have a string of elements *in* H, and multiply it by the empty string (in H) then we must leave the original string unchanged, so the only sensible interpretation for $\phi_1(1)$ is that it is the identity element of H.

So far we have used ϕ to construct a nicely behaved $\phi_1\colon W(X) \to H$. Is ϕ_1 compatible with the equivalence relation we have defined on $W(X)$?

6.2 Lemma. 1. *If* $w = w_1 a^0 w_2$ *and* $w' = w_1 w_2$ *then* $\phi_1(w) = \phi_1(w')$.

2. *If* $w = w_1 a^p a^q w_2$ *and* $w' = w_1 a^{p+q} w_2$ *then* $\phi_1(w) = \phi_1(w')$.

We leave the proof to you. It just uses the fact that the laws of indices hold in H. As an immediate consequence of this lemma, we get that if u, v are in $W(X)$ and $u \sim v$ then $\phi_1(u) = \phi_1(v)$. This in turn implies that we can (without fear of getting into difficulties) *define*

$$\phi'[u] = \phi_1(u).$$

This will not depend on which representing u we take for $[u]$. (It is well defined.)

We have taken as operation on these equivalence classes

$$[u][v] = [uv]$$

and we can check that $\phi'([u][v]) = \phi'([u])\phi'([v])$ for all $[u]$, $[v] \in F(X)$ (including the class of the empty string, 1). This completes our checking that ϕ does extend to a suitable ϕ_1. It also shows that if $w = a_1^{n_1} \dots a_r^{n_r}$ then $\phi_1[w] = (\phi(a_1))^{n_1} \dots (\phi(a_r))^{n_r}$.

How about uniqueness? Suppose ψ: $F(X) \to H$ extends ϕ. This means that $\psi[a] = \phi(a)$. If $[w] \in F(X)$, then if $w = a_1^{n_1} \dots a_r^{n_r}$ we have $[w] = [a_1]^{n_1} \dots [a_r]^{n_r}$, and as ψ is a homeomorphism

$$\begin{aligned}\psi[w] &= (\psi[a_1])^{n_1} \dots (\psi[a_r])^{n_r} \\ &= \phi(a_1)^{n_1} \dots \phi(a_r)^{n_r} \\ &= \phi'[w]\end{aligned}$$

by our calculation earlier.

We have proved that $F(X)$ is indeed free with X as basis. We note that we have never actually checked that the mapping $a \mapsto [a]$ from X to $F(X)$ was one–one but if $[a_1] = [a_2]$ for some $a_1, a_2 \in X$ with $a_1 \neq a_2$, then we could pick $H = C_2$, the cyclic group of order 2, and a mapping ϕ: $X \to C_2$ such that $\phi(a_1) \neq \phi(a_2)$. Now to extend to the unique ϕ': $F(X) \to C_2$, we must have $\phi'[a_1] = \phi'[a_2]$ (since $[a_1] = [a_2]$!) but $\phi'[a_1] = \phi(a_1)$ and $\phi'[a_2] = \phi(a_2)$, ... We clearly have to abandon any idea that we can have both $a_1 \neq a_2$ and $[a_1] = [a_2]$.

If two groups F_1 and F_2 are free with bases X_1 and X_2 respectively, then any function f: $X_1 \to X_2$ extends uniquely to a homeomorphism $F(f)$: $F_1 \to F_2$. We use the 'universal' defining property of freeness to check this.

Suppose we write i_1: $X_1 \to F_1$ for the inclusion, $i_1(a) = [a]$ similarly for i_1: $X_2 \to F_2$. The function f: $X_1 \to X_2$ gives us a mapping $\phi = i_2 f$: $X_1 \to F_2$ which extends uniquely by the defining property to a homeomorphism ϕ_1: $F_1 \to F_2$ which satisfies $\phi_1 i_1 = i_2 f$. It is this ϕ_1 that we take as $F(f)$. Thus if X_1 has the same cardinality as X_2, the corresponding free groups F_1 and F_2 are isomorphic. (Check this.) In fact the converse holds: if F_i is free on X_i for $i = 1, 2$ and F_1 and F_2 are isomorphic, then X_1 and X_2 have the same cardinality. This cardinality will be called the *rank* of the free group.

We refer the interested reader to Johnson (1990) where the theory of free groups is explored in more detail. We need free groups primarily for their use within the theory of group presentations. We introduced them earlier as a formal solution to the problem of where the words, that were the 'relations' in a presentation, could live. In detail the way they are used is a result of the following proposition.

6.3. Proposition

(a) *A set $X \subseteq G$ generates G if and only if the homeomorphism from $F(X)$ to G, associated to the inclusion, is an epimorphism.*

(b) *Every group is a homeomorphic image of a free group.*

Proof. First a notational solution to our earlier 'philosophical' problem: if $x \in X$, we will write $\bar{x}$ if we are thinking of it as a generator of $F(X)$ and x, without a bar if as an element of G. This is necessary since we have, in $F(X)$, elements such as $[x_1][x_2]$ and if, as might happen, $x_1x_2 \in X$, we have the possible confusion between $[x_1][x_2]$ and $[x_1x_2]$, which are *different* elements of $F(X)$. (With this notation we have $[\bar{x}_1][\bar{x}_2] = [\bar{x}_1\bar{x}_2]$ whilst the other element is $[\overline{x_1x_2}.]$

The homeomorphism φ associated to the inclusion of X into G is then given by removing the bar above each generator and each square bracket; for instance,

$$[\bar{x}_1\bar{x}_2^{-1}\bar{x}_3\bar{x}_2] \in F(X)$$

will be sent to $x_1x_2^{-1}x_3x_2 \in G$.

Now assume X generates G. Recall that formally this means that any element of G can be written as a product of elements of $X \cup X^{-1}$ where $X^{-1} = \{x^{-1} : x \in X\}$ is the set of inverses of elements in X. Thus given any element $g \in G$, we find an expression of the form

$$g = x_1^{\varepsilon_1} \dots x_n^{\varepsilon_n}$$

where $\varepsilon_i = \pm 1$ and all $x_i \in X$. The word

$$w = [\bar{x}_1^{\varepsilon_1} \dots \bar{x}_n^{\varepsilon_n}] \in F(X)$$

is sent to g by φ, so φ is onto. The converse is simple.

To see (b) we just have to note that any group has a set of generators, for instance one can take X to be the set G itself. ■

6.5 Presentations

Given a set of generators X for a group G we obtain the corresponding epimorphism

$$\varphi\colon F(X) \longrightarrow G.$$

We can now give a precise sense to the term 'relation' and in particular 'defining relation'. The above epimorphism gives us an isomorphism between the quotient group $F(X)/\mathrm{Ker}\,\varphi$ and G (remember the first isomorphism theorem of group theory), so if we knew how to construct this normal subgroup Ker φ, we could construct G from our set of generators which is what we hope to be able to do. Any word w in Ker φ is, of course, 'killed off' by φ. Let us examine this in our example of D_3 with $X = \{R, S\}$. (For the moment we will keep our cumbersome notation $\bar{R}$, $\bar{S}$ etc. to denote the elements of X when they are thought of as basic elements of $F(X)$, but be warned that we shall soon drop the use of the bar and you will have to do a bit of 'doublethink', when looking at presentations.) In $F(X)$ we have, for example, the word

$$w = \bar{S}^{-1}\bar{R}\bar{R}\bar{S}\bar{R}^{-1}$$

which is 'reduced', that is it cannot be made shorter using elementary contractions or expansions. This word w is in the kernel of φ, however, since in D_3 we have

$$S^2 = 1 \quad (\text{so } S^{-1} = S)$$
$$R^3 = 1 \quad (\text{so } R^{-1} = R^2)$$

and $(RS)^2 = 1$ (so $RS = S^{-1}R^{-1} = SR^2$). Hence

$$\begin{aligned}\varphi(w) &= S^{-1}R^2SR^{-1}\\ &= SR^2SR^2\\ &= RSRS\\ &= 1,\end{aligned}$$

as stated. Thus our word w expresses a relationship between the chosen generators of D_3; it is a relation.

Given a set of generators X of a group G with $\varphi\colon F(X) \to G$ the associated epimorphism, an element $w \in \mathrm{Ker}\varphi$ is called a *relation* (between the given generators) and Ker φ the *relation subgroup*.

Except when it is trivial, Ker φ will always be an infinite group, so it is clearly out of the question to specify all relations; however, we could specify a generating set of relations. Let us return to our example of D_3 with generators R and S. We have R^3, S^2 and $(RS)^2$ are relations; do they generate the relation subgroup? Suppose we conjugate R^3 by S to get SR^3S^{-1}; this is in Ker φ but it is not clear whether or not it can be written as a product of R^3, S^2 and $(RS)^2$. How are we to tell? As usual the answer is—with difficulty. You might try all sorts of combinations of copies of R^3, S^2 and $(RS)^2$ in various orders for several days, and yet not hit on one which gave the result, but clearly that will not do as a proof that it cannot be done. The fact is that Ker φ in this case is a free group of rank 7 (why this is true will be explained later) and because of this it cannot be generated by only three elements (again this must wait for justification later). However, it *is* generated by these three elements together with their conjugates. This shows a point of great importance: we need only specify a set of 'normal generators' for Ker φ as it is a *normal* subgroup of $F(X)$. We need to examine this in slightly more detail. First some observations.

1. Given any group G and a collection of subgroups $S_\lambda \subset G$, $\lambda \in \Lambda$, some indexing set, then $\cap\{S_\lambda: \lambda \in \Lambda\}$ is also a subgroup of G. (Check this for yourselves if you have not met the result before.)

2. If $X \subset G$ and X generates a subgroup, H, say, of G (i.e. the elements of H are exactly the products of elements from X and X^{-1}) then

$$H = \cap\{S\colon\ S \text{ a subgroup of } G,\ X \subseteq S\}.$$

3. If all the subgroups S_λ in observation 1 are normal in G, so is their intersection.

4. If $X \subset G$, then

$$H = \cap\{S\colon S \text{ a normal subgroup of } G,\ X \subseteq S\}$$

is the smallest normal subgroup of G that contains X. (It is often called the *normal closure* of X in G.)

If we denote by X^{-1} the set of inverses of elements of X, so $X^{-1} = \{x^{-1}\colon x \in X\}$, then the subgroup generated by X, denoted $\langle X\rangle$, consists of all products of elements from X and X^{-1}. It is the image of the natural homeomorphism from $F(X)$ to G corresponding to the inclusion of X into G. Can we give a similar description of the normal closure of X in G?

Let us denote this by $\langle\langle X\rangle\rangle$. Clearly since $\langle\langle X\rangle\rangle$ is normal in G, it must contain not only each $x \in X$ but all the conjugates gxg^{-1} of x by elements of G, together with all the inverses and products of all these elements. If $g \in G$, $x \in X$, we will write ${}^g x$ for gxg^{-1}. A *consequence* of X will be an element of G that can be written as a product of elements of the form ${}^g x$ and their inverses, so y is a consequence of X if and only if

$$y = ({}^{g_1}x_1)^{\varepsilon_1}({}^{g_2}x_2)^{\varepsilon_2} \dots ({}^{g_n}x_n)^{\varepsilon_n}$$

for some $g_1, \dots, g_n \in G$, $x_1, \dots, x_n \in X$ and where each ε_i is ± 1. The consequences of X in G form a normal subgroup. (Prove it!) In fact the normal subgroup is the normal closure of X in G. (Again prove it.)

Definition

(a) *Given a group G and a set X of generators, a subset R of $F(X)$ is called a set of defining relations for G if* $\langle\langle R\rangle\rangle =$ Ker $(\varphi\colon F(X) \to G)$.

(b) *A presentation of a group G consists of a set X and a subset R of $F(X)$ so that $F(X)/\langle\langle R\rangle\rangle$ is isomorphic to G. We write* $G = (X : R)$.

Examples

Let $X = \{x, y\}$, $R = \{x^3, y^2, (xy)^2\}$; then $(X : R)$ is the presentation of D_3 that we have been looking to for 'guidance'. Usually we write things slightly less formally and it is quite usual to write alternative versions such as

$$(x, y : x^3 = y^2 = (xy)^2 = 1)$$

or possibly, using that $(xy)(xy) = 1$ is the same as $xy = y^{-1}x^{-1}$,

$$(x, y : x^3 = y^2 = 1, xy = y^{-1}x^{-1}).$$

The point is that now we have set up the formal structure of presentations, we can play around with them with more confidence. This 'playing around' often involves substituting some consequence of one relation. For example, consider the presentation

$$(x, y, z : x^3 = y^2 = 1, xyz^{-1} = 1, z^2 = 1).$$

The third relation tells us that $z = xy$, so we do not need z as a generator—it is already there in terms of x and y. We can substitute xy for every occurrence of z in other relations and delete z from the list of generators without changing the group. This sort of process is so useful that it will be used time and time again later on in the discussion of presentations that arise from knots. Because of this, it is important to have a list of acceptable 'transformations' of presentations that do not change the group.

6.6 Tietze transformations

As usual $G = (X : R)$. The following transformations do not change the group G:

T1: Adding a superfluous relation

$$(X : R) \text{ becomes } (X : R') \text{ where } R' = R \cup \{r\}$$

and $r \in \langle\langle R \rangle\rangle$, that is a consequence of R.

T2: Removing a superfluous relation

$$(X : R) \text{ becomes } (X : R') \text{ where } R' = R \backslash \{r\}$$

and r is a consequence of R'.

T3: Adding a superfluous generator

$$(X : R) \text{ becomes } (X' : R') \text{ where } X' = X \cup \{g\}$$

(g is a new symbol not in X) and $R' = R \cup \{wg^{-1}\}$ where w is a word in the other generators, that is w is in the image of the inclusion of $F(X)$ into $F(X')$.

T4: Removing a superfluous generator

$$(X : R) \text{ becomes } (X' : R') \text{ where } X' = X \backslash \{g\}$$

and $R' = R\backslash\{wg^{-1}\}$, with $w \in F(X')$ and $wg^{-1} \in R$ and no other member of R' involves g.

These four operations are called Tietze transformations, after H. Tietze, who did pioneering work on group presentations (see Tietze 1908; Magnus and Chandler 1982).

We normally carry out more than one change at a time and it is advisable to work things out informally first before launching into formal manipulations.

Example

To show $(x, y : x^l = y^m = (xy)^n = 1)$ is isomorphic to $(a, b : a^n = b^m = (ab)^l = 1)$.

Formally, start:

$$(x, y : x^l = y^m = (xy)^n = 1).$$

Apply **T3** twice to introduce a and b:

$$(x, y, a, b : x^l = y^m = (xy)^n = (xya^{-1}) = y^{-1}b^{-1} = 1).$$

Apply **T1** three times to introduce superfluous relations:

$$(x, y, a, b : x^l = y^m = (xy)^n = xya^{-1} = y^{-1}b^{-1} = a^n = b^m = (ab)^l = 1).$$

Really we should justify that $a^n = 1$, $b^m = 1$ and $(ab)^l = 1$ are consequences of the other relations. This step is often not easy to do. Here we can write

$$a^n = (xya^{-1})^{-1} \prod_{i=1}^{n-1} (xy)^i (xya^{-1})^{-1} (xy)^{-i} (xy)^n$$

so a^n *is* a product of conjugates of the other relations, similarly for b^m and $(ab)^l$. The formula above may be represented in a Van Kampen diagram. For $n = 3$, the appropriate Van Kampen diagram is as shown in Fig. 6.5.

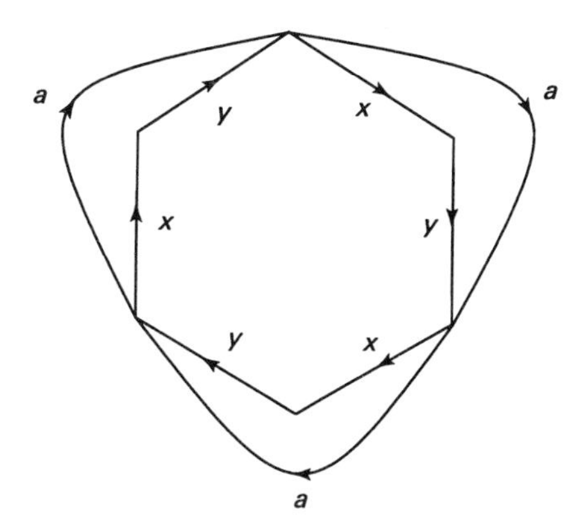

Fig. 6.5

A circuit around the inner hexagon gives $(xy)^3$ whilst the three-sided faces all have circuit xya^{-1}. Note that a circuit around the outside of the diagram gives one a^3. The 'secret' is to break up this outer circuit into a 'sum' of the various faces it 'encircles'. A full account of the theory underlying the use of these diagrams may be found in Johnson (1990).

Returning to the presentation we can treat xya^{-1}, $y^{-1}b^{-1}$, a^n, b^m and $(ab)^l$ as 'basic relations' now and prove that x^l, y^m, and $(xy)^n$ are superfluous relations, so three uses of **T2** reduce our presentation to

$$(x, y, a, b : xya^{-1} = y^{-1}b^{-1} = a^n = b^m = (ab)^l = 1).$$

Next we use $y^{-1}b^{-1}$ and xya^{-1} or rather its conjugate $ya^{-1}x$ to deduce $b^{-1}a^{-1}x$ and hence $x^{-1}ab$ and its conjugate abx^{-1}. Now we have

$$(x, y, a, b : xya^{-1} = y^{-1}b^{-1} = a^n = b^m = (ab)^l = abx^{-1} = 1).$$

The relation xya^{-1} is superfluous and can be deleted; $y^{-1}b^{-1}$ gives us $b^{-1}y^{-1}$, so with abx^{-1} we can apply **T4** twice to get the desired presentation

$$(a, b : a^n = b^m = (ab)^l = 1).$$

This formal method is usually very lengthy, and hence it is usual that people use the informal 'substitution' method described earlier. What is important is to realize that the formality is possible and if any step in the informal sequence of transformations is doubtful, the formal method is there to act as final arbiter. Informally, we start with $(x, y : x^l = y^m = (xy)^n = 1)$. We try to define a and b in terms of x and y to obtain $a^n = b^m = (ab)^l = 1$. The obvious choice is $a = xy$; then $a^n = (xy)^n = 1$.

Next, since we want $(ab)^l = 1$, it seems a good idea to pick b so that $ab = x$, but this means $b = y^{-1}$. We try this out to see if it works. The only relation left to be checked (as we have used two already) is $b^m = 1$, but $b^m = y^{-m} = 1$.

Working the other way: given $(a, b : a^n = b^m = (ab)^l = 1)$, we must define x and y in terms of a and b. Trying $x = ab$ and $y = b^{-1}$ clearly works.

This process of transforming a set of relations to get a simpler equivalent one is very similar to the process we have seen in use on surface symbols. In fact the transformation of surface symbols by operations is completely analogous to the Tietze transformation process, and formally these processes are the same. The intuitive resemblance can be made rigorous via the fundamental group of the surface. This group, which we will meet in detail in Chapter 11, has generators the edge letters and a single relation $A = 1$ where A is the symbol. Transforming the symbol corresponds to applying a Tietze transformation to the presentation.

Sometimes the result of Tietze transformations can be quite surprising. Consider

$$(a, b, c, d : ab = c, bc = d, cd = a, da = b).$$

We can try to build a Van Kampen diagram by sticking copies of the triangles (Fig. 6.6) or merely by 'playing' around with the relations to see what results:

$$\left.\begin{array}{l} ab = c \text{ implies } b = a^{-1}c \\ cd = a \text{ implies } d = c^{-1}a = b^{-1} \end{array}\right\} \Rightarrow d = b^{-1}$$

We can, therefore, reduce the presentation to the following:

$$(a, b, c : ab = c, bc = b^{-1}, b^{-1}a = b).$$

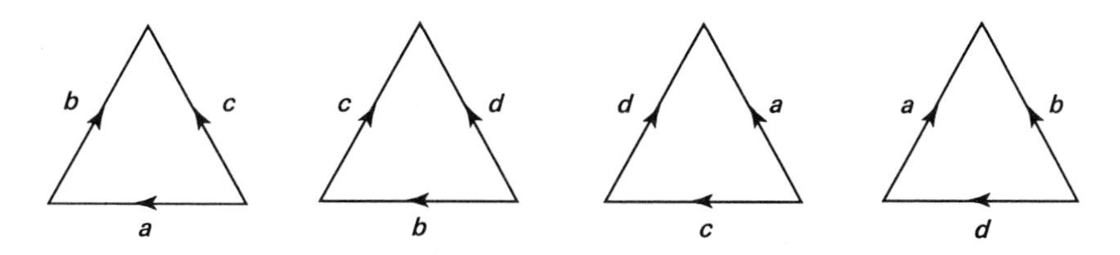

Fig. 6.6

The second and third of these yield $c = b^{-2}$, $a = b^2$, so $ab = c$ implies $b^3 = b^{-2}$ and $b^5 = 1$. Elimination of superfluous generators then gives that the group is isomorphic to C_5.

The success of Tietze transformations is complete. We note the following theorem, proved by Tietze (1908).

6.4 Theorem. *Given two finite presentations of the same group, one can be obtained from the other by a finite sequence of Tietze transformations.*

Proof. This proof is developed in a structured sequence of exercises. We begin with some discussion. Presentations are a combinatorial way of coding up information on groups. When we study groups, we constantly use homeomorphisms as a way of comparing them. Natural curiosity should make us ask if there is a notion of *mapping of presentation* that will reflect information about homeomorphisms. Suppose we have

(1) a presentation $P = (X : R)$ of a group G, yielding a homeomorphism

$$\varphi_P\colon F(X) \longrightarrow G \text{ with } \langle\langle R\rangle\rangle = \operatorname{Ker} \varphi_P;$$

(2) a presentation $Q = (Y : S)$ of a group H, yielding a homeomorphism

$$\varphi_Q\colon F(Y) \longrightarrow H \text{ with } \langle\langle S\rangle\rangle = \operatorname{Ker} \varphi_Q;$$

(3) a homeomorphism $\theta\colon G \longrightarrow H$.

Prove there is a homeomorphism $f\colon F(X) \to F(Y)$ so that in the diagram

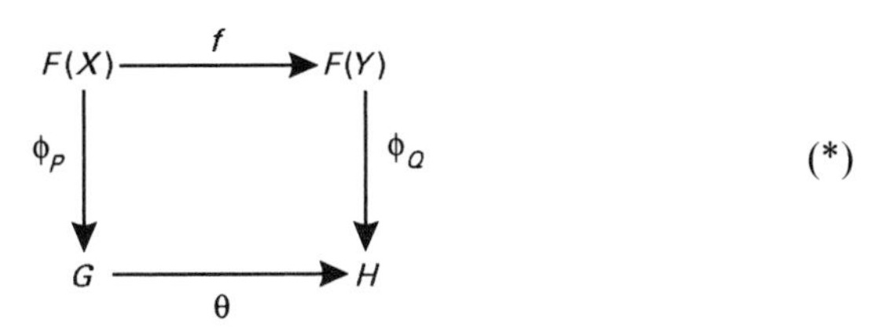

(*)

we have that the two composites $\theta\varphi_P$ and $\varphi_Q f$ are equal. (The square 'commutes'.) If you want a hint, think of the universal property of free groups. What does f do to elements in R? (Remember $\langle\langle R\rangle\rangle = \text{Ker } \varphi$.)

Whenever we show that something exists in mathematics, we should have learnt the reflex of asking if it is unique with the required property. If it is not unique, a variant of the question is to ask how close to unique it is. How close to unique is f for a given θ? If f_1 and f_2 both fit the bill, how can we compare f_1 and f_2?

A *mapping* f: $(X : R) \to (Y : S)$ *of presentations* is a homeomorphism

$$f\colon F(X) \longrightarrow F(Y)$$

such that for each $r \in R$, $f(r) \in \langle\langle S\rangle\rangle$.

If you have completed the above exercise, you will have proved that given a homeomorphism $\theta\colon G \to H$ and presentations P of G, Q of H, then there is a presentation mapping, $f\colon P \to Q$, which mirrors the homeomorphism θ in as much as the square (*) above commutes ($\varphi_Q f = \theta\varphi_P$).

Exercises 6.6

1. Let $P = (X : R)$ be a presentation of G and $Q = (X : R \cup S)$ a presentation of H, say, with $f\colon P \to Q$ given by $f = \text{id}\colon F(X) \to F(X)$. Show that f is a presentation mapping. What is the corresponding homeomorphism θ from G to H? When is θ an isomorphism?
2. Let Y be a set, y a given element of Y, and set $X = Y\backslash\{y\}$. Suppose $P = (X : R)$ and $Q = (Y : S)$ where, if we write $f\colon F(X) \to F(Y)$ for the homeomorphism induced by the inclusion of X into Y, $f(R) \subseteq S$. (This homeomorphism maps the generator, $x \in X$, of $F(X)$ to the corresponding generator, $x \in Y$, of $F(Y)$. It is the unique homeomorphism making the diagram

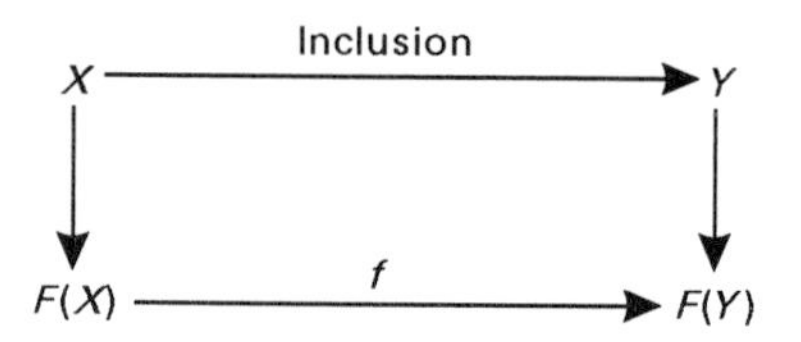

commute. The conditions we have imposed on S make $f\colon P \to Q$ into a presentation mapping.) Describe the corresponding group homeomorphism, θ.

Now suppose $S = f(R) \cup \{yw^{-1}\}$ where w is some word in the image of f. Prove that the homeomorphism θ is an isomorphism. (This is of course the basis for the type **T3** and **T4** Tietze transformations, so do not use those in your proof. If you find it hard to get started, try the following hint: define $g: F(Y) \to F(X)$ by

$$g(x) = x \quad \text{if } x \in X \subset Y$$

$$g(y) = w.$$

Remember that it is only necessary to say where the generators of a free group go to determine where everything else goes. Now prove that g is a presentation mapping from Q to P and that the corresponding group homeomorphism is an inverse for θ.)

3. Let $P = (X : R)$ be a presentation of a group G and let

$$\varphi: F(X) \longrightarrow G$$

be the epimorphism corresponding to the presentation. Suppose Z is a finite set disjoint from X ($X \cap Z = \varnothing$) and pick a homeomorphism

$$\theta: F(X \cup Z) \longrightarrow F(X)$$

so that $\theta(x) = x$ for all $x \in X$. Try to describe a set of normal generators for Ker $\varphi\theta$. Of course, $R \subset \text{Ker } \varphi\theta$, but what other elements do we need? The case when $Z = \{y\}$ is very close to the previous example/exercise, so is a good place to start.

4. Now we are ready to approach the task of proving Tietze's theorem. Suppose we have two finite presentations $P = (X : R)$ $Q = (Y : S)$ of a group G. We have to show that we can get from P to Q using only Tietze transformations. First we lift the identity map on G to give presentation mappings $f: P \to Q$ and $g: Q \to P$. We have to find a sequence of Tietze transformations giving f. The plan of action is to assume first of all that X and Y do not overlap, that is they are disjoint. As we want all of the Y-generators in, we can do this one by one using Tietze transformations. At this stage we have a presentation of the form $(X \cup Y : ?)$. We then reverse the process and go to $(Y : S)$ by getting rid of the X-generators. That much is clear, but how can we do it in detail?

Using Exercise 6.6.3 above we can look at

$$F(X) \rightleftarrows F(X \cup Y)$$

and describe a presentation of G of the form $(X \cup Y : R \cup A)$ say.

Similarly, again using Exercise 6.6.3 we get

$$F(Y) \leftrightarrows F(X \cup Y)$$

and a presentation $(X \cup Y : S \cup B)$. Clearly (but check it), the normal closures of $R \cup A$ and $S \cup B$ are the same.

We now look at the obvious presentation mappings:

$$(X \cup Y : R \cup A) \rightleftarrows (X \cup Y : R \cup S \cup A \cup B) \rightleftarrows (X \cup U : S \cup B),$$

and show that (i) they induce isomorphisms, and (ii) they can be realized by Tietze transformations (of types **T1** and **T2**). You can now look at the various mappings to see if you can get f as the composite (by choosing the 'retractions' from $F(X \cup Y)$ to $F(X)$ etc. in some cunning way)!

You now have the building blocks and the plan. Write out a proof up to this point, with the assumption of $X \cap Y = \varnothing$.

If we are unfortunate and $X \cap Y \neq \varnothing$, we replace X by X_1, Y by Y_1 so that X and X_1 are in one–one correspondence, Y and Y_1 are in one–one correspondence, and $X_1 \cap Y_1 = \varnothing$, $X \cap X_1 = \varnothing$, $Y \cap Y_1 = \varnothing$.

Now the isomorphism $F(X) \xrightarrow{\cong} F(X_1)$, induces a presentation mapping $(X : R) \to (X_1 : R_1)$ and so on. Can you complete the proof of Tietze's theorem? You may need to introduce some notation for the various mappings. Choose it carefully.

Proofs of the theorem can be found in Crowell and Fox (1967) and Johnson (1990).

6.7 Quotient groups

Suppose we have $G = (X : R)$ and we know H is a quotient group of G, say $H \cong G/N$. Is there a presentation of H easily derived from the given one for G? The first point to make is to note why this might be expected. We have formed G as a quotient $F(X)/\langle\langle R\rangle\rangle$ and so have an epimorphism

$$F(X) \xrightarrow{\varphi} G.$$

The normal subgroup $N \lhd G$ corresponds to some normal subgroup M, say, of $F(X)$, that is, the kernel of the composite epimorphism

$$F(X) \xrightarrow{\varphi} G \longrightarrow G/N \cong H$$

so $H \cong F(X)/M$. We know that $\langle\langle R\rangle\rangle \subset M$, so it is reasonable to expect that H has a presentation of the form $(X : R \cup S)$ where $\varphi(S)$ normally generates N in G.

The above argument can be made rigorous so as to prove the following theorem. We leave the proof as an exercise as it is quite simple.

6.5 Von Dyck's theorem. *If R and S are subsets of the free group F on a set X and $R \subseteq S$ then there is an epimorphism*

$$\theta: (X : R) \longrightarrow (X : S).$$

The kernel of θ is just the normal closure of $S \backslash R$ as a subset of $(X : R)$.

As an example, consider a group G with presentation $(a, b : a^2 = b^n)$ for n odd and greater than 2. (This is the knot group of a $(2, n)$-torus knot.) For which values of n does there exist an epimorphism from G to S_3? The group S_3 has a presentation $(x, y : x^2 = y^3 = (xy)^2 = 1)$, so it is clear that we can find an epimorphism from G to S_3 if n is divisible by 3. In fact adding the relations $a^2 = 1$, $b^3 = 1$ and $(ab)^2 = 1$ into the relations for the given presentation of G yields a presentation equivalent to that we have given for S_3. This condition $3 \mid n$ is also necessary, but we leave this to you to check. The method is to assume the existence of an epimorphism $\varphi: G \to S_3$ and then to see what possibilities there are for $\varphi(a)$ and $\varphi(b)$.

Exercise 6.7

Prove Von Dyck's theorem.

6.8 Abelianization

We saw earlier in this chapter that there was an epimorphism from any group G to its Abelianization G_{ab}. Recall that G_{ab} was the quotient of G by the commutator subgroup $[G, G]$ of G. We also saw that

$$[ab, c] = a[b, c]a^{-1}[a, c]$$

and $[a, b]^{-1} = [b, a]$. These facts together with Von Dyck's theorem (6.5) enable us to prove the following theorem.

6.6 Theorem *If G has a presentation $(X : R)$ then G_{ab} has a presentation*

$$(X : R \cup \{[x, x'] : x, x' \in X\}).$$

We know that the kernel of $\varphi: G \to G_{ab}$ is the commutator subgroup and is thus generated by all $[g, g']$, $g, g' \in G$. We also have a commutative diagram

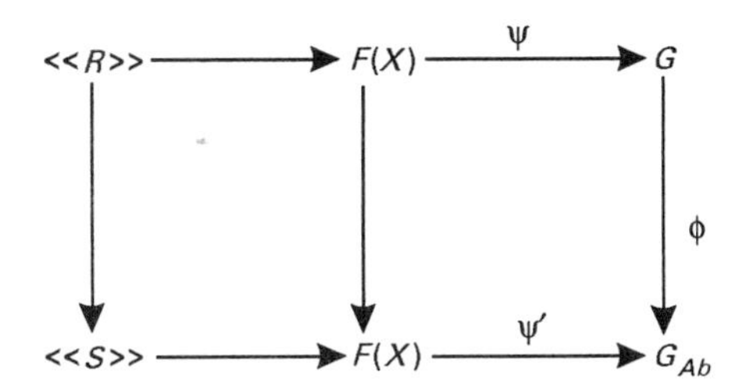

and $\langle\langle R\rangle\rangle$ is a subgroup of $\langle\langle S\rangle\rangle$. Moreover, the kernel of φ is generated by the images of all $[w, w']$, $w, w' \in F[X]$ (since any commutator in G is the image of a commutator in $F(X)$), so we could take $S = R \cup [F(X), F(X)]$. What we have to do is to show that the smaller set of relations

$$R \cup \{[x, x'] : x, x' \in X\} \overset{\text{notation}}{=} R \cup [X, X]$$

has the same normal closure as $R \cup [F(X), F(X)]$. Clearly $R \cup [X, X] \subset R \cup [F(X), F(X)]$ and so all we need to do is to show any $[w, w']$, $w, w' \in F(X)$ can be written as a product of conjugates of elements from $R \cup [X, X']$. (In fact R is not needed here as we really only use $[X, X]$.) As the proof is an induction on the lengths of w and w', we will leave the formalities to you and merely sketch the inductive step.

Suppose we know $[w, w']$ can be written as a product of conjugates of the various $[x, x']$, provided $\ell(w), \ell(w') \leqslant n - 1$, and we look at $[w, w']$ where $\ell(w), \ell(w') \leqslant n$. We assume $\ell(w) = n$; if not we can go to the next step. Then $w = w_1x$ or $w = w_1x^{-1}$ for some w_1 of length $n - 1$ and some $x \in X$. Now use

$$[w, w'] = [w_1x, w'] = w_1[x, w']w_1^{-1} \cdot [w_1, w']$$

(similarly for w_1x^{-1}) to reduce one level in the first variable. If $\ell(w') \leqslant n - 1$, this ends the inductive step; if not write $w' = w_1'y$ (or $w' = w_1'y^{-1}$) and use the expression for $[a, bc]$ to simplify. We have not given all the cases: what about w_1x^{-1}? This gives us a term $[x^{-1}, w']$, not one of the form $[x', w]$ or $[w, x']$ — how can we get around this? We leave it to you to finish this proof.

This result gives us a powerful tool for calculating Abelianizations. We used it earlier to look at $[D_n, D_n]$ and thus at $(D_n)_{ab}$. This tool becomes even more powerful once we have the results on finitely generated Abelian groups that follow.

6.9 Finitely generated Abelian groups

A group G is said to be finitely generated (f.g.) if it has a presentation of the form $(X : R)$ with X a finite set. Most of the groups that we will meet are finitely generated. If G is f.g., so is any quotient group of G, and so in particular if G is finitely generated then so is G_{ab}. Finitely generated Abelian groups are easy to classify up to isomorphism, and so the comparison of Abelianizations of f.g. groups is a useful way of checking for non-isomorphism; that is if G and H

are f.g. groups and we can show that G_{ab} and H_{ab} are not isomorphic, then G and H are not isomorphic.

Presentations of Abelian groups are often written additively, and so when considered as a group C_2 has presentation $(x : x^2)$ but traditionally when considered as an Abelian group one writes $(x : 2x = 0)$. A more complicated example is

$$D_n = (x, y : x^n = y^2 = (xy)^2 = 1)$$
$$(D_n)_{ab} = (x, y : x^n = y^2 = (xy)^2 = [x, y] = 1)$$

as a group but as an Abelian group we might replace

$$x^n = 1 \text{ by } nx = 0$$
$$y^2 = 1 \text{ by } 2y = 0$$
$$(xy)^2 = 1 \text{ by } 2x + 2y = 0$$

and of course, since the group is Abelian,

$$[x,y] \text{ by no relation.}$$

It is now simple to see what $(D_n)_{ab}$ is. We have $2y = 0$ and $2x + 2y = 0$, so $2x = 0$; as $nx = 0$, if n is odd we obtain $x = 0$, so $(D_n)_{ab} \cong C_2$ generated by y, but if n is even $nx = 0$ is implied by $2x = 0$, so we have $C_2 \oplus C_2$ in this case ($\oplus$ indicates direct sum, see below).

Often manipulation of the (additive) relations using common sense suffices to find the Abelian group up to isomorphism. However, the theory of additive presentations is quite easy, so it pays to have the additional tools provided by the method at your disposal. (Proofs will be omitted as they are readily available in many group theory texts.) We start by making precise what we mean by direct sum.

If A_1 and A_2 are subgroups of an Abelian group A, then we can form a new subgroup

$$A_1 + A_2 = \{a_1 + a_2 : a_1 \in A_1, a_2 \in A_2\};$$

$A_1 + A_2$ is called the *sum* of A_1 and A_2. You should check to your satisfaction that this *is* a subgroup. It is easy to extend this to any finite family of subgroups $A_1, \ldots, A_n$ of A. (We will not need to extend it to infinite families here.)

We say A has a *direct sum decomposition* if there are subgroups $A_1, \ldots, A_s$ such that

(1) $A = A_1 + \ldots + A_s$;

(2) for each i, $1 \leqslant i \leqslant s$, $A_i \cap \sum_{i \neq j} A_j = 0$.

(In (2) we have written $\Sigma_{i \neq j} A_j$ for the sum of all the subgroups except A_i.) We write $A = A_1 \oplus A_2 \oplus \ldots \oplus A_s$.

For example, look at $C_6 = \{0, 1, 2, 3, 4, 5\}$ with operation 'addition modulo 6'. Let $A_1 = \{0, 2, 4\}$ which is a subgroup isomorphic to $C_3 = \{0, 1, 2\}$ (addition being mod 3) and let $A_2 = \{0, 3\}$ which is isomorphic to C_2 (i.e. $\{0,1\}$ with addition mod 2). Check these statements yourself—do not just take our word for them! We claim $C_6 \cong A_1 \oplus A_2$. Clearly $A_1 \cap A_2 = \{0\}$ so (2) is satisfied and to prove (1), that is $C_6 = A_1 + A_2$, involves the extra hard calculation: $5 \equiv 2 + 3$ mod 6 and $1 = 4 + 3$ mod 6. Note that $A_1 \cup A_2$ is not a subgroup as it is $\{0, 2, 3, 4\}$ and so is not closed under addition (mod 6).

This shows that C_6 is an *internal direct sum* of a copy of C_2 and one of C_3. It is also useful to be able to construct *external* direct sums starting with a family of groups and forming a big group containing copies of the members of the family. For instance, if the family consists of two Abelian groups A and B, we can form

$$A \oplus B = \{(a, b) : a \in A, b \in B\}$$

with addition performed componentwise. This group $A \oplus B$ has a direct sum decomposition involving its subgroups:

$$A' = \{(a, 0): a \in A\} \quad \text{isomorphic to } A,$$

$$B' = \{(0, b): b \in B\} \quad \text{isomorphic to } B.$$

One can interchange the two forms of direct sum as they are really equivalent notions. One usually drops the 'internal' and 'external' as these are clear from the context. One writes, for instance, '$C_6 \cong C_2 \oplus C_3$' and 'C_6 is the direct sum of subgroups $\{0, 3\}$ and $\{0, 2, 4\}$'.

In fact, if m, n are coprime then $C_{mn} \cong C_m \oplus C_n$. The proof uses the fact that there are $a, b \in \mathbb{Z}$ such that $1 = am + bn$. We leave it as an exercise.

This decomposition as a direct sum of simpler parts is the idea behind the two main theorems on finitely generated Abelian groups.

6.7 Primary decomposition theorem. *Let A be a finitely generated Abelian group; then A has a direct sum decomposition*

$$A = B_1 \oplus \ldots \oplus B_s \oplus B_{s+1} \oplus \ldots \oplus B_{s+t}$$

where (i) B_i is a non-trivial cyclic group of prime power order $p_i^{\alpha_i}$ for $i = 1, \ldots, s$ (the p_i are not necessarily distinct), and (ii) *B_i, for $i = s + 1, \ldots, s + t$, is an infinite cyclic group.*

The integer t is called the *torsion free rank* of A and the prime powers $p_i^{\alpha_i}$ are called the *primary invariants* of A. These are uniquely determined up to order by A and themselves determine A up to isomorphism.

6.8 Cyclic decomposition theorem. *Let A be a finitely generated Abelian group; then A has a direct sum decomposition*

$$A = A_1 \oplus A_2 \oplus \ldots \oplus A_r \oplus A_{r+1} \oplus \ldots \oplus A_{r+t}$$

where (i) A_i is a non-trivial finite cyclic group of order n_i, for $i = 1, \ldots, r$, (ii) A_i is an infinite cyclic group for $i = r+1, \ldots, r+t$, and (iii) $n_1 | n_2 | \ldots | n_r$ (the vertical line is 'divides').

The integer t is the torsion free rank (as in the previous theorem) and $n_1, \ldots, n_r$, the *torsion invariants*, are uniquely determined by A and determine A up to isomorphism.

These two results are usually illustrated by lists of isomorphism classes of Abelian groups of some given finite order. More useful to us is an example showing how to derive torsion or primary invariants and the torsion free rank of an Abelian group A given by an (additive) presentation. Suppose A has torsion free rank t and torsion invariants $n_1, \ldots, n_r$; then it has a presentation

$$(a_1, \ldots, a_{r+t} : n_1 a_1 = \ldots = n_r a_r = 0)$$

(remember additive presentation). A general presentation of A will have the form

$$(x_1, \ldots, x_n : f_1 = \ldots = f_s = 0)$$

where $f_j = \Sigma_{i=1}^{n} r_{ij} x_i$ is some linear expression with integer coefficients. The relations are thus specified by a matrix and Tietze transformations correspond to integer row and column operations on this matrix. The aim of the transformations is to diagonalize this matrix.

Example

$A = (a, b, c, d : 2a + 5b + 8c + 7d = 5a - 4b - 7c + 4d = 8a + 5b + 2c + 7d = 0)$

has matrix

$$R = \begin{pmatrix} 2 & 5 & 8 \\ 5 & -4 & 5 \\ 8 & -7 & 2 \\ 7 & 4 & 7 \end{pmatrix}$$

The resulting diagonal matrix must satisfy $n_1 | n_2 | n_3 \ldots$, so the method must aim to get the highest common factor of the entries of R into the top left corner. In this example the h.c.f is 1:

$$\begin{pmatrix} 2 & 5 & 8 \\ 5 & -4 & 5 \\ 8 & -7 & 2 \\ 7 & 4 & 7 \end{pmatrix} \xrightarrow[R4-R3]{R2-2R1} \begin{pmatrix} 2 & 5 & 8 \\ 1 & -14 & -11 \\ 8 & -7 & 2 \\ -1 & 11 & 5 \end{pmatrix} \xrightarrow{R1 \leftrightarrow R2} \begin{pmatrix} 1 & -14 & -11 \\ 2 & 5 & 8 \\ 8 & -7 & 2 \\ -1 & 11 & 5 \end{pmatrix}$$

$$\longrightarrow \begin{pmatrix} 1 & -14 & -11 \\ 0 & 33 & 30 \\ 0 & 105 & 90 \\ 0 & -3 & -6 \end{pmatrix} \xrightarrow[C3+11C1]{C2+14C1} \begin{pmatrix} 1 & 0 & 0 \\ 0 & 33 & 30 \\ 0 & 105 & 90 \\ 0 & -3 & -6 \end{pmatrix}.$$

Next look for the h.c.f. of the remaining submatrix. As this is 3, we will do a row interchange and multiply by -1 to get

$$\begin{pmatrix} 1 & 0 & 0 \\ 0 & 3 & 6 \\ 0 & 105 & 90 \\ 0 & 33 & 30 \end{pmatrix} \longrightarrow \begin{pmatrix} 1 & 0 & 0 \\ 0 & 3 & 6 \\ 0 & 0 & -120 \\ 0 & 0 & -36 \end{pmatrix} \longrightarrow \begin{pmatrix} 1 & 0 & 0 \\ 0 & 3 & 0 \\ 0 & 0 & -120 \\ 0 & 0 & -36 \end{pmatrix}$$

$$\begin{pmatrix} 1 & 0 & 0 \\ 0 & 3 & 0 \\ 0 & 0 & 36 \\ 0 & 0 & 120 \end{pmatrix} \xrightarrow{R4-3R3} \begin{pmatrix} 1 & 0 & 0 \\ 0 & 3 & 0 \\ 0 & 0 & 36 \\ 0 & 0 & 12 \end{pmatrix} \xrightarrow[\text{followed by } R4-3R3]{R3 \leftrightarrow R4} \begin{pmatrix} 1 & 0 & 0 \\ 0 & 3 & 0 \\ 0 & 0 & 12 \\ 0 & 0 & 0 \end{pmatrix}.$$

So $A \cong C_3 \oplus C_{12} \oplus C_\infty$ since in the new generators a', b', c', d' we have a presentation

$$(a', b', c', d' : a' = 0, 3a' = 0, 12c' = 0).$$

If one wants to find a', b', c', and d' in terms of the original generators, one applies the row operations used in reverse order to a unit matrix of size $s \times s$ where s is the number of original generators (in the example $s = 4$). We leave you to do this in the example.

To find the primary invariants, note that 3 is prime whilst $12 = 2^2 \times 3$, so $C_{12} \cong C_4 \oplus C_3$ and the primary decomposition is $A \cong C_4 \oplus C_3 \oplus C_3 \oplus C_\infty$. (Readers who have not studied this theory in detail can find treatments in Johnson (1990) or Rotman (1984).)

6.10 Pushouts of groups

There is a configuration that will arise many times in various forms in the later parts of this book. The idea is simple. We have two groups G and H and want to form a group that contains both G and H. We saw that even in as simple a case as C_2 and C_3 considered as subgroups of C_6, the setwise union $C_2 \cup C_3$ was not a group. In that example we had a bigger group in which we could work. Now if we are given G and H, how should we multiply $g \in G$ and $h \in H$? We have seen something like this before. When developing the ideas that led to free groups, we formed 'words' in the elements of the set of generators. Here we could clearly form words in the elements of $G \cup H$ (as sets) and then 'remember' the multiplication in G and in H. We can try this out: for instance, if we take the words g_1hg_2 and $g_3h'g_4h''$ and 'multiply' them together we get

$$(g_1hg_2)(g_3h'g_4h'') = g_1h(g_2g_3)h'g_4h'';$$

that is, put the words next to each other (remove the brackets) and multiply the end elements together if this is possible. This is vague. It can be made more exact, but as we have a limited need for these ideas and we have the machinery of presentations available to us, we will take a short cut.

We first note a distinctive property of the union of two sets that is linked to the gluing lemma of Chapter 3. Suppose X, Y, and Z are sets with $f: X \to Z$, $g: Y \to Z$ two functions that agree on $X \cap Y$, that is if $x \in X \cap Y$ then $f(x) = g(x)$, then there is a *unique* function $h: X \cup Y \to Z$ such that if $x \in X$, $h(x) = f(x)$, and if $y \in Y$, $h(y) = g(y)$. This is so obvious as to be almost unnecessary to state. The role of $X \cap Y$ is important here; if f and g did not agree on $X \cap Y$ then no joint extension h could exist.

This sort of situation arises in numerous contexts and is abstracted in the notion of a *pushout diagram*. Suppose we have a diagram

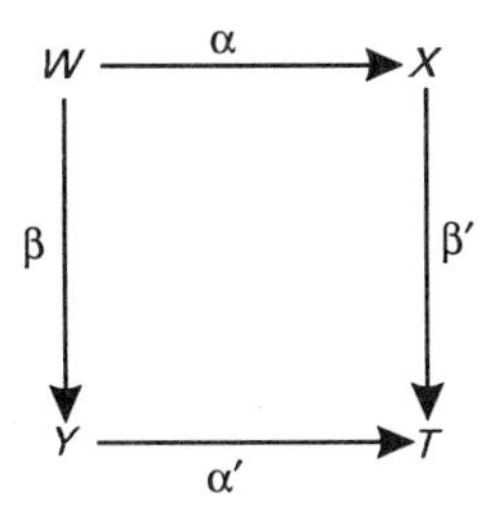

of sets and functions. It is called a pushout diagram (of sets) if it is commutative (i.e. if the two composites $\beta'\alpha$ and $\beta\alpha'$ are equal) and it has the universal property that, given functions $f: X \to Z$, $g: Y \to Z$ satisfying $f\alpha = g\beta$, there is a *unique* function $h: T \to Z$ such that $h\beta' = f$ and $h\alpha' = g$.

(The uniqueness of h with this property is very important and should be thought of in the example we looked at where $W = X \cap Y$ and α and β were inclusions. Note in general that α and β are not assumed to be inclusions.)

We leave it to you to show that given any pair of functions $\alpha: W \to X$, $\beta: W \to Y$ there is a pushout diagram as above containing them. (You form T by taking a disjoint union of X and Y and dividing out by a certain equivalence relation. The disjoint union of two sets can be obtained by taking two new sets X_1, Y_2 consisting of pairs $(x, 1)$, $x \in X$, or $(y, 2)$, $y \in Y$ and then forming $X_1 \cup Y_2$. Even if $X \cap Y$ is non-empty, $X_1 \cap Y_2$ is empty, so X_1 and Y_2 are disjoint copies of X and Y; we write $X \amalg Y$ for this disjoint union.)

We are not really going to use pushouts of sets but a similar notion can be obtained for groups by asking X, Y, W, T, Z to be groups and α, β, α', β', f, and g to be homeomorphisms. We could also use another analogous situation with all the sets replaced by topological spaces and all the functions by continuous maps.

If we are given a 'top left-hand corner' of groups

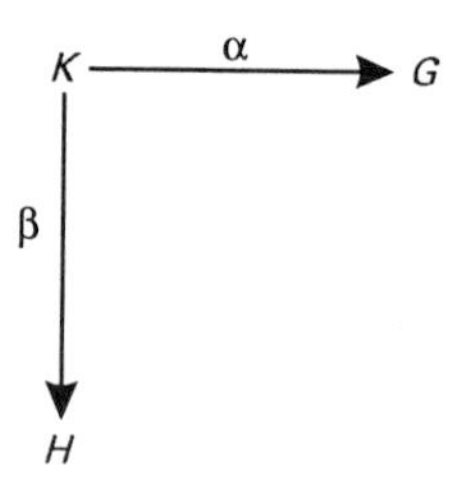

can we complete it to a pushout diagram

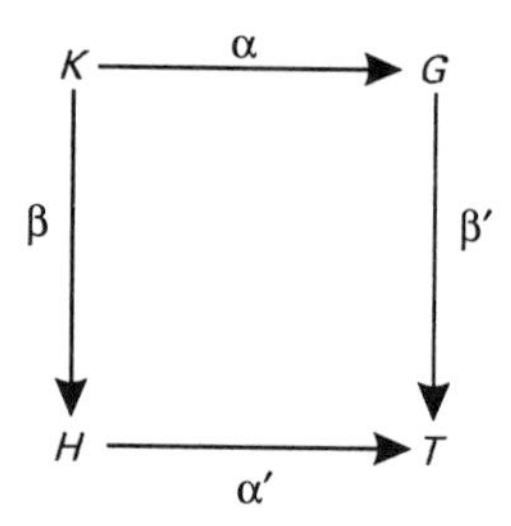

of groups? The case $K = \{1\}$ is the simplest case and is that with which we started this section.

6.9 Proposition. *If $G = (X : R)$, $H = (Y : S)$ then the group $T = (X \cup Y : R \cup S)$ with the obvious inclusions forms a pushout diagram*

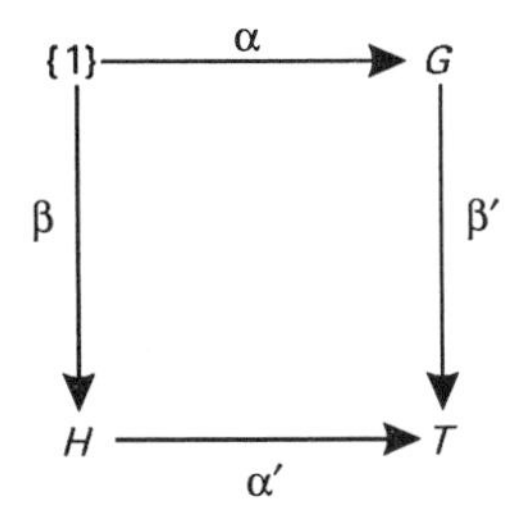

and T is uniquely determined up to isomorphism by this property.

Proof. Suppose we have a group Z and homeomorphisms $f_1: G \to Z$ $f_2: H \to Z$. Of course the commutativity of the square is automatic, since $\{1\}$ only contains the identity element.

We note that $\alpha'(y) = y$, $\beta'(x) = x$ (with some abuse of notation here!). We define $f: T \to Z$ by defining f on generators: if $a \in X \cup Y$ then $a \in X$ or $a \in Y$ but not in both; if $a \in X$, then $f(a) = f_1(a)$, and if $a \in Y$ then $f(a) = f_2(a)$. We have to check that if $w \in R \cup S$ then $f(w) = 1$, but if $w \in R \cup S$ then $w \in R$ or $w \in S$, but not both, so we can check if $w \in R \subset F(X)$ then w only involves the elements of X and $f(w) = f_1(w) = 1$, similarly if $w \in S$. (We will look at the uniqueness of T in the general case later.) ■

This group T will usually be written $G * H$ and it is called the *free product* of G and H. Even if G and H are finite groups, then $G * H$ is infinite unless one of G or H is the trivial group.

The general case is only slightly more difficult to describe.

6.10 Proposition. *If in a corner $H \overset{\beta}{\leftarrow} K \overset{\alpha}{\rightarrow} G$, $G = (X : R)$, $H = (Y : S)$ and K is generated by a set Z then $T = (X \cup Y : R \cup S, \{\alpha(z)\beta(z)^{-1} : z \in Z\})$, with the obvious homeomorphisms, forms a pushout square*

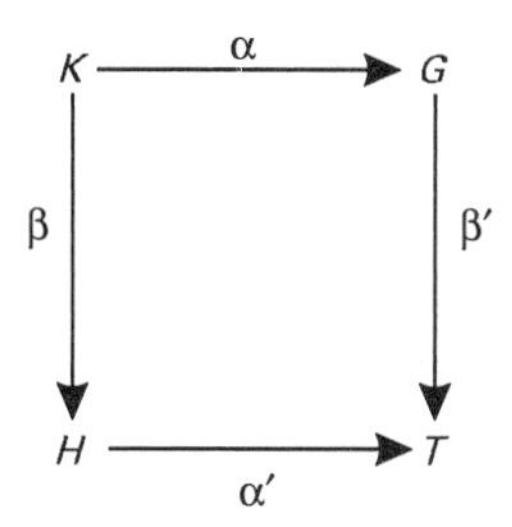

and T is uniquely determined up to isomorphism by this property.

Proof. Firstly notice that the addition of $\alpha(z)\,\beta(z)^{-1} = 1$ for all $z \in Z$ into the relations guarantees that the square commutes.

Again we assume that given $f_1: G \to Z$, $f_2: H \to Z$ such that $f_1\alpha = f_2\beta$ and define $f: T \to Z$ by the same method as before. Again we have to check that f of any relation is trivial but if $w \in R \cup S$ we can use the argument we used before and if w is of the form $\alpha(z)\beta(z)^{-1}$, then $f(w) = f\alpha(z)f\beta(z)^{-1} = f_1\alpha(z)f_2\beta(z)^{-1} = 1$. (We have systematically abused notation in the above. The idea of the proof is simple but to make the distinction between words in $F(X)$ and $F(X \cup Y)$ say, which strictly speaking needs to be done, would seem to complicate and obscure that simple idea.)

We now turn to uniqueness. Firstly we should explain what is meant here. Suppose someone presents us with another square

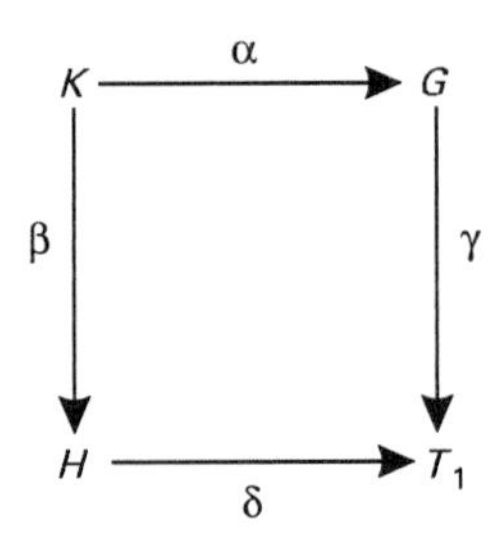

with the same *universal* property: given any $f_1: G \to Z$ and $f_2: H \to Z$ such that $f_1\alpha = f_2\beta$ then there is a unique $f: T_1 \to Z$ satisfying $f_1 = f\gamma, f_2 = f\delta$. If T is a pushout as claimed in the proposition then $T \cong T_1$. To see this, we look at the two diagrams

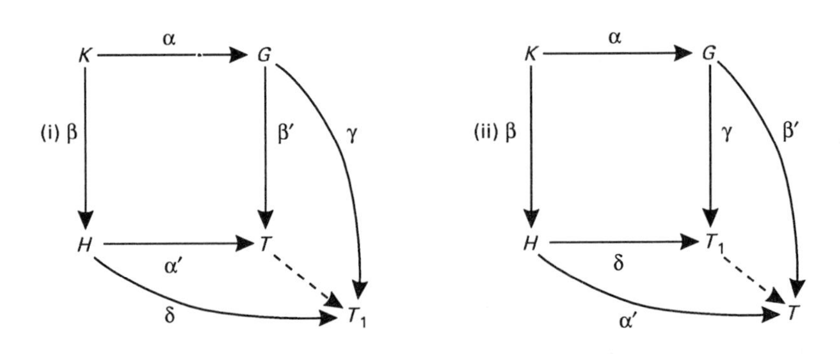

In (i), taking $f_1 = \gamma: G \to T_1, f_2 = \delta: H \to T_1$ gives us a unique $\phi: T \to T_1$ satisfying $\beta' = \gamma$, $\alpha' = \delta$. In (ii), taking $f_1 = \beta', f_2 = \alpha'$ gives us a unique $\psi: T_1 \to T$ satisfying $\psi\gamma = \beta'$ and $\psi\delta = \alpha'$. Now we note that

$$\psi\phi\beta' = \psi\gamma = \beta' \quad \text{and} \quad \psi\phi\alpha' = \psi\alpha = \alpha'$$

whilst

$$\phi\psi\gamma = \phi\beta' = \gamma \quad \text{and} \quad \phi\psi\delta = \phi\alpha' = \delta.$$

Looking at the diagram

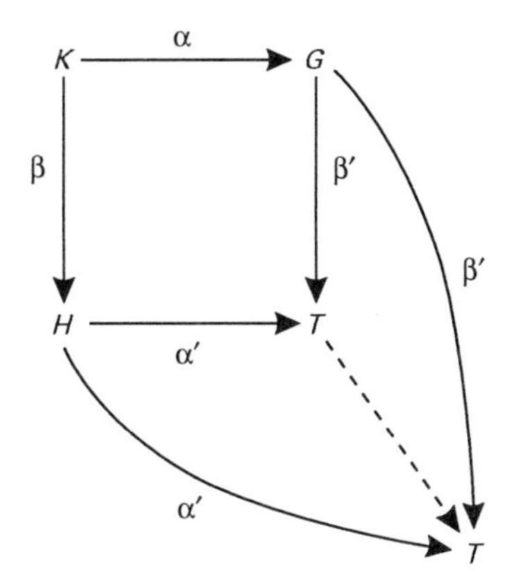

where f_1 is taken to be β' and f_2 to be α' gives a unique map $f\colon T \to T$ satisfying $f\alpha' = \alpha'$, $f\beta' = \beta'$. We have two candidates for f, the identity map on T and $\psi\phi$. Both satisfy the equations and so they must be equal since we know there is a *unique* map satisfying these equations. A similar argument shows that $\phi\psi$ is the identity on T_1; hence ϕ and ψ are isomorphisms.

(We have given this argument in quite a lot of detail as it illustrates the power of definitions involving universal properties. A similar result is true for the universal property defining free groups.)

To complete the proof we have to return to our original T and $f{:}T \to Z$. Why is f unique with the property that $f\beta' = f_1$, $f\alpha' = f_2$? Any homeomorphism h: $T \to Z$ is determined by what it does to generators and any generator is really either an $\alpha'(y)$ or a $\beta'(x)$, so $h\alpha'(y) = f_2(y)$ whilst $h\beta'(x) = f_1(x)$, that is h is the same as f! (Here, to some extent, we must avoid our earlier deliberate abuse and simplification of notation. We wanted previously to think of $F(X)$ as sitting inside $F(X \cup Y)$ and we used the letter x equally for a generator in X or in the X-part of $X \cup Y$; now we need to make a distinction between them, so we call $x \in X$ by the name $\beta'(x)$ if we are considering it as an element of $X \cup Y$.) ■

When $K = \{1\}$ we of course get back exactly to our earlier description of the free product. If α and β are monomorphisms so that G and H contain K as a subgroup, then the group T is called the free product of G and H with K as amalgamated subgroup. Its construction can be thought of as being obtained from $G * H$ by gluing together the two copies of K, one in G and one in H.

The above proof, although an abstract one, contains all the information to enable us to describe the elements of this free product with amalgamated subgroup. Suppose $g \in G$, $h \in H$, $k \in K$; then $g\alpha(k) \in G$, $\beta(k)h \in H$, and we can form the words $(g\alpha(k))h$ and $g(\beta(k)h)$ within T. As $k \in K$, it can be written as a word in the generators, Z, and for each $z \in Z$, $\alpha(z) = \beta(z)$ within T; as a result we find that

$$(g\alpha(k))h = g(\beta(k)h).$$

This example shows what the elements of T look like. They are strings of the g and the h as in $G * H$, but now any element in $\alpha(K)$ can be identified with the corresponding one in $\beta(K)$.

These pushouts of groups are very useful. They also include other, often used constructions.

Exercise 6.10

Suppose in a corner

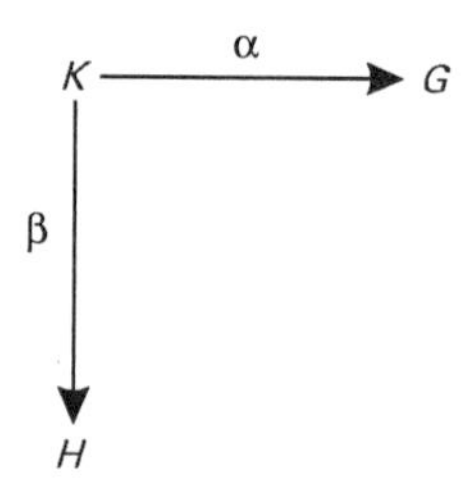

that K is a normal subgroup of G with α the inclusion and that $H = \{1\}$. Prove that the group T that completes the pushout square is isomorphic to G/K with β' the usual quotient map. What happens if K is not assumed to be normal in G?

6.11 Knot groups

One of our main purposes for spending a fair amount of time on group presentations is that they are extremely useful when working with knot groups. The full geometric significance of the definitions and constructions we will give will not be clear until later on, but the combinatorial methods used make calculations very easy to do. We will look at two alternative methods of obtaining a presentation of a group from a knot diagram, and then examine the independence or otherwise of these groups with regard to Reidemeister moves before looking at various links with other ideas we have met.

Wirtinger presentations

The naïve idea behind the Wirtinger presentation is that if you want to explore a knot, just going for a walk along the knot itself is not much help as it is only a circle, but examining the space around the knot will tell you a lot more. You can think of that space as being a three-dimensional labyrinth. Taking the advice of Ariadne, we follow Theseus in taking with us in our exploration a ball of string, letting it out as we go so as to know where we have passed already. Examination of the possible paths yields some generating paths and the crossings give us relations.

(Before we proceed further, it may help if the reader takes a piece of moderately stiff wire and bends it into a trefoil knot, joining the ends together to complete the 'circle'. A piece of strong cord will also be useful!)

Take a picture of the knot, orient it, and label each arc. These labels will be the generators of the knot group, so use sensible distinct labels, possibly labelling around the knot in some order. We will illustrate using the cinquefoil in Fig. 6.7.

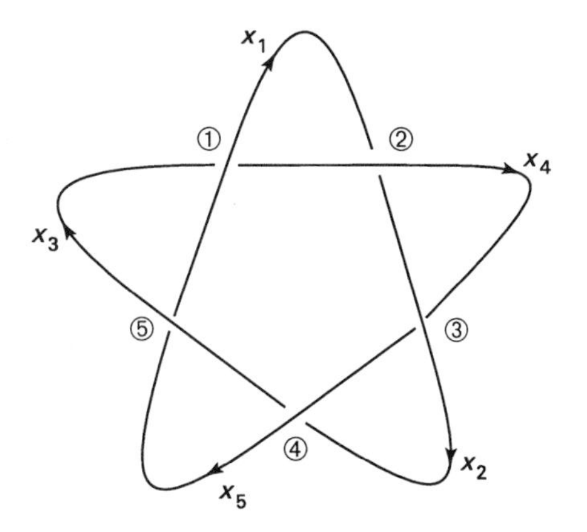

Fig. 6.7

The knot group will have a presentation

$$G(K) = (x_1, \ldots, x_n : r_1, \ldots, r_m)$$

where n is the number of arcs in the picture and m is the number of crossings. (It helps to number the crossings so as to avoid missing any out of the calculation.)

The rule for writing it down is: starting at some point near the crossing move anticlockwise around a small rectangle; at each arc write down the corresponding generator with exponent $+1$ if the arc is

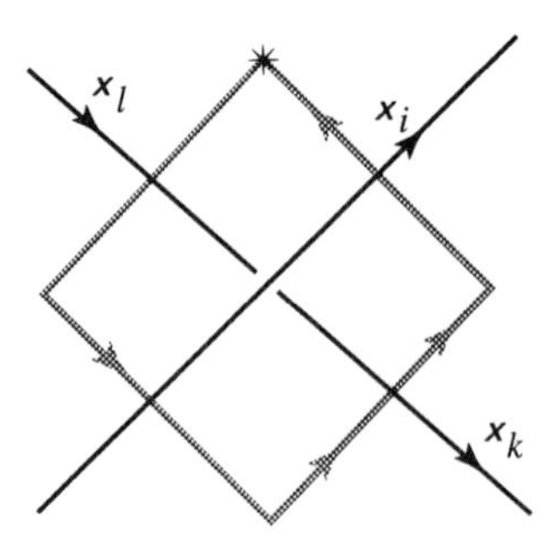

Fig. 6.8

entering the crossing and -1 if it is leaving it. Hence starting at the top gives $x_\ell x_i x_k^{-1} x_i^{-1}$. Clearly starting at a different position produces a conjugate of this relation.

Let us assume that at the first crossing we had the form shown in Fig. 6.8. This corresponds to the relation $r_1 : x_\ell x_i x_k^{-1} x_i^{-1}$.

There is a second possible type of crossing of the form shown in Fig. 6.9 which gives $x_\ell x_i^{-1} x_k^{-1} x_i = 1$.

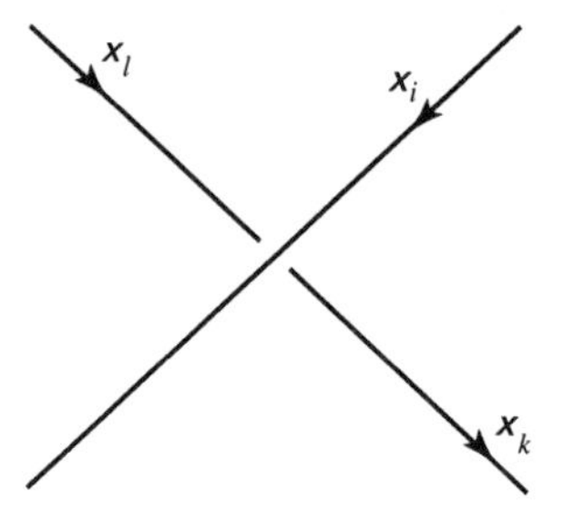

Fig. 6.9

For our example of the cinquefoil knot, we have five generators and the following five relations:

$$
\begin{array}{llll}
r_1 : x_5 x_1 x_4^{-1} x_1^{-1} = 1 & \text{or} & x_4 = x_1^{-1} x_5 x_1 \\
r_2 : x_1 x_4 x_2^{-1} x_4 = 1 & \text{or} & x_2 = x_4^{-1} x_1 x_4 \\
r_3 : x_4 x_2 x_3^{-1} x_2^{-1} = 1 & \text{or} & x_3 = x_2^{-1} x_4 x_2 \\
r_4 : x_2 x_3 x_5^{-1} x_3^{-1} = 1 & \text{or} & x_5 = x_3^{-1} x_2 x_3 \\
r_5 : x_3 x_5 x_1^{-1} x_5^{-1} = 1 & \text{or} & x_1 = x_5^{-1} x_3 x_5.
\end{array}
$$

There are several things to notice here. Firstly each relation expresses a generator (corresponding to the arc leaving from under the overpass) in terms of the 'incoming' generator conjugated by the 'overpass' generator or its inverse. This observation allows one to set up inductive machines to calculate the generators leaving a series of twists in terms of the incoming arcs (Fig. 6.10).

Fig. 6.10

Secondly the cinquefoil is extremely symmetric and this is reflected in the relations. In fact one can write them in a 'string'

$$x_1x_4 = x_5x_1 = x_3x_5 = x_2x_3 = x_4x_2.$$

This can be used to take short cuts to simplify the presentation. You are advised to use special structures such as symmetry to help in calculations. At the same time be warned that as not all knots are symmetric, we need basic simplification techniques that work regardless of any special features. We will apply simple Tietze transformations to the Wirtinger presentation so as to simplify it and will not use the symmetry explicitly.

The presentation is

$$G(K) = (x_1, x_2, x_3, x_4, x_5 : x_1 = x_5^{-1}x_3x_5,\ x_2 = x_4^{-1}x_1x_4,\ x_3 = x_2^{-1}x_4x_2,\ x_4 = x_1^{-1}x_5x_1,\ x_5 = x_3^{-1}x_2x_3).$$

We eliminate (any) one generator, say x_5, by substituting for x_5 the expression $x_3^{-1}x_2x_3$:

$x_1 = x_5^{-1}x_3x_5$	becomes	$x_1 = x_3^{-1}x_2^{-1}x_3x_2x_3$
$x_2 = x_4^{-1}x_1x_4$	does not change	
$x_3 = x_2^{-1}x_4x_2$	does not change	
$x_4 = x_1^{-1}x_3x_1$	becomes	$x_4 = x_1^{-1}x_3^{-1}x_2x_3x_1$

and we delete x_5 from the list of generators. Here strategic thought has to come in. Substitutions of complicated expressions are more likely to lead to slips than are substitutions of simple ones. It therefore is better to proceed by substituting for x_2 or x_3, not x_1 or x_4. We choose to substitute for x_3 next. We will write down the results:

$$G(K) \cong (x_1, x_2, x_4 : x_1 = x_2^{-1}x_4^{-1}x_2x_4x_2^{-1}x_4x_2,\ x_2 = x_4^{-1}x_1x_4,\ x_4 = x_1^{-1}x_2^{-1}x_4^{-1}x_2x_4x_2x_1).$$

Finally, using the expression for x_2 gives

$$G(K) \cong (x_1, x_4 : x_1 = x_4^{-1}x_1^{-1}x_4^{-1}x_1^{-1}x_4x_1x_4x_1x_4,$$
$$x_4 = x_1^{-1}x_4^{-1}x_1^{-1}x_4^{-1}x_1x_4x_1x_4x_1).$$

We note that the two relations are essentially the same:

$$x_1x_4x_1x_4x_1 = x_4x_1x_4x_1x_4.$$

In Wirtinger presentations, any one relation is a consequence of the others. (The reason is that each of the relations corresponds to a path or a loop of string that can be pulled off the knot. Any single such loop can be pulled until it involves not that chosen crossing but all the others.) Keeping all the relations acts as a check on your calculations, so do not just throw a relation away. As a final step we simplify one stage further.

Write $a = x_1x_4$, $b = x_1x_4x_1x_4x_1$; then a few more Tietze transformations give

$$G(K) \cong (a, b : a^5 = b^2).$$

The cinquefoil is a (2, 5)-torus knot. Here are the 2 and the 5 coming out from the calculation of $G(K)$. We will be studying torus knots and their groups more closely later on in Chapter 11.

Dehn presentations

You may recall the way in which the Alexander polynomial was originally introduced by Alexander. That approach also leads to a group presentation introduced by Dehn.

Orient the knot as before but now label the faces of the resulting picture. These face-labels will be the generators. As in the Wirtinger presentation we get a relation for each crossing. A typical crossing might be as shown in Fig. 6.11 and corresponding to this we write down the relation $x_ix_j^{-1}x_kx_l^{-1}$. Finally we add a relation $x_m = 1$ for

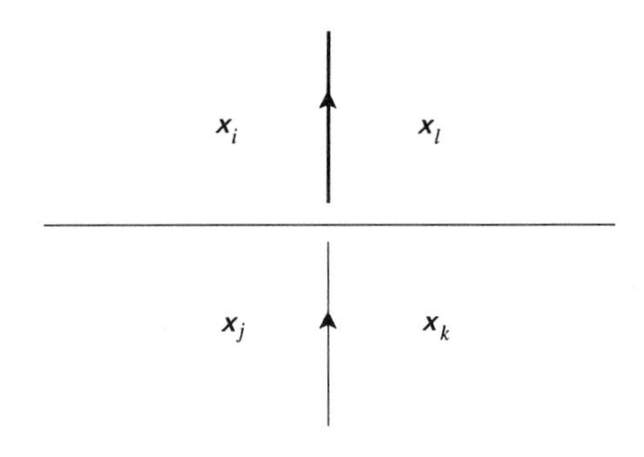

Fig. 6.11

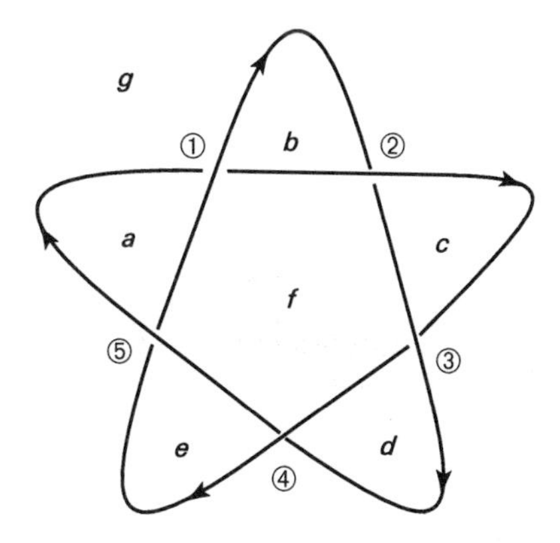

Fig. 6.12

some generator/face-label x_m ~ anyone will do. This last relation effectively deletes x_m from the presentation.

Again we will examine the cinquefoil (Fig. 6.12).

The generators will be a, b, c, d, e, f, and g. The relators are

$$bg^{-1}af^{-1},\quad cg^{-1}bf^{-1},\quad dg^{-1}cf^{-1},\quad eg^{-1}df^{-1},\quad ag^{-1}ef^{-1}$$

and we will choose to 'kill off' g as this seems to cause a lot of simplification to the presentation. This gives

$$(a, b, c, d, e, f : baf^{-1} = cbf^{-1} = dcf^{-1} = edf^{-1} = aef^{-1} = 1).$$

Eliminating f (since $f = ba = cb = dc = ed = ae$) gives us

$$(a, b, c, d, e : ba = cd = dc = ed = ae).$$

Glancing back at the symmetric relations we had for the Wirtinger presentation suggests that this is the same group and so should have a presentation of the form $(x, y : x^5 = y^2)$.

Knowing where we are going shows a short cut. This common element f that we have actually got rid of, if repeated, gives an interesting pattern

$$f^5 = (ed)(cb)(ae)(dc)(ba) = (edcba)^2$$

so we take $y = edcba$ and, using a few Tietze transformations, get

$$(f, y : f^5 = y^2).$$

We leave it to you to try to prove, using the knowledge that you have at this point in the book, that the Dehn presentation is a presentation of the same group as the Wirtinger presentation. We also suggest that you repeat the exercises above for the Dehn method and also examine why it does not matter which of the original face-label generators you kill off.

Invariance

To be useful these knot groups, obtained either by the Dehn or Wirtinger method, should be invariants, that is should not depend on which picture we use to calculate them. To put this more formally we hope that if K and L are equivalent knots then $G(K)$ and $G(L)$ will be isomorphic groups.

Later on we will show that $G(K)$ is the *fundamental group of the knot complement* $\mathbb{R}^3 \backslash K$. This provides a neat use of the fundamental group calculations that we will meet in Chapter 11 and proves invariance in an elegant way. However, there is much to be said for a bare-hands approach. We know equivalences can be broken down into Reidemeister moves, so it would seem a good idea to see what happens to the knot group if we change a knot by a Reidemeister move. We look at Wirtinger presentations, and leave the Dehı versions as an exercise.

Move I (Fig. 6.13)

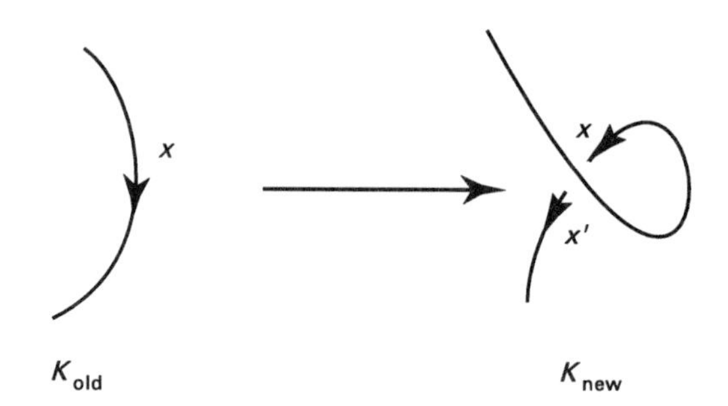

Fig. 6.13

The presentation for K_{new} will have one more generator, x', and some of the relations involving x in $G(K_{\text{old}})$ will be replaced by relations involving x'. In addition we have a relation for the new crossing, namely $xxx^{-1}x'^{-1} = 1$, which, of course, gives $x' = x$. This in turn means that we can substitute x for x' in all relations, eliminating x' and reducing our presentation to that of $G(K_{\text{old}})$. Thus in this case the two groups are isomorphic.

Move II (Fig. 6.14)

Again the change produces new generators and some changes in other relations. It also gives relations

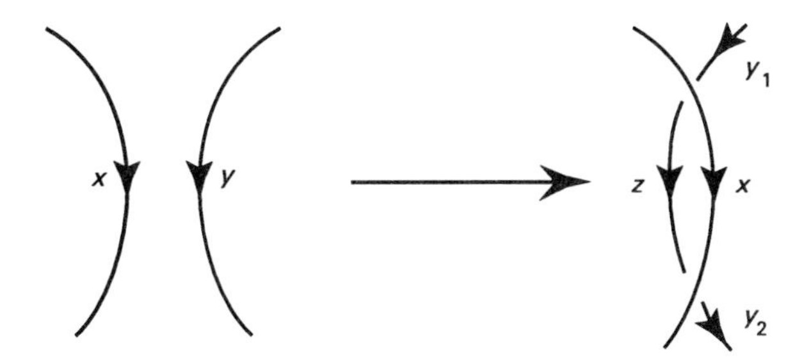

Fig. 6.14

$$z^{-1}x^{-1}y_1x \quad \text{and} \quad y_2^{-1}xzx^{-1}$$

at the two new crossings. The first gives

$$z = x^{-1}y_1x$$

meaning that we can eliminate z from all other relations, but of course it can only occur in the second of these relations. Substituting this in the second relation gives $y_2 = y_1$, so we can again reduce to the old presentation.

We should also look at the move with one changed orientation (Fig. 6.15) but this causes no problems. Check that you can do the adaptation of the argument above to this case.

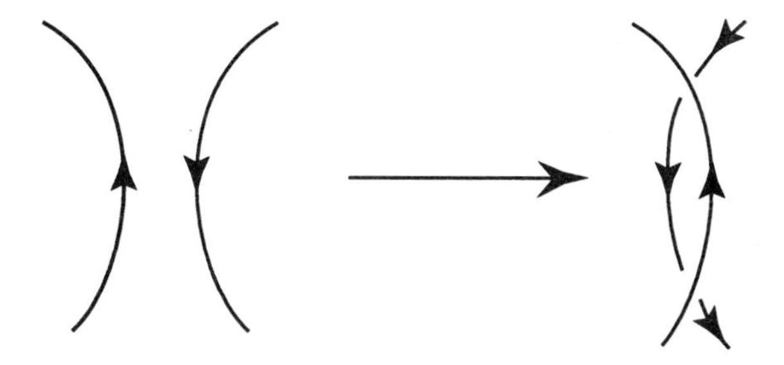

Fig. 6.15

Move III

This causes more bother and again we will only look at one of the numerous different subcases caused by relative positions and orientations.

In Fig. 6.16 the relations are $a^{-1}cbc^{-1}$, $e^{-1}b^{-1}db$, and $f^{-1}cec^{-1}$, whilst in Fig. 6.17 the relations are $a^{-1}cbc^{-1}$, $g^{-1}cdc^{-1}$, and $g^{-1}a^{-1}ga$. (We can safely leave all labels other than e and g

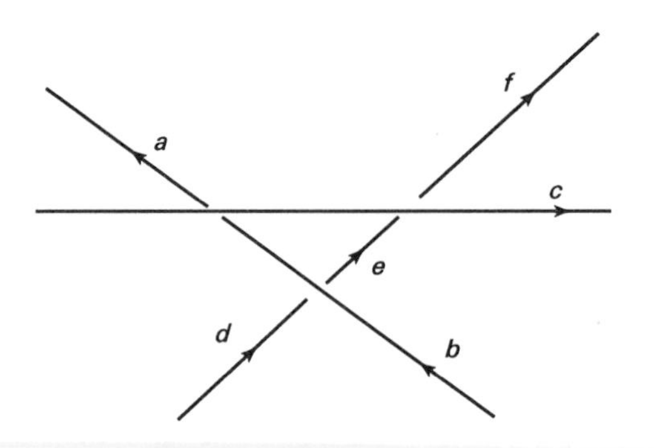

Fig. 6.16

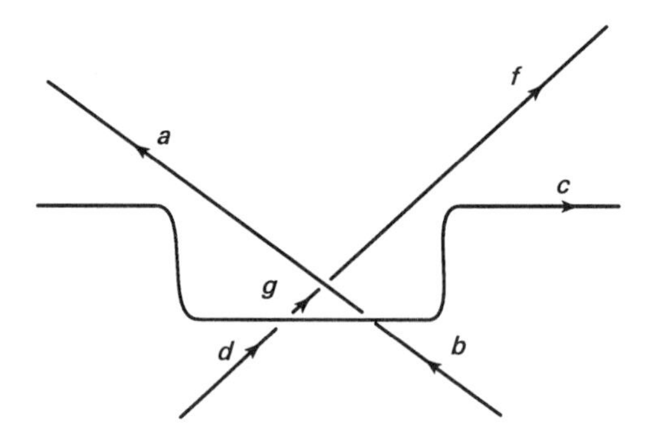

Fig. 6.17

unchanged as we are only changing the knot in the vicinity of this triple crossing.)

Simplifying the old presentation eliminating e we get

$$a^{-1}cbc^{-1} \quad \text{and} \quad f^{-1}a^{-1}cdc^{-1}a.$$

Comparing these we find that the old one has cb^{-1} where the new one has $a^{-1}c$, but the other unchanging relation gives $a^{-1}c = cb^{-1}$, so the two presentations are equivalent and the two knot groups $G(K_{\text{old}})$ and $G(K_{\text{new}})$ are isomorphic.

Exercises 6.11.1

1. Calculate the Wirtinger presentation of (i) the unknot, (ii) the trefoil, and (iii) the figure-eight knot.
2. The $(2, n)$-torus knots can all be drawn in the form shown in Fig. 6.18 where n, the number of crossings, is odd. Using the labelling

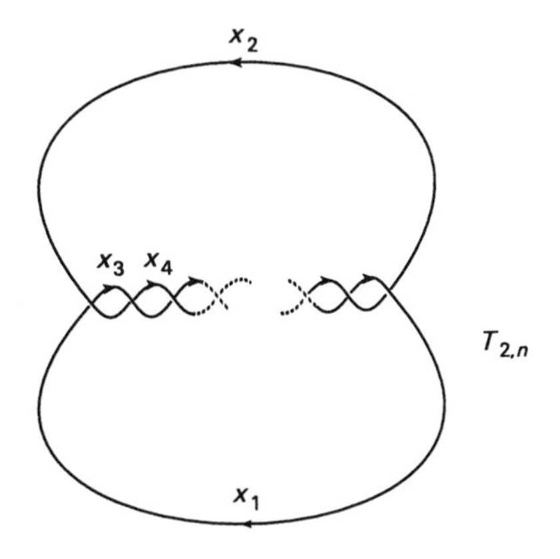

Fig. 6.18

suggested, work out a formula expressing the generator x_k of the Wirtinger presentation in terms of the generators x_1 and x_2. Using the last two crossings (the furthest to the right in the diagram), write x_1 and x_2 in terms of themselves and hence write down a presentation of the group of this knot. Simplify to show the knot group is isomorphic to

$$(a, b : a^n = b^2).$$

3. There is another family of knots that might be called the generalized figure-eight knots; their general form is as shown in Fig. 6.19 with a total of $n + 2$ crossings for the nth knot in the series. For $n = 1$, we have a trefoil and for $n = 2$, a figure-eight. Using similar methods to the preceding exercise, find a presentation for the knot group of the nth knot in the family. Investigate if it can always be presented with just two generators. Is there an

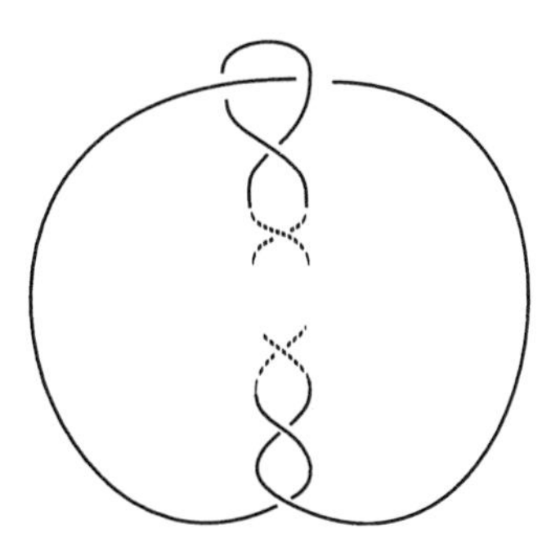

Fig. 6.19

'elegant' way of presenting this group in any way analogous to the $a^n = b^2$ relation form of $G(T_{2,n})$?

4. Prove that $G(T_{2,3})$ is not isomorphic to G(unknot). (You may not be able to complete the proof now but try to understand what the problem is.)
5. If K is any knot, show that $G(K)_{ab}$, the Abelianized knot group, is infinite cyclic.
6. Examine how the Dehn presentation changes under Reidemeister moves and check that the results are equivalent presentations, that is giving isomorphic groups.

Knot groups and the sum of knots

What is $G(K + L)$ in terms of $G(K)$ and $G(L)$? For a change we will examine this for the Dehn presentation but suggest you examine the case of Wirtinger presentations for yourselves.

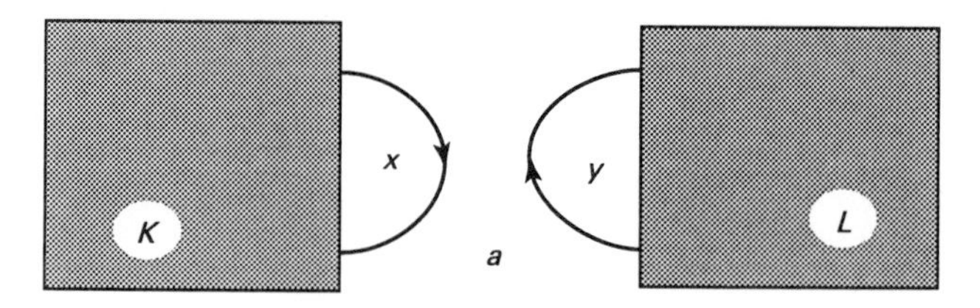

Fig. 6.20

Orient the two knots and label the faces as in Fig. 6.20. Use the same label for the unbounded region, a say, and in both presentations use a as the generator to kill off. (If you have not checked that the choice of which generator to kill is of no consequence to the end result, now would be a good time to pause and think about the problem.)

Now look at $K + L$; this has picture shown in Fig. 6.21 where z is assumed to be a face-label not used in K or L. It should be clear that if

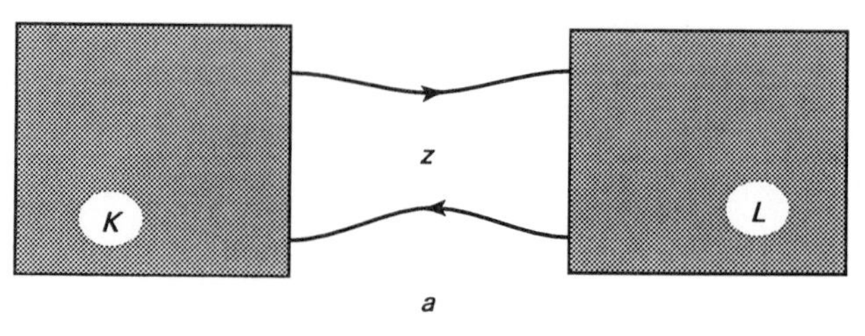

Fig. 6.21

$$G(K) = (X : R) \quad X \text{ a set of generators containing } x$$

and

$$G(L) = (Y : S) \quad Y \text{ a set of generators containing } y$$

with $X \cap Y = \varnothing$, then

$$G(K + L) \cong (X, Y : R, S, x = y)$$

Thus the rule for forming $G(K + L)$ is that of forming the pushout

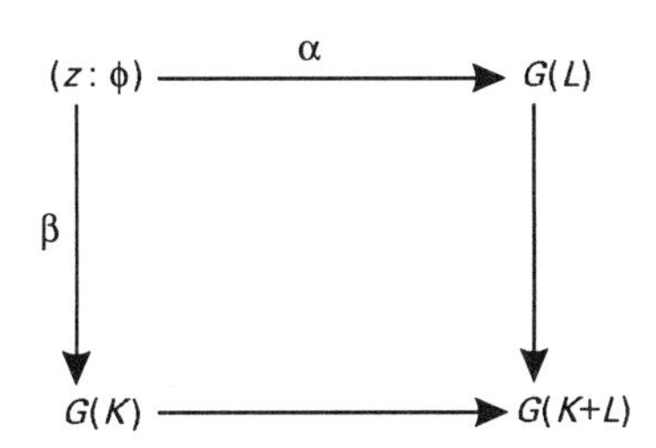

where $\alpha(z) = y$, $\beta(z) = x$.

Knot groups and 3-colourability

Suppose we have a 3-colourable knot, K, coloured by the colours a, b, c. We will construct a group G with generators a, b, c and the relations so that colouring K, that is attaching a colour to each arc and hence to each generator of $G(K)$, defines a map on generators from $G(K)$ to G compatible with the relations, thus giving an epimorphism from $G(K)$ onto G.

The idea is that if x_i and x_j are generators both coloured a then we add in the relation $x_i = x_j$ and call the amalgamated generator thus formed a, and so on.

A typical relation in a Wirtinger presentation is of form $xyx^{-1}z^{-1}$ or $x^{-1}yxz^{-1}$. If this occurs at a crossing where x, y, and z are all coloured the same, the relation will vanish in G, that is we need not worry about compatibility. If x, y, z are coloured a, b, c in some order then mapping arcs to colours maps this relation to one of the following pairs of relations:

(i)	$aba^{-1}c^{-1}$, $a^{-1}bac^{-1}$	(ii)	$bab^{-1}c^{-1}$, $b^{-1}abc^{-1}$	(iii)	$cac^{-1}b^{-1}$, $c^{-1}acb^{-1}$
(iv)	$aca^{-1}b^{-1}$, $a^{-1}cab^{-1}$	(v)	$bcb^{-1}a^{-1}$, $b^{-1}cba^{-1}$	(iv)	$cbc^{-1}a^{-1}$, $c^{-1}bcb^{-1}$.

Half of these are consequences of others, for instance if $aba^{-1}c^{-1} = 1$ then $aba^{-1} = c$, so $b = a^{-1}ca$, that is $a^{-1}cab^{-1} = 1$. The six relations in pairs (iv) to (vi) can therefore be omitted. This leaves us to look for a group with three generators and relations (i), (ii), and (iii). We therefore take

$$G = (a, b, c : aba^{-1}c^{-1} = a^{-1}bac^{-1} = bab^{-1}c^{-1} = b^{-1}abc^{-1} = cac^{-1}b^{-1} = c^{-1}acb^{-1} = 1)$$

and we have a universal quotient of all groups of 3-colourable knots. This group G is rather unfriendly. It is infinite and its structure is not immediately recognizable, but it does have a well-known quotient group.

Thinking of the transpositions (12), (13), (23) in S_3 we note that $(12)(13)(12)^{-1} = (23)$ etc.; this is similar to the type of relation we have above except that the generators in S_3 are of order 2. If we add the relations $a^2 = b^2 = c^2 = 1$ into the presentation of G we get a quotient group

$$G_1 = (a, b, c : aba = c, bcb = a, cac = b, a^2 = b^2 = c^2 = 1)$$

and G_1 mimics much of the behaviour of S_3. We do not need to work out whether or not G_1 is or is not S_3, since by mapping a to (12), b to (13), and c to (23), we can construct a homeomorphism from G_1 onto S_3. What we have done is to use the colouring of K to map $G(K)$ onto S_3, that is we have shown that the geometric and combinatorial notion of 3-colouring translates to some extent to the existence of an epimorphism from $G(K)$ onto S_3. The interest of S_3 rather than G or G_1 is that it is a well-known group and so such a homeomorphism helps our understanding of $G(K)$ itself. The converse of this result holds: if $G(K)$ maps epimorphically to S_3 then K is 3-colourable.

Exercises 6.11.2

1. Prove that if $G(K)$ maps epimorphically to S_3 then K is 3-colourable. (This is quite hard.)
2. Prove, using this method, that the $(2, n)$-torus knot $t_{2,n}$ is 3-colourable if and only if n is divisible by 3. (Some of the parts necessary to this are handled earlier in this chapter.)
3. Investigate the possibility of an analogous result for n-colourability. (This has been done by Crowell but we will not tell you which target group he used.)

7
Graphs and trees

The theory of graphs impinges on the study of knots and surfaces in four significant ways. We shall see how knots can be studied by graph theory applied to their diagrams, and how graphs can be studied by embedding them in surfaces. We have already met (in Chapter 6) the Cayley graph of a group presentation, which produces a geometric representation of a group from sets of generators and defining relations, and later on we shall meet covering graphs that give a proof, via Van Kampen's theorem, of a fundamental result about free groups. It is not our aim to compress a complete introduction to graph theory into this single chapter; rather, we select the essential facts and leave the reader to pursue the wider subject in texts intended for that purpose. We mention Wilson (1985) and White (1984) as being particularly suitable.

7.1 Graphs, quivers, and trees

An initial difficulty in graph theory is that of terminology. No accepted standard has arisen to settle the question of what a graph should be. We shall stretch the term to its fullest extent, but the reader must be warned to consult other authors' definitions when referring to other sources.

A choice presents itself as to whether the edges in a graph come with a preferred direction or not. Our main interest happens to be in *directed graphs*, but many of the important concepts are most easily formulated for graphs without a preferred direction for edges. We shall develop the ideas for directed and undirected graphs side by side, thereby avoiding a choice altogether. An undirected graph we call simply a *graph*; a directed graph could be called just that, or a *digraph*, but we prefer the more picturesque term *quiver*.

A *graph* Γ consists of a set $V\Gamma$ of *vertices* and a collection $E\Gamma$ of pairs of elements of $V\Gamma$. A pair $\{u, v\}$ is called an *edge* of Γ; we allow pairs of the form $\{u,u\}$ and for pairs to be repeated in $E\Gamma$. It is extremely useful to represent a graph Γ (at least when $V\Gamma$ is small) by a diagram in which the elements of $V\Gamma$ are represented by points and an edge $e = \{u, v\}$ is represented by a line joining the points u and v. For example, the graph with $V\Gamma = \{1, 2, 3, 4, 5, 6, 7\}$ and

$$E\Gamma = \{\{1, 1\}, \{1, 2\}, \{1, 3\}, \{1, 4\}, \{3, 4\}, \{3, 4\}, \{3, 4\}, \{3, 5\}, \{5, 4\}, \{5, 4\}, \{6, 7\}\}$$

could be represented by the diagram in Fig. 7.1.

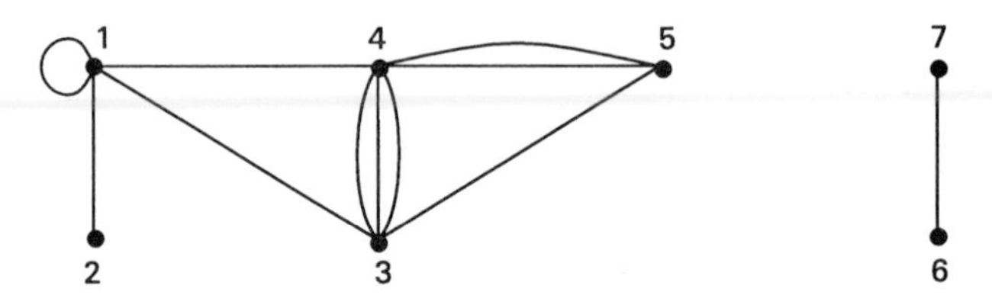

Fig. 7.1

In drawing such a diagram, the essential requirement is correctly to represent the disposition of edges joining vertices; the actual placement of the points, and whether the lines are straight, curved, or have to cross one another, is irrelevant. So the two diagrams below both represent the same graph, as in Fig. 7.2.

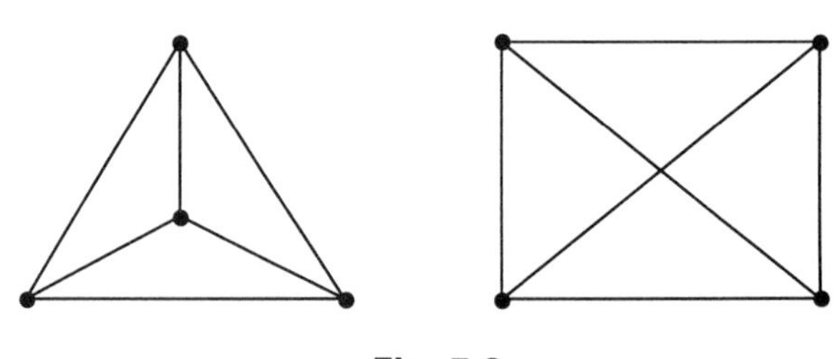

Fig. 7.2

Of course, the fact that we *can* draw a diagram of a graph Γ without edges intersecting is a significant property of Γ, and we shall consider it in detail later in the chapter.

Different choices of labelling for the vertices and edges of a graph will not affect its graph theoretic properties, so we say that two graphs Γ and Γ' are *isomorphic* if there exists a bijection $f\colon V\Gamma \to V\Gamma'$ such that mapping $\{u, v\} \in E\Gamma$ to $\{f(u), f(v)\} \in E\Gamma'$ is a bijection $E\Gamma \to E\Gamma'$. The bijection f is then an *isomorphism* between Γ and Γ'.

It is evident that the notion of a graph embodies a basic idea of everyday life: that two things are connected in some way. Examples include the stations on a railway network, the components of an electrical circuit, students who attend a lecture together, or tennis players who meet in a tournament. Sometimes the direction of connection is important, and sometimes it is irrelevant.

A *quiver* Q consists of a set VQ of vertices and a collection EQ of *ordered* pairs of elements of VQ. A pair $(u, v) \in VQ$ is an edge *from* u *to* v. We allow pairs (u, u) and for pairs to be repeated in EQ. Evidently we can represent a quiver by the same sort of diagram as for graphs, with an arrow on the line representing (u, v) pointing from u to v (Fig. 7.3).

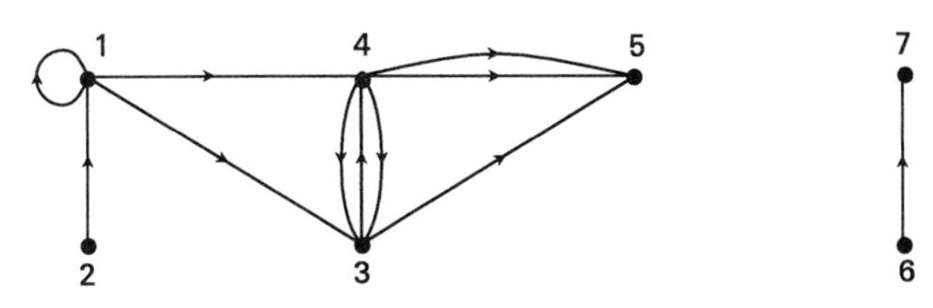

Fig. 7.3

Since every edge in a quiver has a well-defined initial vertex, or *source*, and a terminal vertex, or *target*, a neat definition of a quiver with some conceptual advantages is as follows. A quiver Q consists of two sets VQ and EQ and two functions s, t: $EQ \to VQ$. That is all we have to say. An element of EQ is an *edge* of Q and an element of VQ is a *vertex* of Q. The functions s and t give the *source* and *target* of an edge. Given two quivers Q and Q' with source and target functions s, t and s', t' a *map* of quivers f: $Q \to Q'$ is a pair of functions f_0: $VQ \to VQ'$ and f_1: $EQ \to EQ'$ such that $s'f_1 = f_0 s$ and $t'f_1 = f_0 t$. If each of f_0 and f_1 is a bijection then f is an isomorphism and Q and Q' are isomorphic. The equivalence of the two definitions of quiver should be clear: an edge $e = (u, v)$ has $s(e) = u$ and $t(e) = v$. We shall generally prefer the second definition, but the first makes plain the fact that a quiver is a graph with extra structure.

Any quiver has a graph associated to it, obtained by forgetting about the ordering of pairs in EQ, or by ignoring the arrows in a diagram. We call this associated graph the *underlying graph* of Q and denote it by $|Q|$. It follows at once that $V|Q| = VQ$ and that $E|Q| = \{\{s(e), t(e)\} \mid e \in EQ\}$.

The notion of a *subgraph* of a given graph is straightforward: in terms of diagrams, a subgraph is represented by some subset of the vertices and some subset of the edges joining these vertices. More precisely, if Γ is a graph, then a *subgraph* of Γ consists of a subset W of $V\Gamma$ and a set F of pairs of elements of W such that $F \subseteq E\Gamma$. It is often convenient to regard a graph Γ' as a subgraph of Γ merely if Γ' is isomorphic to a subgraph of Γ in the sense just described.

Let Γ be a graph. An edge $e = \{u, v\} \in E\Gamma$ is said to be *incident* with u and v. A *walk* in Γ is a finite sequence of vertices and edges

$$v_0, e_1, v_1, e_2, \ldots, v_{l-1}, e_l, v_l$$

($v_i \in V\Gamma$, $e_i \in E\Gamma$) such that e_i is incident with v_i and v_{i-1}. Successive edges are therefore incident with a common vertex. The vertex v_0 is the *initial* vertex of the walk and v_1 is its *final* vertex. A walk is *closed* if its initial and final vertices are equal. The graph Γ is *connected* if, given $a, b \in V\Gamma$, there exists a walk in Γ with initial vertex a and final vertex b.

A walk $v_0, e_1, v_1, e_2, \ldots, v_{l-1}, e_l, v_l$ is a *path* if it contains no subsequence u, e, v, e or e, v, e, u with $u, v \in V\Gamma$, $u \neq v$, and $e \in E\Gamma$. Such a subsequence introduces back-tracking into the walk. It is convenient to regard two walks as related if they differ only by such back-tracking, and so we call two walks *equivalent* if one can be obtained from the other from some finite succession of insertions or deletions of back-tracking subsequences u, e, v, e or e, v, e, u with $u, v \in V\Gamma$ and $u \neq v$. Clearly any walk is equivalent to a path, and if a walk is closed then so is any other walk equivalent to it. This idea of equivalence will be generalized to the setting of topological spaces in Chapter 9, to give the important notion of *homotopy*, and we shall return to it in Chapter 12 in order to apply graphs to the study of free groups. A walk in a quiver Q is defined to be a walk in the underlying graph $|Q|$. This definition implies that a walk in a quiver may include an edge that is traversed in the 'wrong' direction; that is, a walk may include a subsequence of the form $t(e), e, s(e)$. If an edge is traversed one way and then immediately the other way, then the walk contains a subsequence $s(e), e, t(e), e$ or $t(e), e, s(e), e$ which may be deleted in an equivalence.

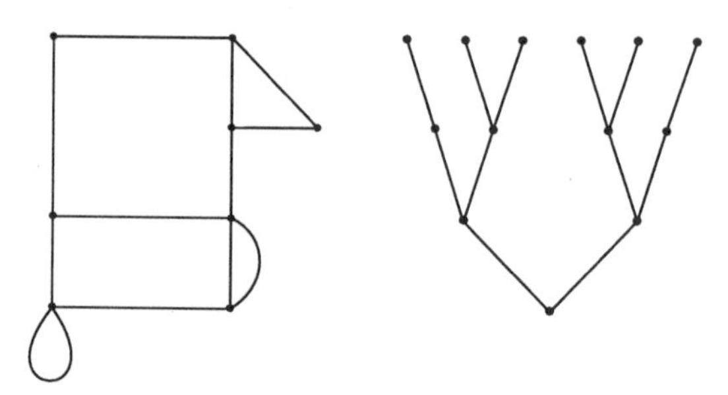

Fig. 7.4

A closed path in a graph Γ is called a *circuit*. The graph on the left in Fig. 7.4. contains many circuits, whilst that on the right contains none.

A graph is a *tree* if it is connected and contains no circuits.

Exercises 7.1

1. Find which of the three graphs in Fig. 7.5 are mutually isomorphic.

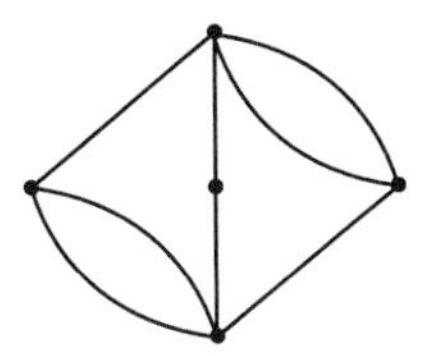
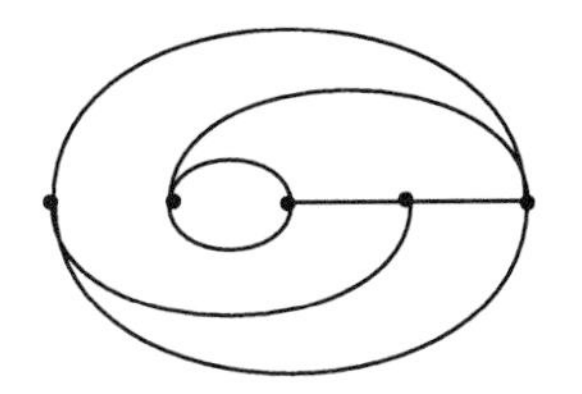

Fig. 7.5

2. Show that the graphs in Fig. 7.6 are not isomorphic.

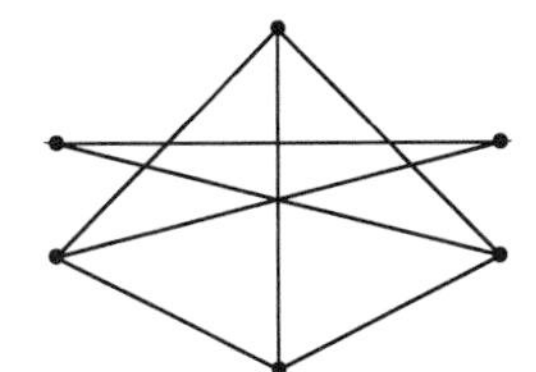
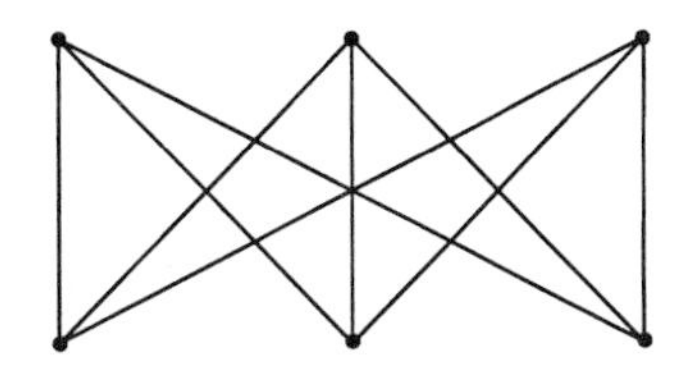

Fig. 7.6

3. Sort out the graphs in Fig. 7.7 into families of isomorphic graphs.
4. A *simple* graph has at most one edge joining a given pair of vertices. How many non-isomorphic simple graphs are there with at most four vertices?
5. Translate the definition of isomorphism of quivers from the terms of the second definition of quiver to those of the first.
6. Show that the two quivers in Fig. 7.8 are not isomorphic.
7. Let V be a set and $\sim$ a relation on V. Define a quiver Q as follows: the vertex set VQ is V and $EQ = \{(a, b) \mid a, b \in Q \text{ and } a \sim b\}$. Hence, given any $a, b \in VQ$, there exists at most one edge in EQ from a to b. A quiver with this property is called *relational*. What properties does Q possess if $\sim$ is (i) reflexive, (ii) symmetric, (iii) transitive, (iv) antisymmetric?
8. Formulate the definition of a *subquiver* of a quiver.

Fig. 7.7

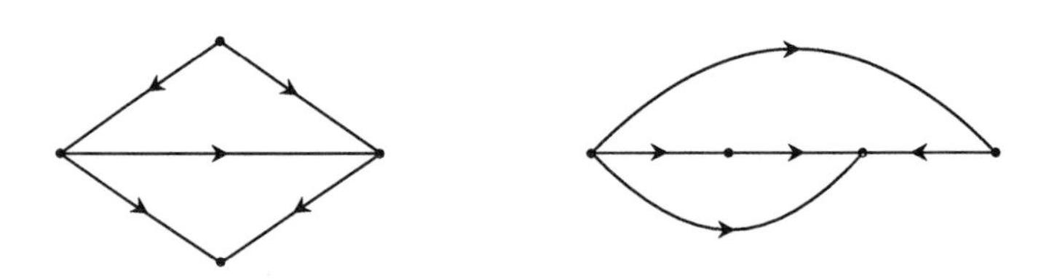

Fig. 7.8

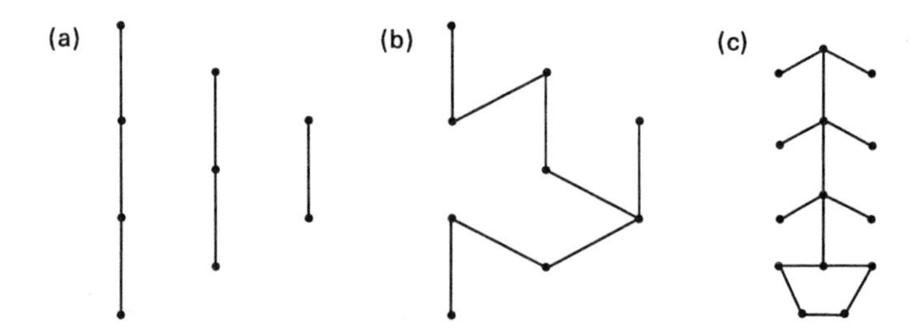

Fig. 7.9

9. Which of the graphs in Fig. 7.9 are trees?
10. Show that a tree is a simple graph. Prove that a graph is a tree if and only if there exists a unique path between any two vertices.
11. How many non-isomorphic trees are there with at most four vertices?

7.2 Examples of graphs

(a) The n-leafed rose X_n

X_n is the graph (unique up to isomorphism) with one vertex and n edges (Fig. 7.10).

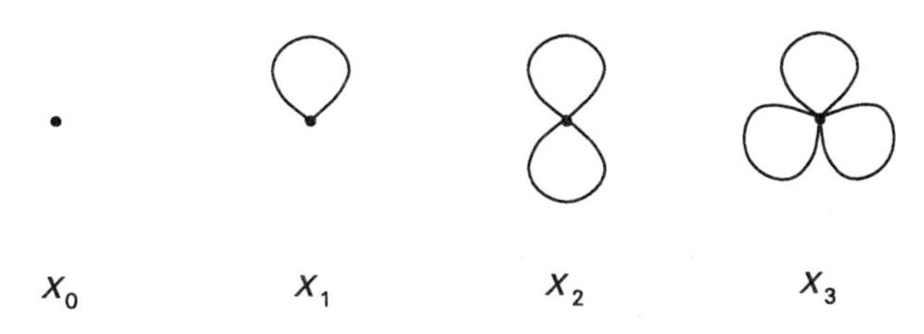

Fig. 7.10

(b) The complete graph K_n

The complete graph K_n has n vertices and $n(n-1)/2$ edges, with exactly one edge joining any two vertices (Fig. 7.11).

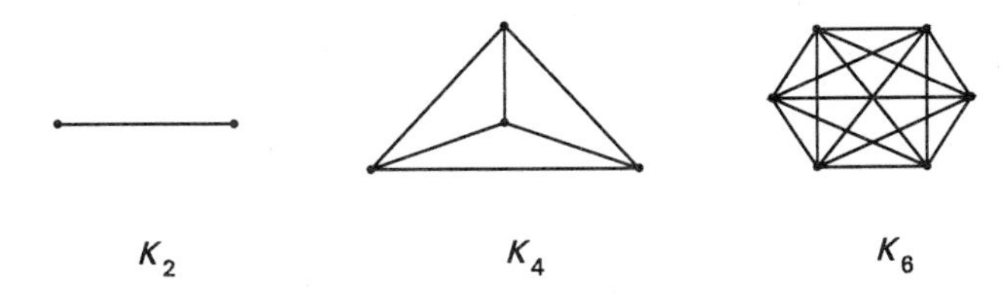

Fig. 7.11

(c) The complete bipartite graph $K_{m,n}$

A graph Γ is *bipartite* if $V\Gamma$ is the union of two non-empty disjoint subsets V_1 and V_2 such that every edge of Γ is incident with one vertex of V_1 and one vertex of V_2 (Fig. 7.12).

If every vertex of V_1 is joined to every vertex of V_2 by exactly one edge then Γ is a *complete bipartite graph*, and is determined up to

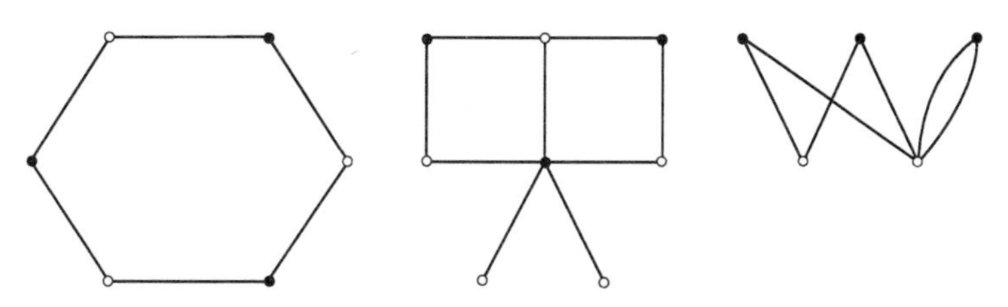

Fig. 7.12

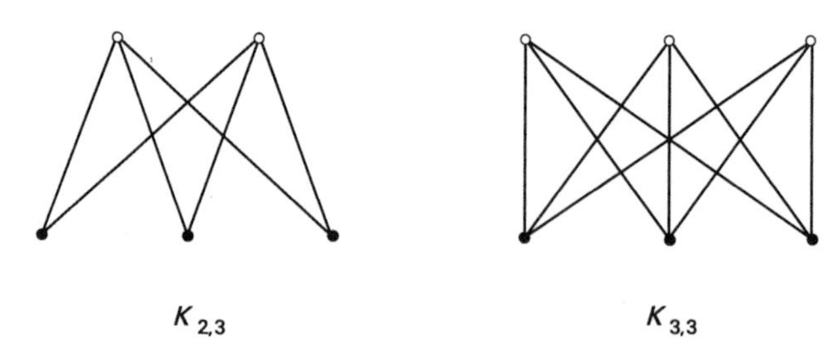

Fig. 7.13

isomorphism by the sizes of the sets V_1 and V_2. If $|V_1| = m$ and $|V_2| = n$ we denote the complete bipartite graph with vertex set $V = V_1 \cup V_2$ by $K_{m,n}$ (Fig. 7.13).

(d) Cayley quivers

The Cayley quiver of a group presentation appeared in Chapter 6. Cayley quivers have the feature that their edge sets are partitioned into subsets, with one subset for each generator in the presentation. The next result explains a method of encoding the partition of edges, and their direction, in a graph (and not a quiver!).

7.1 Lemma. *To each Cayley quiver C there corresponds a graph ΓC from which C may be reconstructed.*

Proof. Fix an ordering of the generators in the presentation, and for each $K \geqslant 1$ do the following. If x is the kth generator, replace each edge e in C labelled by x by a path of three edges and two new vertices e' and e'':

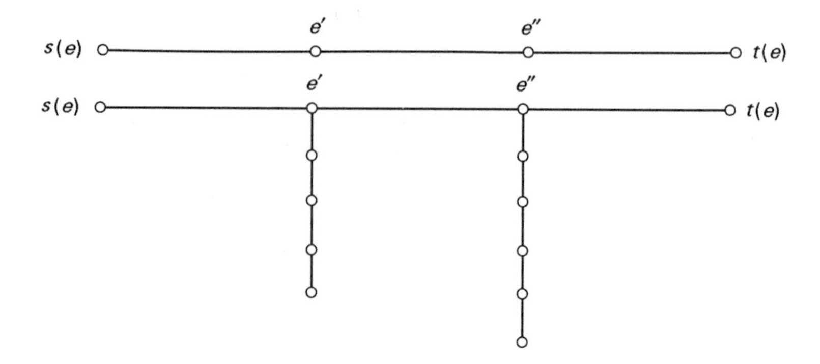

Add a path of length $2k - 2$, with initial vertex e', but otherwise consisting of new edges and vertices, and a similar path of length $2k - 1$ with initial vertex e''.
The new graph is ΓC, and it is easy to see that it has the property claimed in the lemma. ■

7.3 Planar graphs

A graph Γ is *planar* if it has a diagram that may be drawn in the plane without edges intersecting (except at the vertices!). We have seen several examples of planar graphs, and some that you may suspect are not planar. The basic fact about planar graphs is that they satisfy Euler's formula, which dates from 1750 and relates the number of vertices and edges in the graph to the number of regions or faces into which a diagram of the graph divides the plane.

For the remainder of this chapter we shall assume that graphs are finite; that is, for any Γ we assume that $V\Gamma$ and $E\Gamma$ are finite sets. Recall that a graph is connected if there exists a walk between any two vertices. Now any graph consists of a collection of connected graphs called its connected components, so a connected graph has just one connected component.

7.2 Euler's formula for planar graphs. *Let Γ be a planar graph with v vertices, e edges, and having c connected components. Suppose that a diagram of Γ divides the plane into f faces. Then*

$$v - e + f = c + 1.$$

Proof. We prove the formula by induction on e. If $e = 0$ then Γ has as many components as vertices so that $v = c$, and clearly $f = 1$. Therefore the formula holds.

Now suppose that Γ has $e > 0$ edges, and let Γ' be obtained from Γ by removing one edge. Certainly Γ' is planar, and has v vertices and $e - 1$ edges. Further, Γ' has c' connected components, where either $c' = c$ or $c' = c + 1$. From a diagram of Γ drawn in the plane we obtain a diagram of Γ' dividing the plane into f' faces, so that by inductive assumption, $v - e + 1 + f' = c' + 1$, that is $v - e + f' = c'$. First suppose that $c' = c$; that is, the edge removed from Γ does not separate a connected component of Γ into two components of Γ'. This edge either joins two distinct vertices of Γ, or else is an edge of the form $\{u, u\}$, and in either case its removal from Γ reduces the number of faces in the plane by one; that is, $f' = f - 1$ and $v - e + f = c + 1$ as required. If the edge removed does separate a component of Γ into two components of Γ', so that $c' = c + 1$, then this edge cannot be on the boundary of a finite face in the diagram of Γ and so occurs on the boundary of the infinite face. Its deletion does not then change the number of faces, whence $f' = f$ and again $v - e + f = c + 1$. ■

Examples

We may verify the application of Euler's formula to the two graphs shown in Fig. 7.14.

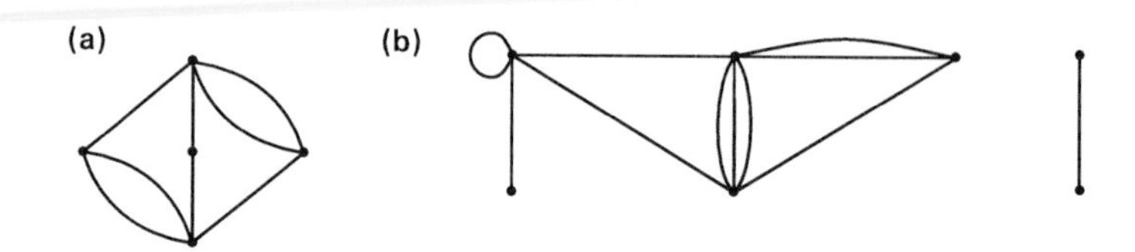

Fig. 7.14

In Fig. 7.14 (a), $v = 5$, $e = 7$, $f = 5$, and $c = 1$. In Fig. 7.14. (b), $v = 7$, $e = 11$, $f = 7$, and $c = 2$.

7.3 Corollary. *Let Γ be a connected simple graph with v vertices and e edges, where $v \geqslant 3$ and is finite. If Γ is planar then $e \leqslant 3v - 6$.*

Proof. In any diagram of Γ, every face has at least three edges in its boundary, and every edge occurs in the boundary of at most two faces. Therefore $3f \leqslant 2e$ and this inequality combines with Euler's formula to give the result. ■

Exercises 7.3

1. The most important case of Euler's formula is that for $c = 1$. Show that a graph is planar if and only if each of its connected components is planar.
2. Prove that K_5 is not planar.
3. Prove that $K_{3,3}$ is not planar.
4. The most celebrated result about the planarity of graphs is *Kuratowski's theorem* (Kuratowski 1930). We say that two graphs Γ and Γ' are *isomorphic modulo vertices of degree 2* if Γ is isomorphic to a graph Γ'' obtained from Γ' by the addition or deletion of vertices with just two edges incident:

Give a precise formulation of this notion in terms of $V\Gamma'$, $V\Gamma''$, $E\Gamma'$ and $E\Gamma''$. Kuratowski's theorem is as follows.

7.4 Theorem. *Let Γ be a finite graph. Then Γ is planar if and only if it contains no subgraph isomorphic modulo vertices of degree 2 to K_5 or $K_{3,3}$.*
We shall not prove this theorem here, but you are urged to consult an extended discussion, such as that in Wilson (1985).

7.4 Embedding graphs in surfaces

The embedding of a graph in the plane is just one part of the more general procedure of embedding graphs in surfaces, as the following observation shows.

7.5 Proposition *A graph Γ is planar if and only if it can be embedded in the sphere S^2.*

Proof. Stereographic projection (see Fig. 7.15) carries plane embeddings to embeddings in the sphere, and vice versa.

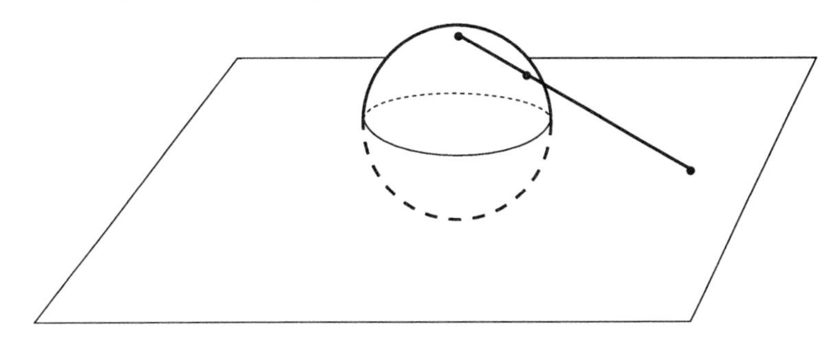

Fig. 7.15

It is now a natural step to consider the embeddings of graphs in other surfaces. If a graph Γ is not planar, how complicated a surface must we take in order to be able to embed Γ? That is, how large must the genus be? Clearly, the larger the genus required, the more complicated is the combinatorial structure of Γ. Note that for the moment it suffices to discuss only simple graphs, for if we are given a Γ that is not simple, then we may construct Γ_0 by replacing multiple edges with single edges, so that Γ_0 is simple, and embeds in a surface S if and only if Γ does.

Denote by S_g an orientable surface without boundary and of genus g. We may picture S_g as a sphere with g handles attached. A graph

Γ with g edges then embeds in S_g; draw one edge around each handle. So K_5 embeds in S_{10}. This is a rather extravagant waste of genus, for we shall see that K_5 may be embedded in S_1, the torus.

A couple of questions now arise, answers to which would shape the theory of graphs embedded in surfaces quite nicely:

1. Which graphs embed in S_g for fixed g?
2. Given a graph Γ, what is the smallest g for which Γ embeds in S_g?

If Γ embeds in S_g then it certainly embeds in S_h for any $h \geqslant g$, so that an answer to question 2 seems more fundamental. An ideal answer to question 1 would be a version of Kuratowski's theorem (see Exercise 4 of Section 7.3) that characterized the graphs embeddable in S_g for each $g \geqslant 0$. Yet far from having ideal answers, we have only partial ones, continually improved and updated by research work. The books by Gross and Tucker (1987) and White (1984) are excellent places to look for more detailed information.

Question 2 does have a theoretical solution, as a consequence of a general method for constructing an embedding of a graph in a surface. This method is known as the *Edmonds algorithm*, named after J. Edmonds who described it in 1960.

Let Γ be a connected simple graph with finitely many vertices. For notational convenience we shall suppose that $V\Gamma = \{1, 2, \ldots, n\}$. Now for each $i \in V\Gamma$, set $V(i) = \{j \in V\Gamma \mid \{i, j\} \in E\Gamma\}$; that is, $V(i)$ is the set of vertices connected to vertex i. Let $|V(i)| = n_i$ and for each i choose an n_i-cycle p_i permuting the set $V(i)$. For example if $V(1) = \{4, 6, 7\}$, we could take $p_1 = (6\ 4\ 7)$. Now let $D = \{(i, j) \mid \{i, j\} \in E\Gamma\}$; hence D is the set of all *ordered* pairs of vertices connected by edges in Γ. The choices of the permutations $p_1, \ldots, p_n$ determine a permutation P of D defined by $P(i, j) = (j, p_j(i))$. If $(i, j) \in D$ then $i \in V(j)$ and so $p_j(i) \in V(j)$. Therefore $\{j, p_j(i)\} \in E\Gamma$ and the orbits of P consist of edges of Γ that fit together into circuits in Γ. We represent such a circuit by drawing it around a polygon; a vertex of such a polygon corresponds to a vertex of Γ and so can be labelled with an element of $\{1, 2, \ldots, n\}$; the edge from vertex j to vertex $p_j(i)$ is labelled with the *ordered* pair $(j, p_j(i))$. Having labelled all the polygons in this way, we can paste them together by matching an edge (a, b) with an edge (b, a) to form a surface in which Γ is embedded.

Example

We show that $K_{3,3}$ may be embedded on the torus S_1 (Fig. 7.16).

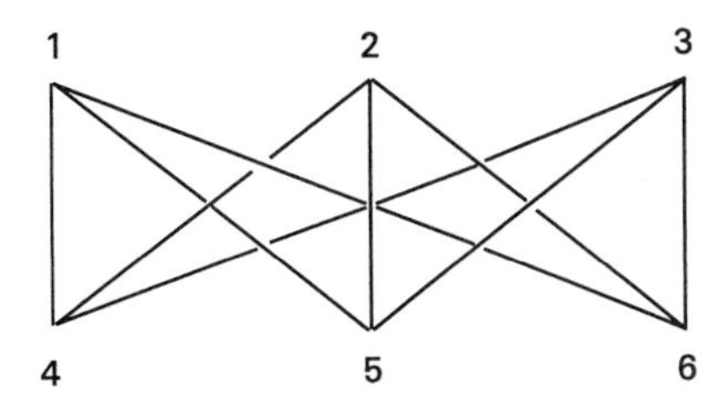

Fig. 7.16

$$V(K_{3,3}) = \{1, 2, 3, 4, 5, 6\}$$
$$V(1) = V(2) = V(3) = \{4, 5, 6\}$$
$$V(4) = V(5) = V(6) = \{1, 2, 3\}.$$

We will take the permutations

$$p_1 = p_2 = p_3 = (4\ 5\ 6)$$
$$p_4 = p_5 = p_6 = (1\ 2\ 3).$$

Thus $P(1, 4) = (4, 2)$, $P(4, 2) = (2, 5)$ and continuing the calculation, we find three orbits:

$$(1, 4) \to (4, 2) \to (2, 5) \to (5, 3) \to (3, 6) \to (6, 1).$$
$$(1, 5) \to (5, 2) \to (2, 6) \to (6, 3) \to (3, 4) \to (4, 1)$$
$$(1, 6) \to (6, 2) \to (2, 4) \to (4, 3) \to (3, 5) \to (5, 1).$$

This gives us three hexagons that can be pasted together to form a surface embedding of $K_{3,3}$. At this stage we can work out the genus of the surface, for the embedding of $K_{3,3}$ subdivides the surface into the three regions given by the orbits, with nine edges and six vertices. Therefore the surface has Euler characteristic

$$\chi = 6 - 9 + 3 = 0$$

and genus $g = 1$, and so it is indeed a torus. We can fit the hexagons together to make the usual gluing instructions for the torus, with $K_{3,3}$ embedded:

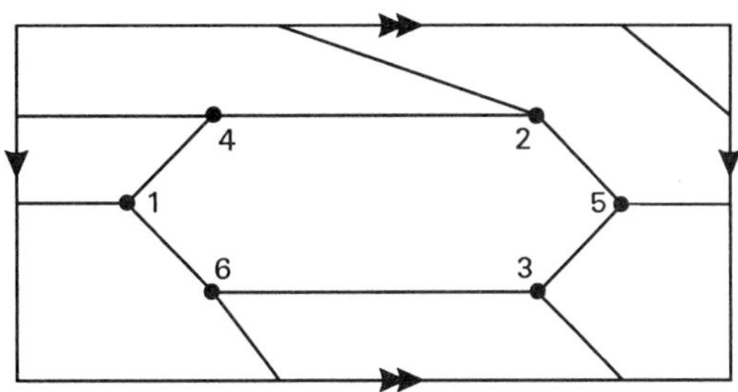

Any embedding of a graph Γ in a surface S constructed by the Edmonds algorithm has the property that $S\backslash\Gamma$ is a union of disjoint regions each homeomorphic to an open disc; these regions arise from the interiors of the polygons around which the circuits are drawn. We call an embedding with this property a *disc embedding*. Now given a disc embedding of Γ in S, the boundaries of the regions of $S\backslash\Gamma$ are circuits in Γ and the succession of the edges in any given circuit can be read as a set of ordered pairs of vertices, that is as a subset of D. Since S is assumed to be orientable, neighbouring regions can be oriented coherently and their boundaries read according to the orientations, partition the set D. The parts of the partition are orbits for a map $P^*: D \to D$ from which permutations p_i for each $i \in V\Gamma$ can be found. These considerations show that *every* disc embedding of a graph Γ in an orientable surface S can be constructed by the Edmonds algorithm.

If a graph Γ can be embedded in an orientable surface of genus g, but not in one of genus $h < g$, we say that Γ has *genus* g. An embedding of Γ into a surface of genus g is then called a *minimal* embedding. We have seen that K_5 has genus 1, as does $K_{3,3}$. The planar graphs are precisely those of genus 0.

7.6 Proposition. *A minimal embedding of a graph Γ in an orientable surface S is a disc embedding.*

Proof. Suppose that some circuit in Γ bounds a region in S *not* homeomorphic to an open disc. We may assume that the region of S contains no edges of Γ (or else we take a circuit containing fewer edges until we find such a region), and we cut S along the circuit. If it happens that S is thereby cut into components, discard the component containing no edges of Γ and attach a disc around the circuit in the other component. This creates a new surface S' of smaller genus than S in which Γ is embedded. Hence the original embedding was not minimal, and this is a contradiction. If S is not cut into two components by cutting along the chosen circuit, then attach disc around each hole caused by the cut, again to form a new surface with smaller genus and with Γ still embedded, and so arrive at the same contradiction. ∎

Given a graph Γ we can theoretically use the Edmonds algorithm to construct all the disc embeddings of Γ into orientable surfaces, and if the smallest genus that arises is g then we know from Proposition 7.6 that Γ has genus g. The practical difficulty is that there a huge number of embeddings constructed by the Edmonds algorithm, corresponding to the choices of permutations to be made, so that it is infeasible to calculate the genus of a graph by this method. However, advanced techniques have been applied to calculate the genus for some special classes of graphs, and we quote two cases here. The symbol $\lceil \chi \rceil$ denotes the smallest integer greater than or equal to χ.

7.7 Theorem. *The complete graph K_n $(n \geqslant 3)$ has genus $\lceil (n-3)(n-4)/12 \rceil$, and the complete bipartite graph $K_{m,n}$ has genus $\lceil (n-2)(m-2)/4 \rceil$.* ■

These results are due to G. Ringel and J.W.T. Youngs; see White (1984) for a fuller discussion.

Exercises 7.4

1. Attempt to embed K_5 in S_1; that is, try to draw a diagram of K_5 on the torus without edges crossing. Represent the torus by the gluing instructions.

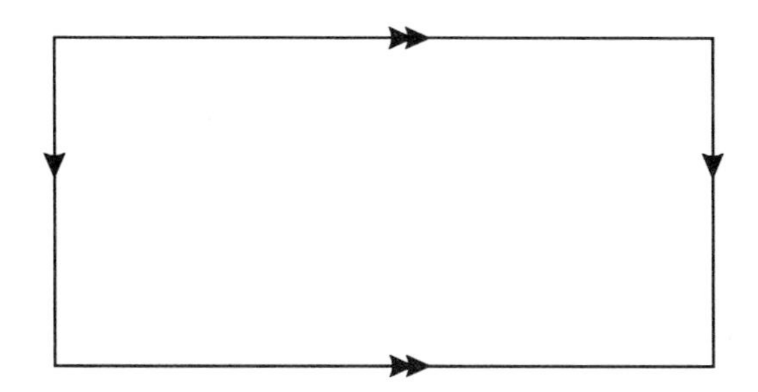

2. Try a different choice of permutations in the example given above for embedding $K_{3,3}$ on the torus. For example, take

$$p_1 = (4\ 5\ 6),\quad p_2 = (4\ 6\ 5) = p_3,\quad p_4 = (1\ 2\ 3)\quad \text{and}\quad p_5 = (1\ 3\ 2) = p_6.$$

3. Use the Edmonds algorithm to construct an embedding of K_5 on a torus.
4. What happens if you apply the Edmonds algorithm to a planar graph? Does an embedding in the sphere always result? Try K_4, $K_{2,2}$ and $K_{2,3}$.
5. Prove the following generalization of Euler's formula. Let Γ be a connected graph with v vertices, e edges (where v and e are finite). Suppose that a disc embedding into a surface S of genus g is such that $S\backslash\Gamma$ is a union of f regions. Then $v - e + f = 2 - 2g$.

8

Alexander matrices and Alexander polynomials

Although the method that we have used to calculate the Alexander polynomial of a knot only uses the theory of determinants, the trouble with it is to know why it works. It is easy to check that the result is invariant under the Reidemeister moves but it would be nice to link it in with other invariants. 'It would be nice' sounds a weak reason, but often when one can do something by a different method, it is possible to see how to 'squeeze' the method a bit to get out more detailed information. What we will do is give a second method of calculating Alexander polynomials, this time using presentations of groups. The payoffs are (i) methods that can be applied to other group presentations which we will use later in Chapter 11; (ii) intermediate stages and related constructions that show more of the fine detail of the situation even in the case of knots; and (iii) new invariants that have important geometric significance although in this book we will not be able to do justice to this aspect.

To do all this will require new algebraic ideas. These in turn have exciting consequences within algebra but we will not be able to explore these fully.

8.1 Group rings

For any group, G, consider the set $\mathbb{Z}G$ (where, as usual, $\mathbb{Z}$ denotes the ring of integers) of all formal sums of the form

$$\sum_{g \in G} n_g g$$

where each $n_g \in \mathbb{Z}$ and only finitely many of the n_g are non-zero. Examples of such expressions are $g_1 - 2g_2$, $4g_3 + 3g_4 - 6g_5$, and so on. The set $\mathbb{Z}G$ has a natural structure of an Abelian group where addition is given by

$$\left(\sum n_g g\right) + \left(\sum m_g g\right) = \sum \left(n_g + m_g\right) g.$$

(As only finitely many of the n_g and m_g are non-zero, the same is true of the $(n_g + m_g)$, so the right-hand side is an element of $\mathbb{Z}G$.) If we think of the *set* $|G|$ of elements of G, that is we forget the multiplication of G, then $\mathbb{Z}G$ is the free Abelian group on this set, $|G|$.

Remembering the multiplication of G, we get a multiplication on $\mathbb{Z}G$, which makes it into a ring. Explicitly we have

$$\left(\sum_{g \in G} n_g g\right)\left(\sum_{h \in G} m_h h\right) = \left(\sum_{g, h \in G} n_g m_h (gh)\right).$$

The result is not quite in the right form but if we write for $k \in G$, $p_k = \Sigma_{g \in G}\, n_g m_{g^{-1}k}$, then we can partition up the sum on the right-hand side to get $\Sigma_{k \in G}\, p_k k$. We leave it to you to check that p_k will be non-trivial only for finitely many k in G. This ring $\mathbb{Z}G$ is called the group ring of G.

Example

One of the most useful examples is when $G = C_\infty$, the infinite cyclic group generated by t, say. A typical element of C_∞ is therefore t^i and a formal sum can be thought of as a polynomial in t and t^{-1} or as a Laurent polynomial

$$\sum_{i=j}^{k} n_i t^i$$

where $j \leqslant k$ are integers. For instance, $t^{-1} - 1 + t$ is an element of $\mathbb{Z}C_\infty$. Another example is when $G = C_2 = \{1, a\}$ with $a^2 = 1$; then $\mathbb{Z}G = \{n_1 + n_2 a : n_1, n_2 \in \mathbb{Z}\}$ where, for instance, $(4a)(1 - a) = 4a - 4a^2 = -4 + 4a$. It is clear that $\mathbb{Z}G$ is, in general, a commutative ring if and only if G is an Abelian group.

The group G can be thought of as sitting inside $\mathbb{Z}G$ via the function $g \mapsto 1g$, that is the element $\Sigma n_h h$ where $n_g = 1$ and $n_h = 0$ if $h \neq g$. Any function $\phi: G \to A$, with A an Abelian group, extends uniquely to an Abelian group homeomorphism from $\mathbb{Z}G$ to A. If $\phi: G \to A$ is such that A is a ring and ϕ preserves multiplication then the extension to $\mathbb{Z}G$ is a ring homeomorphism. Using this we can see that any group homeomorphism $G \to H$ composed with the inclusion of H into $\mathbb{Z}H$ extends uniquely to a ring homeomorphism from $\mathbb{Z}G$ to $\mathbb{Z}H$.

This last result will be useful in two cases in particular:

(1) $\mathscr{A}: G \to G_{ab}$, the Abelianization homeomorphism;

(2) $\varepsilon: G \to \{1\}$, the trivial homeomorphism. The resulting ring homeomorphism sends $\Sigma n_g g$ to Σn_g, and is called the augmentation homeomorphism.

There is a whole class of problems which involve the transfer of properties from G to $\mathbb{Z}G$. For instance, we might ask for what groups G is $\mathbb{Z}G$ an integral domain or a unique factorization domain? (If you have not met integral domains or unique factorization domains, now would be a good time to look up the definitions. You will find them in almost all the usual algebra textbooks.) We will review briefly some ideas of ring theory for convenience:

Let R be a ring with an identity element, 1.

A *unit* in R is an element $u \in R$ with an inverse, that is there is some $v \in R$ with $vu = uv = 1$. For instance, in a group ring the element 1 is the formal sum $n_1 1 = \Sigma n_g g$ where $n_g = 1 \in \mathbb{Z}$ if $g = 1 \in G$ and $n_g = 0$ otherwise, so any element $g \in G$ gives $1g \in \mathbb{Z}G$ and $(1g) \cdot (1g^{-1}) = 1$. Thus any element of G considered as an element of the ring $\mathbb{Z}G$ is a unit. Can there be units other than those of the form g or $(-1)g$ in $\mathbb{Z}G$?

We call two elements a, b in a commutative ring R associated if a divides b (written $a|b$), so there is some $c \in R$ with $b = ac$, and b divides a. If R is an integral domain (so if $ab = 0$, one of a and b must itself be zero), then if a and b are associated, there is a unit u with $a = ub$. (We leave the proof to you.)

An element d of a commutative ring R is called a greatest common divisor (g.c.d.) of a finite set of elements $a_1, \ldots, a_n \in R$ if d divides each a_i, $i = 1, \ldots, n$, and if $c|a_i$ for $i = 1, \ldots, n$ then $c|d$. Any two g.c.d.s of a given set are therefore associated and so if R is an integral domain, the g.c.d. of a set of numbers is determined up to multiplication by units. For instance, both 2 and -2 are g.c.d.s of $\{-6, 4, 24, -26\}$, $-2 = (-1) \times 2$, and -1 is a unit in $\mathbb{Z}$. We say R is a g.c.d. domain if it is an integral domain in which any finite set of elements has a g.c.d.

Recall 'R is a unique factorization domain' means that it is an integral domain with 1 in which every $x \in R$, $x \neq 0$ or 1, has a decomposition as a product of prime elements, that is unique up to choice of order. Any unique factorization domain is a g.c.d. domain as we can use essentially the same method as we might use for finding g.c.d.s in $\mathbb{Z}$ given prime decompositions (e.g. g.c.d. (201, 33) = 3 since $201 = 3 \times 67$ and $3 = 3 \times 11$, so 3 is the only prime power common to both).

If Q is a subring of R, and there is a mapping $\rho: R \to Q$ preserving multiplication and such that a and $\rho(a)$ are associated for any $a \in R$, then

(1) if Q is an integral domain, then so is R;

(2) if Q is a g.c.d. domain, then so is R.

The proofs are quite easy. In each case, to verify a property in R map the test elements back down into Q, verify the property for the images, and then complete since always a and $\rho(a)$ are associated. (Write out a complete proof yourself.)

Before returning to group rings we need to import two results from ring theory, namely (i) that if R is an integral domain, so is the polynomial ring $R[t]$ and the only units in $R[t]$ are those within the subring R (considered a subring via the usual embedding as constant polynomials), and (ii) if R is a unique factorization domain, so is $R[t]$. (Proofs of these results can be found in many standard books on ring theory if you wish to check them.)

The appearance of $R[t]$ in a section on group rings should give a hint as to which group ring is going to be in the limelight for the next few pages. We have already met $\mathbb{Z}C_\infty$ as a ring of Laurent polynomials and clearly $\mathbb{Z}[t]$ is a subring of $\mathbb{Z}C_\infty$. Are we in the situation above? Is there a multiplication preserving mapping $\rho: \mathbb{Z}C_\infty \to \mathbb{Z}[t]$ such that a and $\rho(a)$ are associated for each $a \in \mathbb{Z}C_\infty$? Any such a has the form

$$a = \sum_{n=-\infty}^{\infty} a_n t^n, \qquad a_n \in \mathbb{Z}$$

where only finitely many of the a_n are non-zero. Define, for $a \neq 0$, $\mu(a)$ to be the smallest n such that $a_n \neq 0$, so that for example $\mu(1) = 0$ and $\mu(t^2 - 2t + t^{-24}) = -24$. If $a = 0$ then it will be convenient and consistent with our usage of μ to define $\mu(0) = \infty$. We can easily check (do it) that if $a, b \in \mathbb{Z}C_\infty$,

$$\mu(ab) = \mu(a) + \mu(b).$$

(The case where a or b is zero needs care but is fairly obvious. The other cases are quite simple. Write out a proof.)

Now define $\rho(a) = at^{-\mu(a)}$. This gives a normalization of a, changing it into a polynomial. Of course, a and $\rho(a)$ are associated, since $t^{-\mu(a)}$ is a unit unless $\mu(a) = \infty$. We make the convention that $t^{-\infty} = 0$. You should check that ρ preserves multiplication.

A consequence of this is the following result.

The group ring $\mathbb{Z}C_\infty$ of the infinite cyclic group is a g.c.d. domain.

We can, in fact, say more. The units in $\mathbb{Z}C_\infty$ are all the elements of the form $\pm t^n$, $n \in \mathbb{Z}$. This will be of use when working out Alexander polynomials.

You may wonder why we introduced a general ring R in our earlier discussion. The reason is that a similar argument would have worked for the group ring of any finitely generated free Abelian group. We will not need this in detail, but note the result. You may like to try to investigate this situation and to prove that $\mathbb{Z}(C_\infty \oplus \ldots \oplus C_\infty)$ is always a g.c.d. domain whose only units are group elements and their negatives.

8.2 Derivatives

R. H. Fox, one of the 'founding fathers' of modern knot theory, developed a theory of 'derivatives' (also called derivations) which is a powerful tool in the study of group presentations.

A derivative on a group G is a mapping $D: G \to \mathbb{Z}G$ such that $D(gh) = D(g) + gD(h)$.

Example and Exercise

The augmentation homeomorphism

$$\varepsilon: \mathbb{Z}G \to \mathbb{Z}$$

sends $\Sigma n_g g$ to Σn_g. The kernel of ε contains such elements as $g - 1$. Prove that if G is finite,

(a) Ker ε is a free Abelian group with basis

$$\{g - 1: g \in G\}$$

(b) the mapping from G to $\mathbb{Z}G$ sending g to $g - 1$ is a derivative.

These two properties are very important in the study of the homology groups of a group. In this, one tries to apply methods based on topological ideas to study algebraic objects. At several times in the following chapters we will touch on material related to this area.

Given any derivative $D: G \to \mathbb{Z}G$, there is an extension to a mapping $D: \mathbb{Z}G \to \mathbb{Z}G$ given by $D(\Sigma n_g g) = \Sigma n_g D(g)$. This is additive

$$D(g + h) = D(g) + D(h)$$

but is not a ring homeomorphism.

8.1 Lemma. *If $D: G \to \mathbb{Z}G$ is a derivative, then*

(a) $D(g^{-1}) = -g^{-1}D(g)$.

(b) $D(g^n) = \dfrac{g^n - 1}{g - 1} D(g)$ *for all* $n \in \mathbb{Z}$.

The proofs are by easy algebraic manipulation using the definition $D(gh) = D(g) + gD(h)$ and are left as an exercise.

The case when G is a free group, $F = F(X)$, is especially useful. In this case, to each free generator x of F there corresponds a unique derivative written

$$\frac{\partial}{\partial x}: F \to \mathbb{Z}F$$

defined by

$$\frac{\partial y}{\partial x} = \begin{cases} 1 & \text{if } x = y \\ 0 & \text{if } x \neq y. \end{cases}$$

This defines $\partial/\partial x$ on generators. You should check that there is no difficulty in extending it to the whole of F. You can do this by remembering that elements of F are equivalence classes of words and then checking that the 'obvious' extension to $\partial w/\partial x$ works in such a way that equivalent words have the same image.

Example

We will take $X = \{x, y, z\}$, $F = F(X)$, the free group on X, and will look at $\partial w/\partial x$ where $w = xy^{-1}x^2zx^{-1}$,

$$\frac{\partial}{\partial x}w = \frac{\partial}{\partial x}x + x\frac{\partial}{\partial x}(y^{-1}x^2zx^{-1}) = 1 + x\frac{\partial}{\partial x}y^{-1} + xy^{-1}\frac{\partial}{\partial x}(x^2zx^{-1}).$$

We note that $\partial x^2/\partial x = 1 + x$, so this expression equals

$$\begin{aligned} &1 + xy^{-1}\left((1 + x) + x^2\frac{\partial}{\partial x}zx^{-1}\right) \\ &= 1 + xy^{-1}\left((1 + x) + x^2\frac{\partial}{\partial x}z + x^2z\frac{\partial}{\partial x}x^{-1}\right) \\ &= 1 + xy^{-1}\left(1 + x + x^2z(-x^{-1})\right) \end{aligned}$$

which can now be simplified. (**Caution**: since we are in $\mathbb{Z}F$ and F is free, x^2zx^{-1} is not the same as xz since the generators do not commute.)

Exercises 8.2

1. Working in $F(x, y)$, the free group on $\{x, y\}$, calculate $\partial/\partial x$ and $\partial/\partial y$ of the following words:

(a) $xyxy^{-1}x^{-1}y^{-1}$

(b) $(xy)^n x(xy)^{-n}$, for $n \in \mathbb{N}$

(c) $[x^m, y^n]^p$, for $m, n, p \in \mathbb{N}$.

2. Investigate the following 'short cut' and its generalizations. Suppose we want to calculate $\partial w/\partial x$ and we can decompose w as $w_1 x^n w_2$ where neither w_1 nor w_2 involves the generator x; then

$$\frac{\partial}{\partial x} w = w_1 \left(\frac{x^n - 1}{x - 1} \right).$$

3. Prove that if F is a free group on a set X and D is any derivative, then D is completely determined by the values $D(x)$ for $x \in X$ that it takes on the generators. Knowing these values a formula of the form

$$D(w) = \sum_{x \in X} a_x(w) D(x), \quad a_x(w) \in \mathbb{Z}F$$

can be found. What are these coefficients $a_x(w)$?

4. We suggested earlier that you proved that defining $\partial/\partial x$ on the generators of F was sufficient to define it on all of F. This is a bit like the proof that any mapping defined on the generators extends to a homeomorphism of F. Using the earlier exercise on Ker ε, and the fact that it is generated freely by the set of $x - 1$, $x \in X$, it is possible to find an elegant proof of the 'extension of derivatives' result using just the freeness of F and of Ker ε. Try it out, and see what you can do.

8.3 The Alexander matrix

Suppose we have a group presentation $(X : R)$ of a group $G \cong F/\langle\langle R \rangle\rangle$. We will only need the case when X and R are finite and so will assume this is true throughout. We write $X = \{x_1, \ldots, x_n\}$, $R = \{r_1, \ldots, r_m\}$. Furthermore we will be a bit careful and will suppose all our relations are of the form $r = 1$ (i.e. we rewrite a relation $r = s$ as $rs^{-1} = 1$). We will write

$$\gamma: F \to G$$

for the quotient map and $\gamma: \mathbb{Z}F \to \mathbb{Z}G$ also for its 'extension' to the corresponding group rings. Similarly we write $\mathcal{A}: G \to G_{ab}$ for the natural epimorphism Abelianizing G.

We form a 'Jacobian matrix'

$$J = \left(\frac{\partial r_i}{\partial x_j} \right)$$

with coefficients in $\mathbb{Z}F$. In many ways $\mathbb{Z}F$ is like a Laurent polynomial ring in non-commuting variables $x_1, \ldots, x_n$. Partially to make life simpler (and there, as usual in mathematics, we run the risk of simplifying the interesting detail out of existence) and partially because we are really more interested in G or G_{ab} rather than in F, we use γ and $\mathcal{A}$ to take these coefficients in J down to $\mathbb{Z}G_{ab}$. As this is a commutative ring, we can work more easily here.

We set

$$A = \left(\mathcal{A}\gamma\left(\frac{\partial r_i}{\partial x_j}\right)\right).$$

This is the Alexander matrix of the presentation. The definition perhaps looks slightly technical to you, but the various steps are easily followed through as you will see when we do some examples. The significance of γ is that it allows us to simplify using the relations in R and $\mathcal{A}$ allows us to simplify even further.

Examples

1. $G = S_3 = (a, b : a^3 = b^2 = (ab)^2 = 1)$.

   We take $F = F(a, b)$, the free group on $\{a, b\}$, and relations $r_1 = a^3$, $r_2 = b^2$, $r_3 = (ab)^2$,

$$\frac{\partial}{\partial a} r_1 = 1 + a + a^2 \qquad \frac{\partial}{\partial b} r_1 = 0$$

$$\frac{\partial}{\partial a} r_2 = 0 \qquad \frac{\partial}{\partial b} r_2 = 1 + b$$

$$\frac{\partial}{\partial a} r_3 = [1 + (ab)] \frac{\partial}{\partial a}(ab) \qquad \frac{\partial}{\partial b} r_3 = [1 + (ab)] \frac{\partial}{\partial b}(ab)$$

$$= (1 + ab) \qquad = (1 + ab)a$$

so our 'Jacobian matrix' is

$$\begin{pmatrix} 1 + a + a^2 & 0 \\ 0 & 1 + b \\ 1 + ab & a + aba \end{pmatrix}.$$

Applying γ to the coefficients means that a^3, b^2, and $(ab)^2$ become 1; then applying the Abelianization and writing $\bar{a}$ for $\mathcal{A}(a)$ we find that $\overline{a}\overline{b}\overline{a}\overline{b} = 1$ implies $a^{-2}b^{-2} = 1$; hence $a^{-2} = 1$ and $a^{-3} = 1$, so $\bar{a} = 1$, that is $(S_3)_{ab} \cong C_2 = (t : t^2)$ with generator t corresponding, under the isomorphism, to $\bar{b}$. Thus we obtain an Alexander matrix

$$A = \begin{pmatrix} 3 & 0 \\ 0 & 1+t \\ 1+t & 1+t \end{pmatrix}$$

since $\mathscr{A}(a) = 1$.

Before considering another example, we will look at a computational hint which saves time and diminishes the chance of errors, especially when relations are long.

Consider a relation of the form $r = s$; this normalizes to give $rs^{-1} = 1$ and if x is a generator, we have

$$\frac{\partial}{\partial x}(rs^{-1}) = \frac{\partial r}{\partial x} + r\frac{\partial}{\partial x}(s^{-1})$$

$$= \frac{\partial r}{\partial x} - rs^{-1}\frac{\partial s}{\partial x}.$$

However, when we calculate the Alexander matrix of the presentation we first apply $\gamma\colon F \to F/\langle\!\langle R\rangle\!\rangle$ and in $F/\langle\!\langle R\rangle\!\rangle$, $\gamma(rs^{-1}) = 1$, so

$$\gamma\left(\frac{\partial}{\partial x}(rs^{-1})\right) = \gamma\left(\frac{\partial r}{\partial x} - \frac{\partial s}{\partial x}\right).$$

This implies that we can compute the derivatives of rs^{-1} by using $r - s$ in the group ring $\mathbb{Z}F$. If r and s are of moderate length, rs^{-1} is long and it is easy to make an error in calculating $\partial(rs^{-1})/\partial x$. For instance, in the cinquefoil knot group presentation, we get

$$(x, y : xyxyx = yxyxy)$$

or

$$(x, y : xyxyxy^{-1}x^{-1}y^{-1}x^{-1}y^{-1} = 1).$$

In fact, here the derivatives of this single relation can be handled in several different ways, but one of the best is to calculate the derivatives of $xyxyx - yxyxy$ in $\mathbb{Z}F(x, y)$.

2. $G = (x, y : x^2 = 1, y^3xy^{-2}x^{-1})$.

The Abelianization G_{ab} will have $y = 1$ and we will put $\mathscr{A}\gamma(x) = t$ ($t^2 = 1$ in G_{ab}, which is isomorphic to C_2).

Put $r = x^2$ so

$$\frac{\partial}{\partial x}r = 1 + x \qquad \frac{\partial r}{\partial y} = 0.$$

Put $s = y^3xy^{-2}x^{-1}$; we use the calculation via $y^3x - xy^2$ in $\mathbb{Z}F(x, y)$ to get

$$\mathscr{A}\gamma\left(\frac{\partial}{\partial x}s\right) = \mathscr{A}\gamma\left(\frac{\partial}{\partial x}(y^3x) - \frac{\partial}{\partial x}(xy^2)\right)$$
$$= \mathscr{A}\gamma(y^3 - 1) = 0.$$

Similarly

$$\mathscr{A}\gamma\left(\frac{\partial}{\partial y}s\right) = \mathscr{A}\gamma\Big((1 + y + y^2) - x(1 + y)\Big)$$
$$= 3 - 2t.$$

Exercises 8.3

1. If you have not already done this, check that if G is a knot group, G_{ab} is an infinite cyclic group with one generator, t. (Although you can use either the Dehn or Wirtinger presentations to obtain this, the Wirtinger presentation has the advantage that all the generators map to t under the composite $\mathscr{A}\gamma$.)
2. Check that $(a, b : a^p = b^q)_{ab}$ is infinite cyclic if the highest common factor of p and q is 1. (This exercise is a repeat.)
3. Calculate the Alexander matrices of the following presentations:

(a) $G(\text{trefoil}) = (x, y, z : z = xyx^{-1}, x = yzy^{-1}, y = zxz^{-1})$.

(b) $G(\text{cinquefoil}) = (x, y : (xy)^2x = y(xy)^2)$.

(c) If $n = 2m + 1$, and $t_{2,n}$ is the $(2, n)$-torus knot,

$$G(t_{2,n}) = (x, y : (xy)^m x = y(xy)^m).$$

(d) Again as in (c), $G(t_{2,n}) = (a, b : a^n = b^2)$.

(Beware, you need to do the preceding exercise first.)

8.4 Elementary ideals

The Alexander matrix A of a presentation is not invariant under Tietze transformations since even its size can change if a generator is added or removed. What sorts of quantities could we expect to construct from A that might be invariant? One way to guess the sort of answer to expect to such a question is to see how an Alexander matrix of a given presentation changes as we apply Tietze transformations. (We suggest that you do this investigation yourself. That way you can adjust the complexity of the example you tackle to your present understanding of the question. If you feel you understand quite well, it would be a good idea to consider the completely general question of how the Alexander matrix of $(x_1, \ldots, x_m : r_1, \ldots, r_2)$

changes under the various transformations. Whichever way you go about this, compare the transformations that result with those you know from linear algebra.)

Alexander matrices are matrices over group rings, $\mathbb{Z}G_{ab}$. You are most used to matrices over $\mathbb{R}$ or $\mathbb{C}$ and may also have met, for instance when handling presentations of finitely generated Abelian groups, matrices over $\mathbb{Z}$. To keep things fairly general for the moment, we will start by working over a general commutative ring R. If S is a subset of R, the *ideal E generated by S* is the set of all finite sums $\Sigma a_i s_i$, $a_i \in R$, $s_i \in S$. Formally an ideal behaves like the normal subgroup generated by a subset of a group.

Let A be an $m \times n$ matrix over R. For k, a non-negative integer, the *kth elementary ideal* $E_k(A)$ of A is:

(1) if $0 < n - k \leqslant m$, the ideal generated by all $(n-k) \times (n-k)$ minors of A.

(2) if $(n-k) > m$, $E_k(A) = 0$.

(3) if $0 \geqslant n - k$, $E_k(A) = R$, the whole ring.

Remarks

(a) In case you have not met the term (or have need to recall it) an $r \times r$ minor of A is the determinant of an $r \times r$ submatrix of A.

(b) The slight awkwardness in the definition is needed because if we merely said that $E_k(A)$ is generated by the $(n-k) \times (n-k)$ minors of A then what could be the meaning of, say, a -2×-2 minor or a 10×10 minor in a 4×5 matrix? Because of this, we had to assign meanings to $E_k(A)$ when $r - k$ is small or larger than m.

(c) Since any $(r+1) \times (r+1)$ minor can be expanded as a linear combination (over R) of $r \times r$ minors of A, these ideals are nested:

$$0 \subseteq E_0(A) \subseteq E_1(A) \subseteq \ldots \subseteq E_n(A) = R.$$

Examples

1. $$A = \begin{pmatrix} 3 & 0 \\ 0 & 1+t \\ 1+t & 1+t \end{pmatrix},$$

$E_0(A)$ is generated by 2×2 minors:

$$\begin{vmatrix} 3 & 0 \\ 0 & 1+t \end{vmatrix}, \quad \begin{vmatrix} 0 & 1+t \\ 1+t & 1+t \end{vmatrix}, \quad \begin{vmatrix} 3 & 1+t \\ 1+t & 1+t \end{vmatrix};$$

hence by $3(1+t)$, $-(1+t)^2$, $3(1+t)$.

We have not stated over which ring we are working but A is, in fact, the form of the Alexander matrix of the usual presentation of S_3 and $(S_3)_{ab}$ is cyclic of order 2, so we will assume $R = \mathbb{Z}C_2$ generated by t with $t^2 = 1$. This gives us that $E_0(A)$ is generated by $3(1+t)$ and $-2(1+t)$ (we write $E_0(A) = \langle 3(1+t), -2(1+t)\rangle$.

Since $(1+t) = 3(1+t) - 2(1+t)$, we get

$$E_0(A) = \langle 1 + t\rangle.$$

$E_1(A)$ is generated by 1×1 minors and hence by the entries of A. We have

$$E_1(A) = \langle 3, 1+t\rangle \quad \text{over } \mathbb{Z}C_2.$$

We know

$$0 \cong E_0(A) = \langle 1+t\rangle \subseteq E_1(A) = \langle 3, 1+t\rangle \subseteq E_2(A) = \mathbb{Z}C_2.$$

Are both of the inclusions proper or is one an equality? If $E_0(A) = E_1(A)$ then $3 \in \langle 1+t\rangle$. This would mean there was some $\lambda, \mu \in \mathbb{Z}$ such that $(\lambda + \mu t)(1+t) = 3$. (We must remember that we are working in $\mathbb{Z}C_2$, so all elements have the form $\alpha + \beta t$ for $\alpha, \beta \in \mathbb{Z}$, and on multiplying, we put $t^2 = 1$.) Multiplying out our conjectured equation gives

$$(\lambda + \mu) + (\lambda + \mu)t = 3$$

which is silly since $\lambda + \mu$ cannot be both 3 and 0. This shows that $3 \notin E_0(A)$ and the first inclusion is proper. Similarly one knows that $1 \in \mathbb{Z}C_2$ and we can easily adapt the above to show that $1 \notin E_1(A)$.

2. $G = (x, y : x^2 = 1, y^3x = xy^2)$.

As we saw $G_{ab} \cong C_2$ and the Alexander matrix is

$$A = \begin{pmatrix} 1+t & 0 \\ 0 & 3-2t \end{pmatrix}$$

over $\mathbb{Z}C_2$ as before.

$E_0(A)$ is generated by 2×2 minors, that is by the determinant of A:

$$\begin{aligned}(1+t)(3-2t) &= 3 + 3t - 2t - 2t \\ &= 1 + t.\end{aligned}$$

$E_1(A)$ is generated by 1×1 minors $1+t$ and $3-2t$, but as $(1+t) = (1+t)(3-2t)$, $E_1(A) = \langle 3-2t\rangle$.

We get a chain

$$0 \subseteq \langle 1+t\rangle \subseteq \langle 3-2t\rangle \subseteq \mathbb{Z}C_2.$$

This is a different chain of ideals than that in Example 1 since $3 \notin \langle 3 - 2t \rangle$. (Check for yourself.)

We can now see how we might use these chains of elementary ideals. Let us suppose we can prove that the chain of elementary ideals is an invariant of the group G and does not depend on the presentation used. We have above presentations

$$(a, b : a^3 = b^2 = (ab)^2 = 1)$$

of S_3, and

$$(x, y : x^2 = 1, y^3x = xy^2)$$

of our group G of example (2). Is $G \cong S_3$? We know that $G_{ab} \cong C_2 \cong (S_3)_{ab}$, so Abelianization does not help, but we also know that if $G \cong S_3$, then we could find Tietze transformations linking the two presentations. This in its turn would imply that the two presentations would have the same chains of elementary ideals. This last statement is false, so $G \not\cong S_3$.

In this case, it would be possible to prove that G was not S_3 in many other different (and probably quicker) ways. However, later on we will need this technique to prove that certain groups arising from surfaces are not free groups even though their Abelianization is free Abelian. It is important to note that, as usual, the logic is: the invariants are different, and therefore the groups are non-isomorphic; we cannot conclude that two groups are isomorphic if their elementary ideals are the same.

Our next preoccupation must be to prove invariance of the elementary ideals of a presentation.

8.2 Theorem. *Any two (finite) presentations of a group G have the same chains of elementary ideals.*

Proof. Since by Tietze's theorem, we can link the two presentations by Tietze transformations, the proof reduces to checking that Tietze transformations leave the elementary ideals unchanged.

Clearly we need only look at **T1** (Adding a superfluous relation) and **T3** (Adding a superfluous generator).

T1: Replace $(X : R)$ by $(X : R \cup \{r\})$ where $r \in \langle\langle R \rangle\rangle$. Thus there are elements $r_i \in R$, $u_i \in F(X)$ (possibly with repeats), such that

$$r = \prod_{i=1}^{p} u_i r_i u_i^{-1}.$$

Then if $x \in X$,

$$\frac{\partial r}{\partial x} = \frac{\partial}{\partial x}(u_1 r_1 u_1^{-1}) + u_1 r_1 u_1^{-1} \frac{\partial}{\partial x}(u_2 u_2 u_2^{-1})$$

$$+ \ldots + \prod_{k=1}^{p-1} (u_i r_i u_i^{-1}) \frac{\partial}{\partial x}(u_p r_p u_p^{-1}).$$

However, $\gamma(r_i) = 1$ for $i = 1, \ldots, p$, so

$$\gamma\left(\frac{\partial r}{\partial x}\right) = \sum_{i=1}^{p} \gamma\left(\frac{\partial}{\partial x}\right)(u_i r_i u_i^{-1}).$$

Next we calculate these p individual terms:

$$\frac{\partial}{\partial x}(u_i r_i u_i^{-1}) = \frac{\partial}{\partial x} u_i + u_i \frac{\partial}{\partial x} r_i - u_i r_i u_i^{-1} \frac{\partial}{\partial x} u_i.$$

Again we map this down by γ to get

$$\gamma\left(\frac{\partial}{\partial x}(u_i r_i u_i^{-1})\right) = \gamma(u_i) \frac{\partial}{\partial x}(r_i).$$

Thus the Alexander matrix of $(X : R \cup \{r\})$ is like that of $(X : R)$ except that it has an extra row which is a linear combination of other rows. Hence for any k, the corresponding minors generate the same ideal. (If you do not see this last step immediately try a simple example and then try to *prove* that the statement is true.)

T3: Replace $(X : R)$ by $(X \cup \{g\} : R \cup \{wg^{-1}\})$. We denote the two (isomorphic) groups by G and G', and the two Alexander matrices by A and A' respectively. We assume $R = \{r_1, \ldots, r_n\}$; then as g is a symbol not in X, $\partial r_i / \partial g = 0$ for all i. Of course the last relation $s = wg^{-1}$ will involve some at least of the $x \in X$. The inclusion of X into $X \cup \{g\}$ induces a homeomorphism

$$\alpha : F(X) \longrightarrow F(X \cup \{g\}).$$

Dividing out by relations, this induces

$$\alpha' : G \longrightarrow G' = F(X \cup \{g\}) / \langle\langle R \cup \{wg^{-1}\} \rangle\rangle$$

(which is an isomorphism) and hence an isomorphism

$$\alpha'' : \mathbb{Z}G_{ab} \longrightarrow \mathbb{Z}G'_{ab}.$$

It is clear that the (i, j)-entry of A' is $\alpha''(a_{ij})$ except if the entry is in the last row or column. In fact our calculations show that

$$A' = \begin{pmatrix} \alpha''(A) & \mathbf{0} \\ \boldsymbol{a}' & 1 \end{pmatrix}$$

for some row vector $\boldsymbol{a}'$. We can use a column operation to eliminate $\boldsymbol{a}'$ and get an equivalent matrix with zeros in those positions of the bottom row:

$$A' \sim \begin{pmatrix} \alpha''(A) & \mathbf{0} \\ \mathbf{0} & 1 \end{pmatrix} = A'', \quad \text{say.}$$

Now it follows from elementary arguments of linear algebra that the $((n+1)-k) \times ((n+1)-k)$ minors of A'' are of the form

$$\begin{pmatrix} \alpha''(M_{n-k}) & \mathbf{0} \\ \mathbf{0} & 1 \end{pmatrix}$$

where M_{n-k} is an $(n-k) \times (n-k)$ minor of A, or are an image $\alpha''(M_{(n+1)-k})$ of some $((n+1)-k) \times ((n+1)-k)$ minor, or have determinant zero. All possible M_{n-k} can arise in this way, so $E_k(A'') = E_k(A') = E_k(A)$. ■

Exercise 8.4

Show, using elementary ideals, that

$$G = (a, b : a^3b = ba, b^2a^2 = a^2b)$$

is not isomorphic to

$$H = (x, y : x^2 = 1, y^3x = xy^2).$$

8.5 Alexander polynomials via the Wirtinger presentation

Given a group G and a presentation $(X : R)$, we have found a matrix A over $\mathbb{Z}G_{ab}$ and a chain $\{E_k(A)\}$ of ideals that are invariants of G. As we saw, the ideals $E_k(A)$ need not always be singly generated. If, however, G_{ab} is a free (finitely generated) Abelian group, then $\mathbb{Z}G_{ab}$ is a g.c.d. domain, so if $E_k(A)$ is generated by elements $a_1, \ldots, a_n \in \mathbb{Z}G_{ab}$, letting d be a greatest common divisor of $a_1, \ldots, a_n$, then as each of its generators is in the principal ideal $\langle d \rangle$, $E_k(A) \subseteq \langle d \rangle$ and $\langle d \rangle$ is the smallest principal ideal with this property (prove it!). The various d for the different $E_k(A)$ form an important and calculable set of invariants of G. Of course this holds in particular if $G = G(K)$, a knot group, as $G(K)_{ab} \cong C_\infty$.

Let A be the Alexander matrix of the Wirtinger presentation of the knot group $G(K)$ of a knot K. The *kth knot polynomial* $\Delta_k(t)$ is the greatest common divisor of all $(n-k) \times (n-k)$ minors of A. For $k = 1$, $\Delta_1(t)$ is the Alexander polynomial of K.

Example

Earlier you should have found that the Alexander matrix for the Wirtinger presentation of the trefoil is

$$\begin{pmatrix} t-1 & -1 & 1 \\ 1 & t-1 & -t \end{pmatrix}.$$

This has 2×2 minors $(t-1)^2 + t = t^2 - t + 1$ and $-t^2 + t - 1$ and again $t^2 - t + 1$; thus $\Delta_1(t) = t^2 - t + 1$.

Remark

We see that since g.c.d.s can only be determined up to multiplication by units, the same is true for these polynomials $\Delta_k(t)$. This in part explains why the combinatorial derivation of the Alexander polynomial had to be 'normalized' to be well defined. It also means that 'the' in the definition of $\Delta_k(t)$ is rather too strong as $\Delta_k(t)$ is not completely determined by the definition.

Example

It often pays to simplify a Wirtinger presentation before working out A and hence the $\Delta_k(t)$. An eight-crossing knot with seven generators will have a 7×8 Alexander matrix if the presentation is not reduced in size first; then $\Delta_1(t)$ will involve evaluating all 7×7 minors of A! The example we will look at in detail only involves five crossings and five generators (Fig. 8.1), but even that would be bad enough.

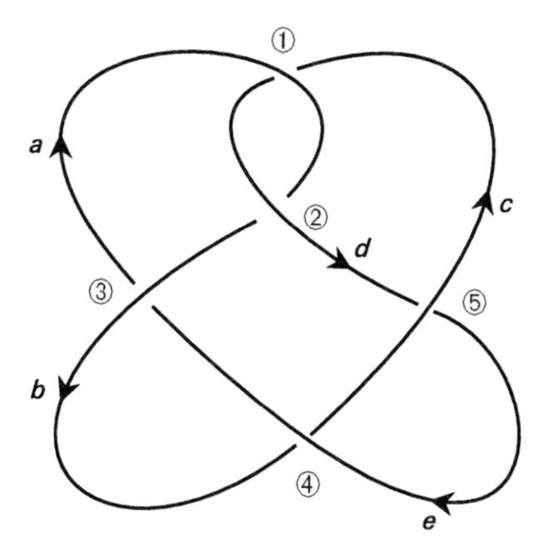

Fig. 8.1

Here $G(K) = (a, b, c, d, e : d = a^{-1}ca, \quad b = d^{-1}ad, \quad a = b^{-1}eb, c = e^{-1}be, \; e = c^{-1}dc)$. Simplifying the presentation, we obtain $(a, b, c, d : d = a^{-1}ca, b = d^{-1}ad, a = b^{-1}c^{-1}dcb, c = c^{-1}d^{-1}cbc^{-1}dc) =$

$$(a, c, d : d = a^{-1}ca, a = d^{-1}a^{-1}dc^{-1}dcd^{-1}ad, c = d^{-1}cd^{-1}adc^{-1}d)$$

$$= (a, c : a = \; c^{-1}a^{-1}cac^{-1}a^{-1}caca^{-1}c^{-1}ac, c = a^{-1}c^{-1}aca^{-1}c^{-1}acac^{-1}a^{-1}ca).$$

Examination of the two relations shows that they are conjugate as one would expect, as we know that any one relation in a Wirtinger presentation is a consequence of the others. (This is a useful check that we have made no errors.) Thus

$$G(K) = (a, c : c^{-1}a^{-1}cac^{-1}a^{-1}c = ac^{-1}a^{-1}cac^{-1}a^{-1}).$$

(We have evened up the two sides of the relation to help with the calculation.)

Take

$$r = c^{-1}a^{-1}cac^{-1}a^{-1}c - ac^{-1}a^{-1}ac^{-1}a^{-1},$$

so

$$\frac{\partial r}{\partial a} = - c^{-1}a^{-1} + c^{-1}a^{-1}c - c^{-1}a^{-1}cac^{-1}a^{-1} - 1 + ac^{-1}a^{-1}$$

$$- ac^{-1}a^{-1}c + ac^{-1}a^{-1}cac^{-1}a^{-1}.$$

Similarly

$$\frac{\partial r}{\partial c} = - c^{-1} + c^{-1}a^{-1} - c^{-1}a^{-1}cac + c^{-1}a^{-1}cac^{-1}a^{-1}$$

$$+ ac^{-1} - ac^{-1}a^{-1} + ac^{-1}a^{-1}cac^{-1}.$$

Applying γ and then $\mathscr{A}$ (in effect substituting $a = c = t$) we get

$$\mathscr{A}\gamma\left(\frac{\partial r}{\partial a}\right) = - 2t^{-2} + 3t^{-1} - 2$$

whilst

$$\mathscr{A}\gamma\left(\frac{\partial r}{\partial c}\right) = 2t^{-2} - 3t^{-1} + 2.$$

Our Alexander matrix is

$$(- 2t^{-2} + 3t^{-1} - 2 \;\; 2t^{-2} - 3t + 2).$$

We will 'normalize' $\Delta_1(t)$ to have no negative powers of t and positive constant term, so

$$\Delta_1(t) = 2t^2 - 3t + 2.$$

Properties of $\Delta_1(t)$

We list the following without proof. They can be useful in checking for errors:

1. For any knot K, $\Delta_1(1) = \pm 1$.
2. For any knot K, there is an integer n such that

$$\Delta_1(t) = t^n \Delta_1(t^{-1}).$$

Thus if $\Delta_1(t) = c_n t^n + \ldots + c_0$, the coefficients pair up $c_i = c_{n-i}$ for $i = 0, \ldots, n$. (Proofs are given in Crowell and Fox (1967, chapter IX).

3. If the knot is alternating, and the genus of K can be calculated, it is useful to note that

$$\text{genus of } K = \tfrac{1}{2} \text{ degree of } \Delta_1(t).$$

Exercises 8.5

1. The group $G(t_{p,q})$ with presentation

$$(x, y : x^p = y^q)$$

is the group of a (p, q)-torus knot where p, q are coprime positive integers. Calculate $\Delta_1(t)$ for this knot and show that

$$\Delta_1(t) = \frac{(t^{pq} - 1)(t - 1)}{(t^p - 1)(t^q - 1)}; \quad E_2 = 1.$$

2. Suppose K is a knot with crossing number c. Investigate the relationships between c and the degree of the normalized Alexander polynomial.
3. Investigate what happens if one applies the Fox calculus, Alexander matrix, elementary ideal analysis to a Dehn presentation of a knot group. Examine in particular the relationship with Alexander's original method of defining the Alexander polynomial.
4. Investigate, using results and exercises from earlier chapters, how the Alexander polynomial reacts to knot sum. (In fact we showed that it changed knot sum to product of polynomials using the 'original' method. Try to see why this is true from the point of view of Wirtinger presentations, Alexander polynomials, etc.

This may seem to you more complicated but it does provide a new type of insight.)

5. Calculate lots of Alexander polynomials. (You choose the knots from anywhere in the book!)
6. Can the Alexander polynomial construction be extended to be

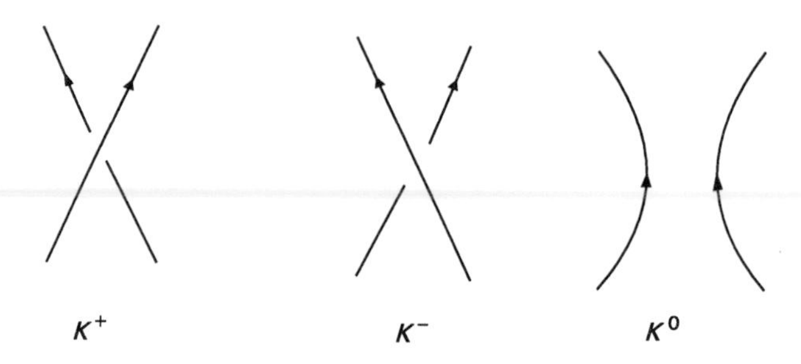

Fig. 8.2

applicable to links with more than one component? See how far you can get. Many of the necessary changes in the theory are hinted at in this chapter. If you are successful in doing such a generalization, then try to adapt Alexander's original method!

7. Using the preceding exercise, find how the Alexander polynomials of knots related by the sequence of changes in Fig. 8.2 are themselves related. There is a simple equation rather like the Homfly defining equation. This should help you to find a relationship between the Alexander polynomial and the Homfly polynomial. Use either formulation of the Alexander polynomial.

9
The fundamental group

The fundamental group is an algebraic label attached to a topological space. Naturally enough, it is a group and is constructed using paths in the space, continuous analogues of the sequences of edges constituting paths in a graph. The more convoluted is the shape of the space, the more variety there will be among the paths in that space and the larger will be the fundamental group. The fundamental group first arose in Poincaré's work on the three-body problem. His idea was to classify the possible evolutions of a system of three bodies moving under mutual gravitational forces as paths in the *phase space* with coordinates given by the bodies' positions and velocities, but to regard nearby paths as the same. This led to the notion of deformation of one path into another, now formalized as *homotopy of paths*. The fundamental group consists of *homotopy classes* of paths in a space. The notion of homotopy is also crucial in complex analysis: Cauchy's theorem on contour integration may be phrased as saying that the integrals of a holomorphic function around two homotopic contours are the same (see e.g. Priestley 1990). Further development of the extensive theory of the fundamental group, beyond that covered here, may be found in Armstrong (1983) and Brown (1988). Our principal application will be to knots: any knot yields a group, namely the fundamental group of its complement in $\mathbb{R}^3$, and algebraic properties of the group reflect properties of the knot.

9.1 Paths

Let X be a topological space. A *path* ω in X is a continuous map $\omega: [0, r] \to X$, where $r \geqslant 0$. The real number r is the *duration* of the path; if $r = 1$ we say that the path is *normalized*. Given any path ω in X, there is a normalized path ω_1 with the same image in X; we simply define $\omega_1(t) = \omega(rt)$ for $0 \leqslant t \leqslant 1$. We can therefore restrict attention to normalized paths if we choose, but it is often more convenient to allow paths of arbitrary duration.

If $\omega: [0, r] \to X$ and $\omega': [0, r'] \to X$ are paths in X, such that $\omega(r) = \omega'(0)$, then ω and ω' have an obvious *product*, the path obtained by first going along the image of ω and then along the image of ω'. We write this product as $\omega \cdot \omega'$; then we have $\omega \cdot \omega': [0, r + r'] \to X$ defined by $\omega \cdot \omega'(t) = \omega(t)$ if $0 \leqslant t \leqslant r$ and $\omega \cdot \omega'(t) = \omega'(t - r)$ if $r \leqslant t \leqslant r + r'$. The duration of $\omega \cdot \omega'$ is $r + r'$ and clearly $\omega \cdot \omega'(0) = \omega(0)$, $\omega \cdot \omega'(r + r') = \omega'(r')$. There is a product of normalized paths as well. If $\omega: [0, 1] \to X$ and $\omega': [0, 1] \to X$ are normalized, then their *normalized product* is the normalized path σ obtained from $\omega \cdot \omega': [0, 2] \to X$. Since $\sigma(t) = \omega \cdot \omega'(2t)$, we have $\sigma(t) = \omega(2t)$ if $0 \leqslant t \leqslant 1/2$ and $\sigma(t) = \omega'(2t - 1)$ if $1/2 \leqslant t \leqslant 1$. Thus σ is obtained by going along ω and ω' in turn, but twice as quickly.

It is important to note that the product of paths is not defined for arbitrary paths ω and ω': it must be that ω ends at the same point at which ω' begins. In particular, the product $\omega \cdot \omega'$ may be defined whilst $\omega' \cdot \omega$ is not.

9.1 Lemma. *Let $\omega: [0, r] \to X$, $\omega': [0, r'] \to X$ and $\omega'': [0, r''] \to X$ be paths in X with $\omega(r) = \omega'(0)$ and $\omega'(r') = \omega''(0)$. Then the products $\omega \cdot (\omega' \cdot \omega'')$ and $(\omega \cdot \omega') \cdot \omega''$ are both defined and are equal.*

We leave the proof of Lemma 9.1 as an exercise. Note that it says that the product of paths is an *associative* operation. This is *not* true of the normalized product: why not?

For each point $x \in X$ there is a constant path $0_x: \{0\} \to X$ of duration 0 defined by $0_x(0) = x$. If $\omega: [0, r] \to X$ is a path then $\omega \cdot 0_{\omega(r)} = \omega$ and $0_{\omega(0)} \cdot \omega = \omega$, so that the paths of duration 0 provide a set of left and right identities for the product operation on paths. However, there are no inverses; for if ω has duration $r > 0$, then for any ω' the path $\omega \cdot \omega'$ (if it is defined) has duration at least r and so cannot be a path 0_x. The best approximation that we have to an inverse for $\omega: [0, r] \to X$ is the reverse path $\bar{\omega}: [0, r] \to X$ defined by $\bar{\omega}(t) = \omega(r - t)$. Both $\omega \cdot \bar{\omega}$ and $\bar{\omega} \cdot \omega$ are defined and are paths of duration $2r$.

We see that the set of paths in X is endowed with quite a reasonable structure, but one which is none the less deficient in certain respects, and is plainly not that of a group. The circumstances are similar to those encountered in the construction of free groups in Chapter 6, when it proved necessary to consider equivalence classes of words on an alphabet. The next step towards the fundamental group is to put an equivalence relation on the set of paths in the space X. It turns out that, with this equivalence relation, it will be sufficient to consider only normalized paths.

9.2 Homotopy

Let X and Y be topological spaces and suppose that f: $Y \to X$ and g: $Y \to X$ are continuous maps. A *homotopy* between f and g is a continuous map h: $Y \times I \to X$ such that, for all $y \in Y$, $h(y, 0) = f(y)$ and $h(y, 1) = g(y)$. The set of functions $\{h(-, t) \mid 0 \leqslant t \leqslant 1\}$ can be thought of as a continuous deformation from f to g. If a homotopy between f and g exists, then f and g are *homotopic* and we write $f \simeq g$.

9.2 Proposition. *Homotopy is an equivalence relation on the set of all maps* $Y \to X$.

There are variations on the notion of homotopy obtained either by considering only a limited class of functions, or by adding extra conditions on the homotopy. *Pointed homotopy* of *closed* paths leads to the fundamental *group*. The properties of paths with which this chapter began are more accurately reflected by *fixed end point* homotopy of paths, and this leads to the fundamental *groupoid*. For a full account of groupoids, see Brown (1988), and for the first steps in the theory of the fundamental groupoid, see Exercises 9.3.2–7 below.

As already remarked, it will suffice to consider only normalized paths in our route to the fundamental group. Let X be a topological space and $x_0 \in X$ any point of X. A *closed path in X at x_0* is a continuous map $\omega : [0, 1] \to X$ with $\omega(0) = x_0 = \omega(1)$. Observe that the normalized product of any two closed paths at x_0 is defined and is again a closed path at x_0. Suppose that α and ω are closed paths in X at x_0. A *pointed homotopy* from α to ω is a continuous map h: $[0, 1] \times [0, 1] \to X$ such that, for all $s \in [0, 1]$, $h(s, 0) = \alpha(s)$ and $h(s, 1) = \omega(s)$, and further such that for all t, $h(0, t) = x_0 = h(1, t)$. In

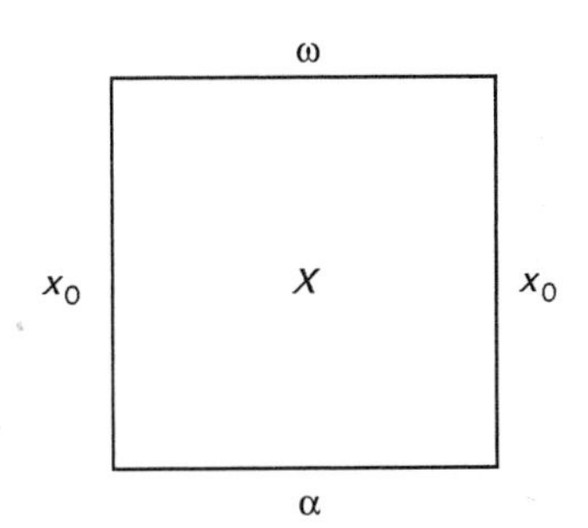

a pointed homotopy, we have a set of closed paths at x_0 that give a continuous deformation from α to ω. A simple but useful picture to bear in mind is the unit square $[0, 1] \times [0, 1]$ adorned with labels to show how it is mapped into X by h.

Any point on the sides of the square is mapped to x_0: the bottom of the square is mapped into X by α, the top by ω, and horizontal slices give the intermediate closed paths $h(-, t)$, $0 \leqslant t \leqslant 1$.

Pointed homotopy is an equivalence relation on the set of all closed paths in X at x_0, and we will use $[\alpha]$ to denote the equivalence class of a closed path α. The set of all the equivalence classes of closed paths in X at x_0 is denoted $\pi_1(X, x_0)$. Paths are composed by juxtaposition, just as were the words in a free group, and this analogy with the construction of free groups leads us to hope that $\pi_1(X, x_0)$ is a group, with multiplication of equivalence classes of paths obtained by taking the product of representative paths.

9.3 Theorem. *For any* $[\alpha]$, $[\alpha'] \in \pi_1(X, x_0)$, *the binary operation* $[\alpha][\alpha'] = [\alpha \cdot \alpha']$ *is well defined and makes* $\pi_1(X, x_0)$ *into a group.*

Proof. For ease of digestion the proof is divided into four steps.

Step 1

We shall show that the operation is *well defined*; that is, it is *independent* of any choices made of one closed path in an equivalence class. Suppose that $[\alpha] = [\omega]$ and that $[\alpha'] = [\omega']$. There are pointed homotopies h from α to ω and h' from α' to ω'. The following picture indicates the

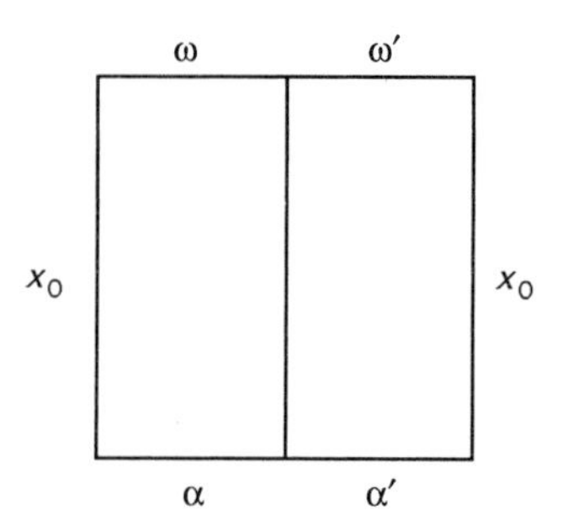

definition of a pointed homotopy h'' from $\alpha \cdot \alpha'$ to $\omega \cdot \omega'$.

Define h'' as follows. We have, for all t, $h''(s, t) = h(2s, t)$ if $0 \leqslant s \leqslant 1/2$, and $h''(s, t) = h'(2s - 1, t)$ if $1/2 < s \leqslant 1$. Then h'' is indeed a pointed homotopy from $\alpha \cdot \alpha'$ to $\omega \cdot \omega'$, and therefore $[\alpha \cdot \alpha'] = [\omega \cdot \omega']$.

Step 2

The operation is associative. Given closed paths α, α', α'' at x_0, we need to construct a homotopy h from $(\alpha \cdot \alpha') \cdot \alpha''$ to $\alpha \cdot (\alpha' \cdot \alpha'')$. Remember that, because we are dealing with normalized paths, $(\alpha \cdot \alpha') \cdot \alpha''$ and $\alpha \cdot (\alpha' \cdot \alpha'')$ need not be *equal* paths. They *are* homotopic, and again a picture indicates what we should do.

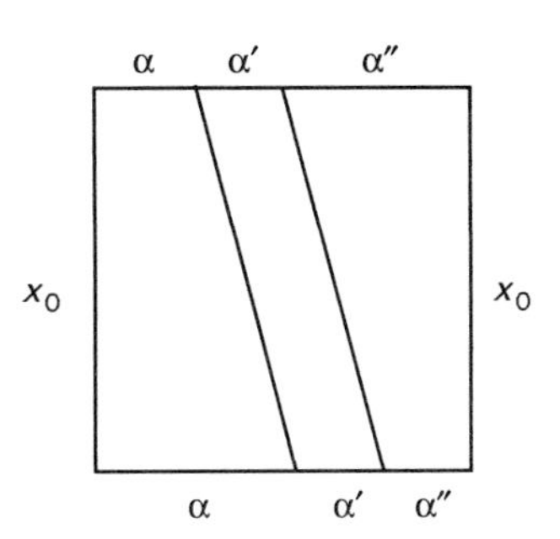

Define $h(s, t) = \alpha(4s/(t + 1))$ if $0 \leqslant s \leqslant (t + 1)/4$, $h(s, t) = \alpha'(4s - t - 1)$ if $(t + 1)/4 \leqslant s \leqslant (t + 2)/4$, and $h(s, t) = \alpha''((4s - t - 2)/(2 - t))$ if $(t + 2)/4 < s \leqslant 1$, and this has the properties we need.

Step 3

Let e: $[0, 1] \to X$ be the constant path at x_0. Then $[e]$ is an identity element in $\pi_1(X, x_0)$; for given any closed path α in X at x_0 we have $\alpha \cdot e \simeq \alpha$ and so $[\alpha][e] = [\alpha]$. A homotopy from $\alpha \cdot e$ to α is indicated by the picture

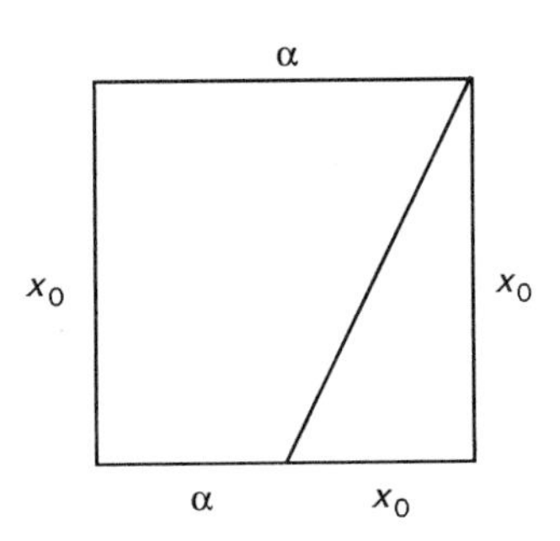

and given by the explicit formulae $h(s, t) = \alpha(2s/(t + 1))$ if $0 \leqslant s \leqslant (t + 1)/2$ and $h(s, t) = x_0$ otherwise.

Step 4

If α is a closed path in X at x_0, its reverse $\bar{\alpha}$ (which you will recall is defined by $\bar{\alpha}(s) = \alpha(1 - s)$) is also a closed path in X at x_0 and $[\bar{\alpha}]$ is the inverse of $[\alpha]$; that is, $[\alpha][\bar{\alpha}] = [e]$. A homotopy from $\alpha \cdot \bar{\alpha}$ to e is indicated by the picture

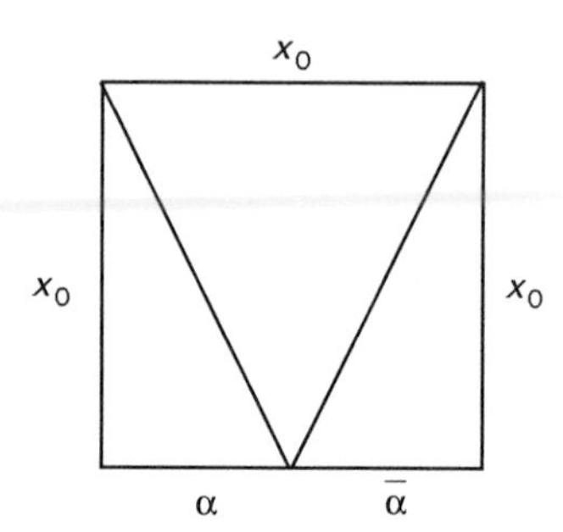

and we leave it as an exercise for the reader to write out an explicit formula for the homotopy. ■

The group $\pi_1(X, x_0)$ is the *fundamental group* of the space X at the point x_0. What happens if we choose another point x_1 instead of x_0 at which to base all our closed paths?

9.4 Proposition. *If X is path connected and x_0, x_1 are two points of X, then there is an isomorphism between $\pi_1(X, x_0)$ and $\pi_1(X, x_1)$.*

Proof. Let ω: $[0, 1] \to X$ be a path in X with $\omega(0) = x_0$ and $\omega(1) = x_1$. Suppose that α is a closed path in X at x_0; then $\omega \cdot \alpha \cdot \bar{\omega}$ is a closed path in X at x_1, and mapping $[\alpha] \mapsto [\omega \cdot \alpha \cdot \bar{\omega}]$ gives an isomorphism $\pi_1(X, x_0) \to \pi_1(X, x_1)$. ■

Exercises 9.2

1. The definition of homotopy will have reminded the reader of the definition of isotopy of knots. If k, k': $S^1 \to \mathbb{R}^3$ are isotopic, then are they homotopic? Why isn't homotopy a useful equivalence relation for knots?
2. Prove Proposition 9.2. [**Hint.** The only difficult part is transitivity. The gluing lemma should help.]
3. The proof of Proposition 9.4 is rather sketchy. Supply the details.
4. Is it reasonable to expect an isomorphism between $\pi_1(X, x_0)$ and $\pi_1(X, x_1)$ if x_0 and x_1 lie in different path components of X?

9.3 Calculations of the fundamental group

The computation of fundamental groups will be the main theme of the remainder of the book, so that examples will be in plentiful supply. We pause here to consider some spaces whose fundamental group is trivial.

A topological space is *simply* connected if it is path connected and if $\pi_1(X, x_0) = 1$ for some (and hence all) $x_0 \in X$.

9.5 Proposition. *If $X \subseteq \mathbb{R}^n$ is a convex set then X is simply connected.*

Proof. Recall that $X \subseteq \mathbb{R}^n$ is *convex* if for all $\mathbf{x}$, $\mathbf{y} \in X$, and all t with $0 \leqslant t \leqslant 1$, we have $t\mathbf{x} + (t - 1)\mathbf{y} \in X$. Let ω: $[0, 1] \to X$ be a closed path at $\mathbf{x}_0 \in X$, and define h: $[0, 1] \times [0, 1] \to X$ by $h(s, t) = t\mathbf{x}_0 + (1 - t)\omega(s)$. Then h is a homotopy from ω to the constant path at $\mathbf{x}_0$. ∎

Our first example of a non-trivial fundamental group is S^1, the circle. It is convenient to regard $\mathbb{R}^2$ as the complex plane and so identify S^1 as $\{z \in \mathbb{C} \mid |z| = 1\}$; we take $1 \in \mathbb{C}$ as basepoint.

9.6 Theorem. *The fundamental group $\pi_1(S^1, 1)$ is infinite cyclic, and is generated by the homotopy class of the closed path ω: $t \mapsto \exp(2\pi i t)$.*

Proof. Before we embark on the detailed proof, it is worth considering the matter in a vaguer fashion. A path in S^1 may wander around in a complicated way, but it is forced to remain within the one-dimensional bounds of the circle. Its wanderings are thus no more than changes of direction, or back-tracking, and these portions of the path can be smoothed out by some suitable homotopies. Any path will be homotopic to one without any back-tracking, and will wind around the centre of the circle a certain number of times. Our claim is that this *winding number* completely determines the homotopy class of the path; that is, that two paths in S^1 are homotopic if and only if they have the same winding number. Since composition of paths induces addition of winding numbers, we will have an isomorphism ϕ: $\pi_1(S^1, 1) \to \mathbb{Z}$ as claimed.

Our first problem is to give a definition of the winding number. You might like to think of ways in which this might be done, and see if you can use your definition to prove the theorem. The solution we shall adopt here may initially seem strange: we consider paths in $\mathbb{R}$ instead of paths in S^1. The path ω: $[0, 1] \to S^1$ given by ω: $t \mapsto \exp(2\pi i t)$ extends to a function $\mathbb{R} \to S^1$ given by the same formula: you may picture ω as winding the real line $\mathbb{R}$ into a spiral and then flattening the spiral onto the unit circle. The crucial fact about ω for our purposes is this: given any path σ: $[0, 1] \to S^1$ there exists a unique path $\tilde{\sigma}$: $[0, 1] \to \mathbb{R}$ with $\omega \circ \tilde{\sigma} = \sigma$. The path $\tilde{\sigma}$ is a *lifting* of σ and is unwound along $\mathbb{R}$; furthermore, since $\tilde{\sigma}(1) = 1$ we must have $\tilde{\sigma}(1) \in \mathbb{Z}$. We can therefore define the *winding number* of σ to be $\tilde{\sigma}(1)$. The homeomorphism ϕ: $\pi_1(S^1, 1) \to \mathbb{Z}$ that we seek will then be defined by $\phi([\sigma]) = \tilde{\sigma}(1)$, and it will only remain to show that ϕ is bijective.

Is the previous sentence true? In fact it disguises the problems that we must still face: it may be difficult to show that ϕ is bijective (what do you think at this stage?), but more fundamentally (pun intended), is our definition of ϕ a workable one? Firstly we have yet to prove that ω does possess the crucial path-lifting property, so that given σ we can get a unique $\tilde{\sigma}$. Secondly, in the definition of ϕ, we use the homotopy class of σ and not σ itself. Does it matter which path in the homotopy class we lift to $\mathbb{R}$? Of course, we would not have come this far unless everything was going to be all right in the end, but it is extremely important to see exactly what it is we still have to do to make the idea of the proof into a watertight argument. We carry out this waterproofing process in four steps:

Step 1: we show that given any path σ: $[0, 1] \to S^1$ there exists a unique path $\tilde{\sigma}$: $[0, 1] \to \mathbb{R}$ with $\omega \circ \tilde{\sigma} = \sigma$.

Step 2: we show that ϕ: $\pi_1(S^1, 1) \to \mathbb{Z}$ is well defined by $\phi([\sigma]) = \tilde{\sigma}(1)$; that the answer we get does not depend on the choice of a path in the homotopy class $[\sigma]$.

Step 3: we show that ϕ is a homeomorphism.

Step 4: we show that ϕ is bijective.

Step 1

The path-lifting lemma. *If σ: $[0, 1] \to S^1$ is a path with $\sigma(0) = 1$ there exists a unique path $\tilde{\sigma}$: $[0, 1] \to \mathbb{R}$ such that $\omega \cdot \tilde{\sigma} = \sigma$ and $\tilde{\sigma}(0) = 0$.*

To prove the path-lifting lemma we set $U = S^1 \backslash \{-1\}$ and $V = S^1 \backslash \{1\}$; then U and V are open and $U \cup V = S^1$. Now $\omega^{-1}(U) = \cup_{n \in \mathbb{Z}} (n - 1/2, n + 1/2)$ and for each $n \in \mathbb{Z}$, $\omega|(n - 1/2, n + 1/2)$ is a homeomorphism onto U. Let α_n: $U \to (n - 1/2, n + 1/2)$ be its inverse. Similarly, we consider $\omega^{-1}(V) = \cup_{n \in \mathbb{Z}}(n, n + 1)$, and we let β_n: $V \to (n, n + 1)$ be the inverse of $\omega|(n, n + 1)$. Each of $\omega^{-1}(U)$ and $\omega^{-1}(V)$ is open in $[0, 1]$ and $[0, 1] = \omega^{-1}(U) \cup \omega^{-1}(V)$. It follows from *Lebesgue's lemma* in the theory of metric spaces that there exist points $0 = t_0 < t_1 < \ldots < t_n = 1$ such that $\omega([t_i, t_{i+1}])$ lies in U or in V. We shall not prove Lebesgue's lemma in this book, but it is easy to find in standard texts on metric spaces, such as Sutherland (1981), or in

Armstrong (1983). Since $\omega(0) \in U$ we have $\omega([0, t_1]) \subseteq U$ and we can define $\tilde{\sigma}$: $[0, t_1] \to \mathbb{R}$ by $\tilde{\sigma}(t) = \alpha_0 \circ \sigma(t)$. If we now assume that we have defined $\tilde{\sigma}$ on the interval $[0, t_k]$, we show that we can extend the definition to $[0, t_{k+1}]$. If $\sigma([t_k, t_{k+1}]) \subseteq U$ then $\tilde{\sigma}(t_k) \in (n - 1/2, n + 1/2)$ for some $n \in \mathbb{Z}$, and we may define $\tilde{\sigma}$: $[t_k, t_{k+1}] \to \mathbb{R}$ by $\tilde{\sigma}(t) = \alpha_n \circ \sigma(t)$. Similarly, if $\sigma([t_k, t_{k+1}]) \subseteq V$ then $\tilde{\sigma}(t_k) \in (n, n + 1)$ for some $n \in \mathbb{Z}$ and we may define $\tilde{\sigma}$: $[t_k, t_{k+1}] \to \mathbb{R}$ by $\tilde{\sigma}(t) = \beta_n \circ \sigma(t)$. Repeating this procedure, we eventually define $\tilde{\sigma}([0, 1])$. Notice that we have no choice for the definition of $\tilde{\sigma}$ at each step if we are to ensure that $\omega \circ \tilde{\sigma} = \sigma$ and $\tilde{\sigma}(0) = 0$, and so $\tilde{\sigma}$ is unique.

Step 2

This is also accomplished by lifting the problem to $\mathbb{R}$. The result we need is as follows.

Homotopy-lifting lemma. *If h: $[0, 1] \times [0, 1] \to S^1$ is a homotopy between the closed paths σ and τ, (so that $h(0, t) = 1 = h(1, t)$ for all $t \in [0, 1]$) there is a unique homotopy $\tilde{h}$: $[0, 1] \times [0, 1] \to \mathbb{R}$ such that $\omega \circ \tilde{h} = h$ and $\tilde{h}(0, t) = 0$ for all t.*

We shall not give a detailed proof of this lemma because the ideas needed are the same as for the path-lifting lemma. The unit square $[0, 1] \times [0, 1]$ must be divided into smaller squares, again by use of Lebesgue's lemma, each of which maps into U or into V, and $\tilde{h}$ then defined square by square.

Now suppose that σ and τ are closed paths in S^1 and are homotopic by means of a homotopy h. The homotopy $\tilde{h}$ given by the homotopy-lifting lemma has the property that $\omega \circ \tilde{h}(s, 0) = h(s, 0) = \sigma(s)$. Furthermore, $\tilde{h}(0, 0) = 0$. It follows by the path-lifting lemma that $\tilde{h}(-, 0) = \tilde{\sigma}$ and so $\varphi(\sigma) = \tilde{\sigma}(1) = \tilde{h}(1, 0)$. Similarly, $\tilde{h}(-, 1) = \tilde{\tau}$ and so $\varphi(\tau) = \tilde{\tau}(1) = \tilde{h}(1, 1)$. Now for all $t \in [0, 1]$ we have $\tilde{h}(1, t) \in \omega^{-1}(1) = \mathbb{Z}$; but the image of the connected set $\{(1, t) \mid t \in [0, 1]\}$ under the continuous map $\tilde{h}$ must be connected, and so is a single integer, n say, and now $\varphi(\sigma) = n = \varphi(\tau)$.

Step 3

This is nice and easy. Given two closed paths σ and τ in S^1, with corresponding paths $\tilde{\sigma}$ and $\tilde{\tau}$, define a path ψ: $[0, 1] \to \mathbb{R}$ by

$$\psi(t) = \tilde{\sigma}(2t), \quad 0 \leqslant t \leqslant 1/2$$

$$= \tilde{\tau}(2t - 1) + \tilde{\sigma}(1), \quad 1/2 \leqslant t \leqslant 1$$

Then $\omega \cdot \psi = \sigma + \tau$ and $\psi(0) = \tilde{\sigma}(0) = 0$. By uniqueness, $\psi = (\sigma + \tau)^\sim$ and $\psi(1) = \tilde{\tau}(1) + \tilde{\sigma}(1)$.

Step 4

It is easy to see that φ is surjective: given $n \in \mathbb{Z}$, let σ_n: $[0, 1] \to S^1$ be defined by $\sigma_n(t) = \exp(2n\pi i t)$. We leave it to the reader to check that $\tilde{\sigma}_n(1) = n$ so that $\varphi(\sigma_n) = n$.

It remains to show that φ is injective; all we need to prove is that if $\varphi([\sigma]) = 0$ then σ is homotopic to the constant path at 1. If $\varphi([\sigma]) = 0$ then $\tilde{\sigma}(1) = 0$, so that $\tilde{\sigma}$ is a closed path in $\mathbb{R}$. Define a homotopy H: $[0, 1] \times [0, 1] \to \mathbb{R}$ by $H(s, t) = (1 - t)\tilde{\sigma}(s)$. Then H is a homotopy from $\tilde{\sigma}$ to the constant path at 0. Set $h = \omega \circ H$. Then $h(s, 0) = \omega \circ \tilde{\sigma}(s) = \sigma(s)$ and $h(s, 1) = \omega(0) = 1$. We see that h is a homotopy from σ to the constant path at 1, and finally the proof is complete. ∎

The complexities of the proof should have convinced you that the computation of fundamental groups from first principles is likely to be a formidable task. Even though we began with a very plausible, and rather simple, candidate for $\pi_1(S^1, 1)$ we still encountered some subtle problems in establishing that our candidate was indeed correct. How are we to compute the fundamental group of a space when no likely candidate presents itself to our imaginations? Fortunately, a powerful tool exists: Van Kampen's theorem. This will be the subject of the following two chapters. For the remainder of this chapter, we develop some more basic theory of the fundamental group.

How are operations on spaces reflected in the calculation of the fundamental group? For products of spaces (with the product topology) the answer is very satisfactory.

9.7 Proposition. *Let X and Y be topological spaces with $x_0 \in X$ and $y_0 \in Y$. Then $\pi_1(X \times Y, (x_0, y_0))$ is isomorphic to the direct product $\pi_1(X, x_0) \times \pi_1(Y, y_0)$.*

Proof. Let α be a closed path in $X \times Y$ at (x_0, y_0). There is an obvious candidate for an isomorphism:

$$\varphi\colon \pi_1(X \times Y, (x_0, y_0)) \to \pi_1(X, x_0) \times \pi_1(Y, y_0);$$

we take $\varphi([\alpha]) = ([p\alpha], [q\alpha])$ where p: $X \times Y \to X$ and q: $X \times Y \to Y$ are the projection maps. The proof of the proposition consists of a verification of the credentials of φ.

Is φ well defined? If $\alpha \simeq \omega$ then do we have $p\alpha \simeq p\omega$ and $q\alpha \simeq q\omega$? Let h be a homotopy from α to ω. Then ph and qh are homotopies from $p\alpha$ to $p\omega$ and from $q\alpha$ to $q\omega$ respectively. Check the details of this assertion.

Is φ a homeomorphism? If α and β are closed paths at (x_0, y_0), check that $p(\alpha \cdot \beta) = p\alpha \cdot p\beta$ as closed paths in X at x_0 and that $q(\alpha \cdot \beta) = q\alpha \cdot q\beta$ as closed paths in Y at y_0.

Is φ injective? Let e, e' be the constant paths at x_0 and y_0 respectively and suppose that h and h' are homotopies from $p\alpha$ to e and from $q\alpha$ to e'. Define H: $[0, 1] \times [0, 1] \to X \times Y$ by $H(s, t) = (h(s, t), h'(s, t))$. Show that H is a homotopy from α to the constant path at (x_0, y_0). Explain why this shows φ to be injective.

Complete the proof by showing that φ is surjective. ∎

Before continuing with the theory of the fundamental group, we return to the properties of paths, and in the following exercises we consider a homotopy notion more closely related to path products than based homotopy.

Exercises 9.3

1. $X \subseteq \mathbb{R}^n$ is *starlike* if it is path connected and if there exists $x_0 \in X$ such that for any $y \in X$ and for all t with $0 \leqslant t \leqslant 1$, we have $tx_0 + (1 - t)y \in X$. Show that starlike subsets of $\mathbb{R}^n$ are simply connected.
2. Let α and ω be paths in X with $\alpha(0) = x_0 = \omega(0)$ and $\alpha(1) = x_1 = \omega(1)$. A *fixed-end-point homotopy* from α to ω is a continuous function h: $[0, 1] \times [0, 1] \to X$ such that, for all $s \in [0, 1]$, $h(s, 0) = \alpha(s)$ and $h(s, 1) = \omega(s)$, and further such that for all $t \in [0, 1]$, $h(0, t) = x_0$ and $h(1, t) = x_1$. If there is a fixed-end-point homotopy from α to ω we write $\alpha \simeq \omega$ (f.e.p.). Show that fixed-end-point homotopy is an equivalence relation on the set of all paths α in X with fixed end points $\alpha(0) = x_0$ and $\alpha(1) = x_1$.
3. Show that if α, β, γ are normalized paths such that the normalized product $\alpha \cdot (\beta \cdot \gamma)$ is defined, then $(\alpha \cdot \beta) \cdot \gamma$ is defined and $\alpha \cdot (\beta \cdot \gamma) \simeq (\alpha \cdot \beta) \cdot \gamma$ (f.e.p.).
4. Show that $\omega \cdot \bar{\omega} \simeq 0_{\omega(0)}$ (f.e.p.) and that $\bar{\omega} \cdot \omega \simeq 0_{\omega(1)}$ (f.e.p.).
5. Let α, ω, α', ω' be paths in x such that $\alpha \simeq \omega$ (f.e.p.), $\alpha' \simeq \omega'$ (f.e.p.), and such that $\alpha \cdot \alpha'$ is defined. Show that $\omega \cdot \omega'$ is defined and that $\alpha \cdot \alpha' \simeq \omega \cdot \omega'$ (f.e.p.).
6. Let ΠX denote the set of fixed-end-point homotopy classes of paths in X (so we take all possible pairs of end points, but within each class the end points are fixed). Show that the product of homotopy classes $[\omega][\omega'] = [\omega \cdot \omega']$, defined when $\omega(1) = \omega'(0)$, is well defined.
7. Let X be a path connected space. Show that X is simply connected if and only if given any two paths α and ω in X with $\alpha(0) = \omega(0)$ and $\alpha(1) = \omega(1)$ then $\alpha \simeq \omega$ (f.e.p.).

9.4 Homotopy equivalence

The fundamental group construction attaches an algebraic label to a topological space that records some information about the shape of the space. A task that we might hope to have made easier by this labelling is that of distinguishing one space from another: spaces with different shapes should carry different labels. On the other hand, homeomorphic spaces might be expected to have the same label, that is to have isomorphic fundamental groups. This is indeed the case, and we arrive at the result by extending the construction of the fundamental group to continuous maps between spaces: a continuous map becomes a homeomorphism of groups.

Let X and Y be topological spaces. To discuss fundamental groups, we need a pair of basepoints, $x_0 \in X$ and $y_0 \in Y$. The space X and the selected point x_0 together form the *pointed space* (X, x_0). Given the pointed spaces (X, x_0) and (Y, y_0), a continuous function $f: Y \to X$ is *pointed* if $x_0 = f(y_0)$. If f is pointed we write $f: (Y, y_0) \to (X, x_0)$. If $\alpha: [0, 1] \to Y$ is a closed path in Y at y_0 and $f: (Y, y_0) \to (X, x_0)$ is a pointed continuous function then $f\alpha$ is a closed path in X at x_0. [**Exercise:** exactly what has just been asserted? Why is it true?] If h is a pointed homotopy from α to ω in Y then fh is a pointed homotopy from $f\alpha$ to $f\omega$, and so given $[\alpha] \in \pi_1(Y, y_0)$ the class $[f\alpha]$ is well defined in $\pi_1(X, x_0)$. From $f: (Y, y_0) \to (X, x_0)$ we have obtained a function $f_*: \pi_1(Y, y_0) \to \pi_1(X, x_0)$ defined by $[\alpha] \mapsto [f\alpha]$. Now if α, β are closed paths in Y at y_0, then $f(\alpha \cdot \beta) = (f\alpha) \cdot (f\beta)$ [**Exercise:** check this!] and therefore f_* is a homeomorphism between the fundamental groups $\pi_1(Y, y_0) \to \pi_1(X, x_0)$.

Since homotopy is clearly a very good idea, we should attempt to perceive its effects wherever possible. What is the effect on the homeomorphism f_* of a homotopy from f to another pointed map g? The answer is that there is no effect at all. Let f, $g: (Y, y_0) \to (X, x_0)$ be pointed maps. A *pointed homotopy* from f to g is a homotopy $H: Y \times [0, 1] \to X$ from f to g such that $H(y_0, t) = x_0$ for all t, with $0 \leqslant t \leqslant 1$. If $\alpha: [0, 1] \to Y$ is a closed path in Y at y_0 then $h: [0, 1] \times [0, 1] \to X$ defined by $h(s, t) = H(\alpha(s), t)$ is a pointed homotopy from $f\alpha$ to $g\alpha$ and so in $\pi_1(X, x_0)$ we have $[f\alpha] = [g\alpha]$; that is, for all $[\alpha] \in \pi_1(Y, y_0)$, $f_*[\alpha] = g_*[\alpha]$ and so $f_* = g_*$. Our conclusions are summarized in the next result.

9.8 Theorem. *A pointed map $f: (Y, y_0) \to (X, x_0)$ induces a homeomorphism $f_*: \pi_1(Y, y_0) \to \pi_1(X, x_0)$ between fundamental groups. If f is homotopic to $g: (Y, y_0) \to (X, x_0)$ by a pointed homotopy then $f_* = g_*$; that is, pointed homotopic maps induce the same homeomorphism on fundamental groups.*

We say that a pointed map $f\colon (Y, y_0) \to (X, x_0)$ is a *homotopy equivalence* if there exists a pointed map $g\colon (X, x_0) \to (Y, y_0)$ and pointed homotopies between gf and id_Y and between fg and id_X. The map g is a *homotopy inverse* for f. If f is a homeomorphism then it is certainly a homotopy equivalence, and f^{-1} is a homotopy inverse for f.

9.9 Proposition. *If* $f\colon (Y, y_0) \to (X, x_0)$ *is a homotopy equivalence then* $f_*\colon \pi_1(Y, y_0) \to \pi_1(X, x_0)$ *is an isomorphism.*

Proof. Let g be a homotopy inverse for f. We leave it to you to show that g_* is an inverse for f_*.

Especially interesting is the setting in which there exists a homotopy equivalence between a space and a subspace. The subspace is then a simpler version of the space that, with our vision blurred by homotopy, is indistinguishable. Let A be a subspace of X and choose a basepoint $x_0 \in A$. Let $i\colon (A, x_0) \to (X, x_0)$ be the inclusion. A *retraction* of X to A is a continuous map $r\colon (X, x_0) \to (A, x_0)$ such that $ri(a) = a$ for all $a \in A$. If there exists a retraction of X to A we say that A is a *retract* of X.

9.10 Proposition. *If A is a retract of X then the retraction* $r\colon (X, x_0) \to (A, x_0)$ *induces a surjection* $r_*\colon \pi_1(X, x_0) \to \pi_1(A, x_0)$ *and* r_*i_* *is the identity on* $\pi_1(A, x_0)$.

Proof. Exercise! ∎

As an application we can prove a famous result of L. E. J. Brouwer.

9.11 The Brouwer fixed point theorem. *Let D be the unit disc in* $\mathbb{R}^2$, *that is* $D = \{\boldsymbol{x} \in \mathbb{R}^2 \mid \|\mathbf{x}\| \leqslant 1\}$. *Then any continuous map* $f\colon D \to D$ *has a fixed point; there is some* $\boldsymbol{x} \in D$ *such that* $f(\boldsymbol{x}) = \boldsymbol{x}$.

Proof. Suppose that no fixed point exists. Then for each $\boldsymbol{x} \in D$ we may define $r(\boldsymbol{x})$ to be the point on the boundary S^1 of D crossed by the line segment from $f(\boldsymbol{x})$ to $\boldsymbol{x}$ (Fig. 9.1).

Then r is a retraction of D to S^1, and we leave you to check this. Therefore r_* is surjective from $\pi_1(D, (1, 0))$ to $\pi_1(S^1, (1, 0))$; now find a contradiction by combining Proposition 9.5 with Theorem 9.6. ∎

The subspace A of the space X is a *deformation retract* of X if there exists a retraction $r\colon (X, x_0) \to (A, x_0)$ and a pointed homotopy h from ir to id_X. The inclusion $i\colon (A, x_0) \to (X, x_0)$ is then a homotopy equivalence.

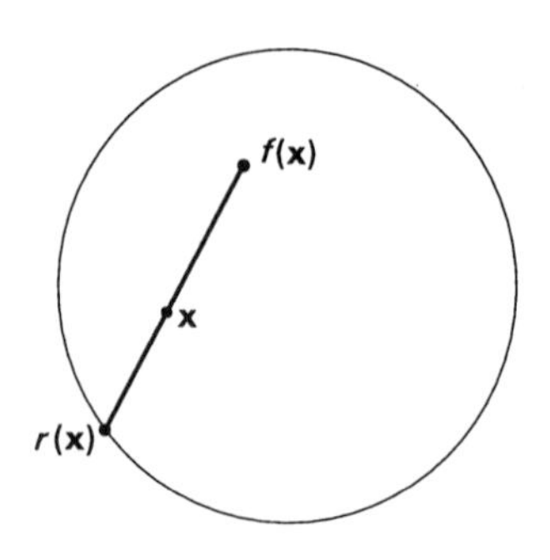

Fig. 9.1

Examples

(a) Let $X = \{(x, y) \mid 1 \leqslant x^2 + y^2 \leqslant 4\}$ and $A = \{(x, y) \mid x^2 + y^2 = 1\}$. Then $r(x, y) = (x, y)/(x^2 + y^2)$ is a retraction and

$$h((x, y),t) = t(x, y) + \frac{1 - t}{x^2 + y^2}(x, y)$$

is a homotopy $ir \simeq \mathrm{id}_X$.

(b) In Fig. 9.2, the subspace A (a line segment joining two circles) is a deformation retract of the space X: visualize the retraction!

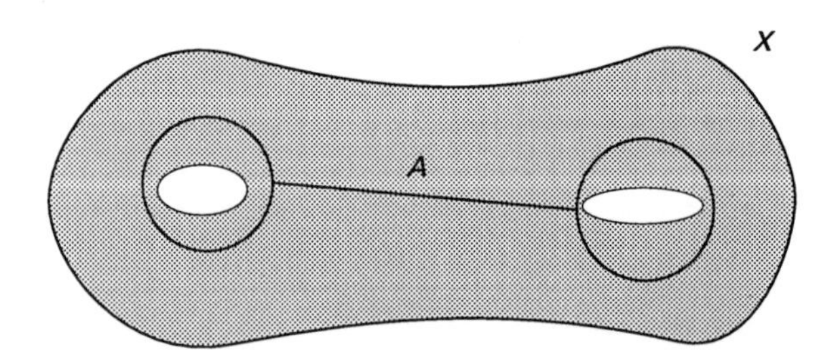

Fig. 9.2

(c) See Fig. 9.3.

Exercises 9.4

1. If $g: (Z, z_0) \to (Y, y_0)$ and $f: (Y, y_0) \to (X, x_0)$ are pointed maps, show that $(fg)_*$ and f_*g_* are equal.
2. Let f, g: $Y \to X$ be homotopic maps (with no assumption that the homotopy is pointed) and let $y_0 \in Y$. Show that there exists an isomorphism $\theta: \pi_1(X, f(y_0)) \to \pi_1(X, g(y_0))$ such that $g_*: \pi_1(Y, y_0) \to \pi_1(X, g(y_0))$ is equal to the composite

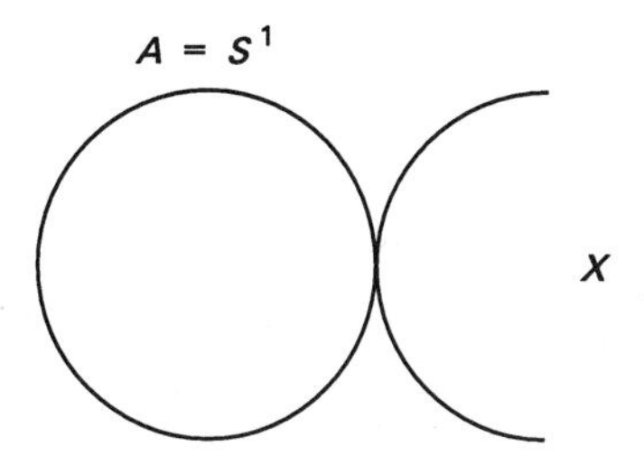

Fig. 9.3

$$\theta f_*: \pi_1(Y, y_0) \to \pi_1(X, f(y_0)) \to \pi_1(X, g(y_0)).$$

3. Let f: $Y \to X$ be a continuous map. We say that f is a homotopy equivalence if there exists a map g: $X \to Y$ such that $fg \simeq \mathrm{id}_X$ and $gf \simeq \mathrm{id}_Y$. If X, Y are path connected spaces and f: $Y \to X$ is a homotopy equivalence, show that for any $y_0 \in Y$ the fundamental groups $\pi_1(Y, y_0)$ and $\pi_1(X, f(y_0))$ are isomorphic.

10
Van Kampen's theorem

In this chapter we will examine the statement and proof of the Van Kampen theorem. If you glance forward to where this theorem is stated in full and in detail, you may wonder why such an apparently technical theorem is needed. Complicated theorems occur at numerous places within mathematics courses at this level and students often ask themselves why such theorems are deemed necessary. Again it is often the case that the lecturers have used the results explicitly or implicitly so many times in research projects that it becomes self-evident to them that these theorems are important and hence it can be extremely difficult to explain their importance to someone meeting them for the first time. With the Van Kampen theorem, we are lucky. We have been developing the theory of the fundamental group and if we pause to take stock of our present knowledge (and ignorance), it should become clearer why theorems of the Van Kampen type are needed. An even closer look will suggest what it should say and will also give us a sketch proof. Writing out a formal proof is another matter and will be dealt with later on; the first priorities are to guess what the result will say (before we meet its detailed statement), to see why the result is true and why certain technical restrictions are necessary; the details of the formal proof are (initially) somewhat less important. Once you understand the idea of the proof the details can be seen merely as the 'justification of the obvious'. The fact that these details are technically non-trivial merely shows that the 'obvious' often does need justification!

Why do we need a Van Kampen theorem? The state of our art at present is that we have calculated a large class of fundamental groups with great ease only to find they are all trivial! Working considerably harder we were able to prove that π_1 of a circle is an infinite cyclic group. We proved a result on π_1 of a product of spaces and so can show $\pi_1(\text{torus}) \cong C_\infty \oplus C_\infty$, but we cannot handle π_1 of an n-leafed rose, a simple enough space you might think, nor π_1 (genus g orientable surface) if $g > 1$. These spaces are good examples as we can understand them well. We can build geometric models of them, we can picture them, and so should be able, one would hope, to compute their fundamental groups. One important thing they have in common is that they can be built up from simpler 'building block' spaces such as lines and discs. The n-leafed rose is got from n circles. The genus g orientable surface can be built up from a $2g$-leafed rose and a disc.

This raises the question: can one calculate π_1 of a space if one knows π_1 of the pieces it is built up from? (If the answer is no, it is not at all clear how to go further in calculating fundamental groups!) To answer this question we will need to examine how loops in a space are related to loops in parts of a space.

Suppose we have a path connected space X divided up into two non-empty parts U and V. We are working with pointed spaces since we want to calculate $\pi_1(X)$; it is therefore best to assume that x_0, the basepoint of X, is in $U \cap V$. (If $U \cap V$ were empty then X would not be pathwise connected.) Let us consider a picture (Fig. 10.1) that will help the discussion (but always beware of arguments that rely too much on pictures!).

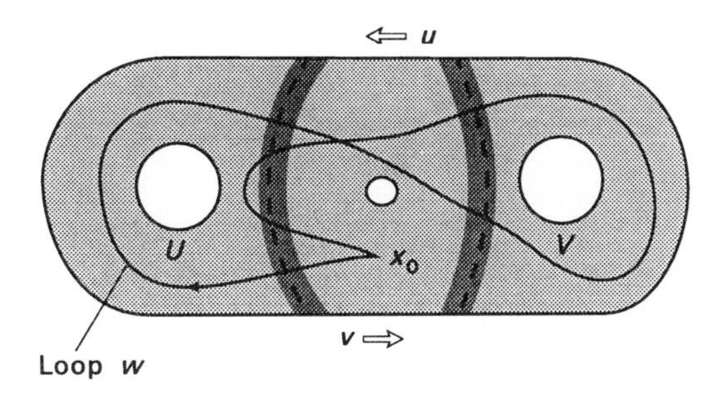

Fig. 10.1

Suppose w: $[0, 1] \to X$ is a normalized closed path at x_0. We hope to decompose w, up to homotopy, into parts, each of which is either wholly in U or wholly in V. In the picture, it is clear that one could go around w until one arrives in $U \cap V$, and then continue in V until one is back in $U \cap V$ and so on. This idea would seem to decompose the path w into *paths* totally within U or totally within V. However, this decomposition would be taking place not within $\pi_1(X)$ but within the fundamental groupoid of X, as we get *paths* in U and V that need not be closed.

Let us be a bit more precise (Fig. 10.2). We will assume that we can find points $t_1, t_2, \ldots, t_{n-1} \in [0, 1]$ so that (putting $t_0 = 0$, $t_n = 1$) each $w|[t_i, t_{i+1}]$ is a path wholly within U or wholly within V. How are we to ensure this is possible? For instance, if we take

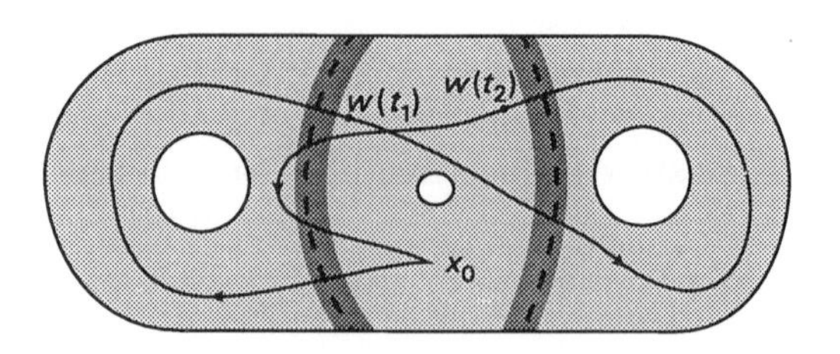

Fig. 10.2

$w: [0, 1] \to \mathbb{R}^2$ defined by $w(t) = (t, t \sin(1/t))$, $U = \{(x, y): y \geqslant 0\}$, $V = \{(x, y): y \leqslant 0\}$ then we will be in difficulties, since we would need to chop [0,1] into infinitely many subintervals to get the sort of property we want. We cannot, however, compose an infinite family of paths in the fundamental groupoid! Our problem is that U and V in this example are not open. If we increased U and V slightly in size, say $U_\varepsilon = \{(x, y): y \geqslant -\varepsilon\}$ and $V_\varepsilon = \{(x, y): y \leqslant \varepsilon\}$ for any small $\varepsilon > 0$, then we would be able to get away with finitely many subpaths. Why?

To handle the general case we need to recall the idea of compactness. First the definition: a space X is *compact* if, given any family $U = \{U_\lambda : \lambda \in \Lambda\}$ of open sets such that $X = \cup\{U_\lambda : \lambda \in \Lambda\}$, there is a finite subfamily $\{U_{\lambda_1}, \ldots, U_{\lambda_n}\}$ of U so that $X = \cup_{i=1}^n U_{\lambda_i}$. (*Every open covering of a compact space has an open subcovering.*) Now if U, V are open sets, so are $w^{-1}(U)$ and $w^{-1}(V)$. The open subsets of [0, 1] are unions of open intervals and [0, 1] is compact. (The Heine–Borel theorem from real analysis can be thought of as saying: $X \subseteq \mathbb{R}^n$ is compact if and only if it is closed and bounded. Thus both $I = [0, 1]$ and $I^2 = I \times I$, which is the closed unit square in $\mathbb{R}^2$, are compact spaces.) We need a further piece of information about compact spaces or, more precisely, compact *metric* spaces, that is spaces where the open sets are defined by a distance function or metric. If Y is any compact metric space and U is any open covering, there is a number $\varepsilon > 0$ (often called the Lebesgue numbering of the covering) such that any subset of Y of diameter less than ε is wholly contained in some set of the covering. (Try to prove this yourself using techniques from analysis, or look it up in a book on metric spaces or on point set topology.)

Applying these ideas to our situation, we find that there is an $\varepsilon > 0$ such that any interval of length less than ε is contained wholly within $w^{-1}(U)$ or $w^{-1}(V)$. We can thus pick points $t_0 = 0, t_1, \ldots, t_{n-1}, t_n = 1$ in I such that each $t_i - t_{i-1} < \varepsilon$ and hence $w[t_{i-1}, t_i]$ is either within U or within V. Now clearly if we write w_i for $w|[t_{i-1}, t_i]$

composed with the linear homeomorphism between $[0, t_i - t_{i-1}]$ and $[t_{i-1}, t_i]$, then we have

$$w = w_1 \cdot \ldots \cdot w_n$$

and so we have written w as a composite of paths wholly within U or within V. This takes place in the fundamental groupoid of X. The fundamental groupoid is useful, but for the moment we are searching for a result on fundamental groups and the paths w_i need not be closed. To get around this difficulty we proceed as follows: pick for each $i = 0, 1, \ldots, n$, a path v_i joining x_0 to $w(t_i)$; these paths are to satisfy

(1) v_0 and v_n are the trivial path at x_0;

(2) if w_i and w_{i+1} are both paths in U, v_i is wholly within U; similarly if they are both in V, v_i is wholly within V;

(3) if one of w_i and w_{i+1} is in U and the other in V, then v_i is a path in $U \cap V$.

This last condition overlaps with condition (2) since if w_i and w_{i+1} are *both* in $U \cap V$ then v_i, under both conditions, has to be within $U \cap V$. In any case we must have U, V, and $U \cap V$ path connected if we are to be able to fulfil these conditions. (This requirement could have been avoided to some extent if we had worked with the fundamental groupoid rather than the fundamental group as our principal tool; we will briefly examine this possibility later.)

We can rewrite the decomposition of w, up to homotopy this time, as

$$w \sim v_0 \cdot w_1 \cdot \bar{v}_1 \cdot v_1 \cdot w_2 \cdot \bar{v}_2 \cdot \ldots \cdot v_{n-1} \cdot w_n \cdot \bar{v}_n,$$

as each $\bar{v}_i v_i$ is homotopic to the trivial path. (At this point, it is useful to think of working with unnormalized paths as otherwise we would need to bother with the fact that composition of normalized paths is not associative. In fact, we can disregard this technical problem as we know that if we work with unnormalized paths and then normalize at the end of a calculation, the end result is homotopic to the end result obtained by consistently normalizing and taking into account the lack of associativity that results.) We can group this decomposition as

$$w \sim u_1 \cdot u_2 \cdot \ldots \cdot u_n$$

within X, where $u_i = v_{i-1} w_i \bar{v}_i$. Each u_i is a loop either in U or in V. We will use the convention that u_i is in $U(i)$ where $U(i) = U$ or V. This allows us to introduce a neat piece of notation: if w is a loop in X we write $[w]_X$ for the corresponding equivalence class in $\pi_1(X)$; similarly with X replaced by U, V or $U \cap V$. This means that the induced homeomorphism

$$i_U: \pi_1(U) \longrightarrow \pi_1(X)$$

is defined by $i_U[w]_U = [w]_X$. We note that in the decomposition above

$$[w]_X = [u_1]_X[u_2]_X \ldots [u_n]_X$$
$$= i_{U(1)}[u_1]_{U(1)} \ldots i_{U(n)}[u_n]_{U(n)}.$$

This gives us a result that we can summarize as follows.

10.1 Lemma. *The images of $\pi_1(U)$ and $\pi_1(V)$ in $\pi_1(X)$ generate that group.*

An alternative and slightly more formal way of putting this same result is as follows. The induced homeomorphisms

$$i_U: \pi_1(U) \longrightarrow \pi_1(X)$$
$$i_V: \pi_1(V) \longrightarrow \pi_1(X)$$

in their turn induce a homeomorphism

$$\pi_1(U) * \pi_1(V) \longrightarrow \pi_1(X)$$

where as before $*$ indicates free product and the lemma states that this homeomorphism is onto $\pi_1(X)$. This is already a positive result. For instance, if we have presentations $(\boldsymbol{x} : \boldsymbol{r})$ of $\pi_1(U)$ and $(\boldsymbol{y} : \boldsymbol{s})$ of $\pi_1(V)$ then it says that $\pi_1(X)$ is a quotient of the group with presentation $(\boldsymbol{x} \cup \boldsymbol{y}: \boldsymbol{r} \cup \boldsymbol{s})$. This means that generators for $\pi_1(U)$ and $\pi_1(V)$ together generate $\pi_1(X)$ and that the relations between these generators still hold in $\pi_1(X)$. What it fails to say is what additional relations are needed, or equivalently what the kernel of the epimorphism looks like. If we try to interpret this geometrically, we can see it as the uncertainty about uniqueness or otherwise of the list $[u_1]_{U(1)}, \ldots, [u_n]_{U(n)}$ due to the choices that had to be made in its formation, and perhaps to questions of invariance of the list under homotopies of w. The general rule should always be that if we make choices we should see how much the end result depends on the choices.

One of the subtlest choices we made was deciding in which open set a given w_i was. This may seem a strange place to start, but it will point out the necessity of thinking clearly. Suppose w was a closed path in $U \cap V$; we would not need to decompose it at all, but if we write $j_U: \pi_1(U \cap V) \to \pi_1(U)$ and $j_V: \pi_1(U \cap V) \to \pi_1(V)$, then it is clear that the two elements $[w]_U = j_U[w]_{U \cap V}$ and $[w]_V = j_V[w]_{U \cap V}$ give the same element of $\pi_1(X)$. In other words this subtle choice is really basic: any closed path in $U \cap V$ 'decomposes' trivially in two different ways giving the same element of $\pi_1(X)$. This first observation suggests the following points:

1. The square

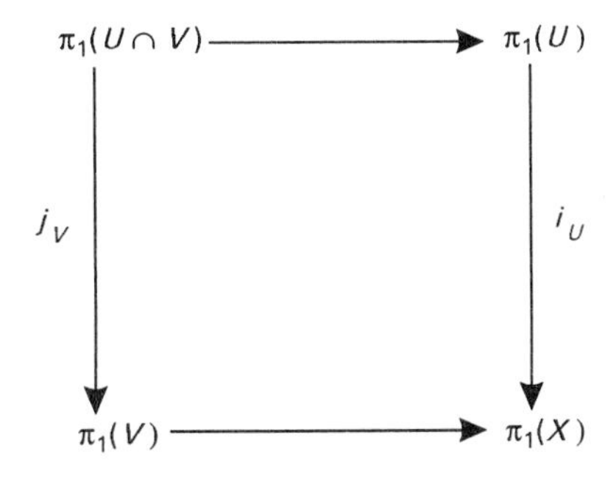

is commutative, so there is an induced homeomorphism, θ say, from the pushout

$$\pi_1(U) \underset{\pi_1(U \cap V)}{*} \pi_1(V)$$

to $\pi_1(X)$.

2. As $i_U[w]_U = i_V[w]_V$ for any $[w]_{U \cap V} \in \pi_1(U \cap V)$, we should 'kill off' within $\pi_1(U) * \pi_1(V)$ those elements of the form $[w]_U[w]_V^{-1}$. We do this by dividing out by the normal subgroup generated by them. An efficient way to do this is to pick a set of generators z, say, for $\pi_1(U \cap V)$ and to kill off the words $j_U z \cdot j_V z^{-1}$ for $z \in Z$ within the presentation $(\boldsymbol{x} \cup \boldsymbol{y} : \boldsymbol{r} \cup \boldsymbol{s})$ of $\pi_1(U) * \pi_1(V)$, giving a group G with presentation $(\boldsymbol{x} \cup \boldsymbol{y} : \boldsymbol{r} \cup \boldsymbol{s} \cup \{(\overline{j_U(z)})(\overline{j_V(z)})^{-1}\})$. Here we have written $\overline{j_U(z)}$ for any element of $F(\boldsymbol{x} \cup \boldsymbol{y})$ that maps down to $j_U(z)$ under the natural composite epimorphism

$$F(\boldsymbol{x} \cup \boldsymbol{y}) \to F(\boldsymbol{x}) \to (\boldsymbol{x} : \boldsymbol{r});$$

similarly for $\overline{j_V(z)}$.

Of course, we saw earlier that G is precisely the pushout

$$\pi_1(U) \underset{\pi_1(U \cap V)}{*} \pi_1(V)$$

so really the two points are the same, that is there is an epimorphism from G onto $\pi_1(X)$ induced by the inclusions of the subspaces. This group G is often much smaller than the free product, so this is real progress to cutting $\pi_1(X)$ 'down to size'.

We next turn to the other choices that were made in determining the decomposition of w. We might expect that, each time, we would find a new geometrically inspired set of relations to add in, further reducing the size of the group we can construct and also that of the kernel of the resulting epimorphism. What actually happens is different from this, and is slightly surprising, making one realize that the end result is a very elegant and pretty theorem indeed (as well as being an extremely useful tool.)

The first step is to investigate what happens when we add another point in the decomposition. The reasoning behind this is the following. Suppose we decompose w using $t_0, t_1, \ldots, t_n$ and also using $t'_0, \ldots, t'_m$; we would seem to get two elements g and g', say, in G. Are these elements the same? Putting the two subdivisions of $[0,1]$ together gives yet another subdivision. If the resulting element g'' of G can be shown to be equal to g and to g', then of course $g = g'$. The advantage of g'' over g' is that the subdivision used can be obtained by adding in finitely many extra points into that used for g and hence if we can show that one extra point added into the subdivision changes nothing we can show that $g = g''$, and also $g' = g''$ as hoped for, simply using proof by induction. You may recognize the basic method here. To prove invariance under choices, prove invariance under basic elementary changes of choice (here adding in or taking out a point, and previously by performing one of the Reidemeister moves or Tietze transformations).

Suppose we take $t' \in [t_{i-1}, t_i]$ and without loss of generality we assume that w maps $[t_{i-1}, t_i]$ into U. In the decomposition before adding t', we have

$$[u_i]_U \in \pi_1(X)$$

and after adding t' we choose a v from x_0 to $w(t')$ and find closed paths $v_{i-1} \cdot w|[t_{i-1}, t'] \cdot \bar{v} = u$, say $v \cdot w|[t_1, t_2] \cdot v_i = u'$, giving $[u]_U \cdot [u']_U$ in place of $[u_i]_U$. However, $[u]_U \cdot [u']_U = [u \cdot u']_U = [u_i]$, so nothing has changed. If the section $w|[t', t_i]$ is in $U \cap V$ and v is chosen to be within $U \cap V$ then u' will be a closed path in $U \cap V$. Given that we have already within the presentation of G allowed for the double identity of such closed paths (within both $\pi_1(U)$ and $\pi_1(V)$) we can, if need be, now consider u' as being a closed path in V and delete t_i from the subdivision (provided $t_i \neq 1$ of course). Thus the independence of the element defined by the decomposition on the choice of subdivision is clear and the mechanism set up to handle the problem of the closed paths with a 'split personality' allows one to pass parts of a closed path from U to V and vice versa.

If we do not change the subdivision but merely the choice of the v_i then again the same idea works (Fig. 10.3).

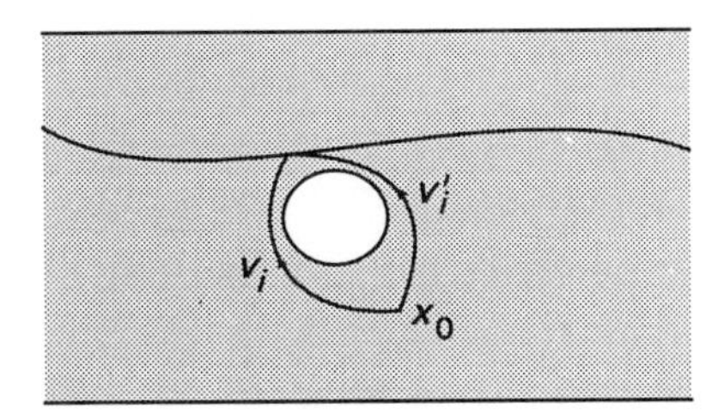

Fig. 10.3

Writing $u_i = v_{i-1}w_i\bar{v}_i$, $u_i' = v_{i-1}w_i\bar{v}_i'$, $u_{i+1} = v_iw_{i+1}\bar{v}_{i+1}$, $u_{i+1}' = v_i'w_{i+1}\bar{v}_{i+1}$, we find that $u_i \cdot u_{i+1} \sim u_i \cdot (v_i \cdot \bar{v}_i') \cdot u_{i+1}' \sim u_i' \cdot u_{i+1}'$ within X. If both u_i and u_{i+1} are in one only of U and V, this is unremarkable as the homotopies happen equally well within U and V, but if, say, u_i is in U and u_{i+1} is in V, then v_i and v_i' are in $U \cap V$ (by the way we chose them) and we obtain within the group G,

$$\begin{aligned} i_U[u_i]_U\, i_V[u_{i+1}]_V &= i_U[u_i]_U\, i_V[v_i \cdot \bar{v}_i']_V\, i_V[u_{i+1}']_V \\ &= i_U[u_i]_U\, i_Vj_V[v_i \cdot \bar{v}_i']_{U\cap V}\, i_V[u_{i+1}']_V \\ &= i_U[u_i]_U\, i_Uj_U[v_i \cdot \bar{v}_i']_{U\cap V}\, i_V[u_{i+1}']_V \\ &= i_U[u_i']_U\, i_V[u_{i+1}']_V, \end{aligned}$$

so the algebraic structure in G faithfully reproduces the geometric structure within $\pi_1(X)$. Note that we have used the '$j_U(\alpha) = j_V(\alpha)$' condition in the form $i_Uj_U = i_Vj_V$.

The final thing we have to check is that if $w_0 \sim w_1$ then the resulting elements of G are equal. We suppose given

$$h\colon [0, 1] \times [0, 1] \longrightarrow X$$

such that $h\colon w_0 \simeq w_1$. Using the compactness of $[0, 1] \times [0, 1]$ we find an $\varepsilon > 0$ such that any set of diameter less than ε is mapped by h either wholly into U or wholly into V. Next we subdivide the square $[0, 1] \times [0, 1]$ into rectangles small enough that the diagonal length is

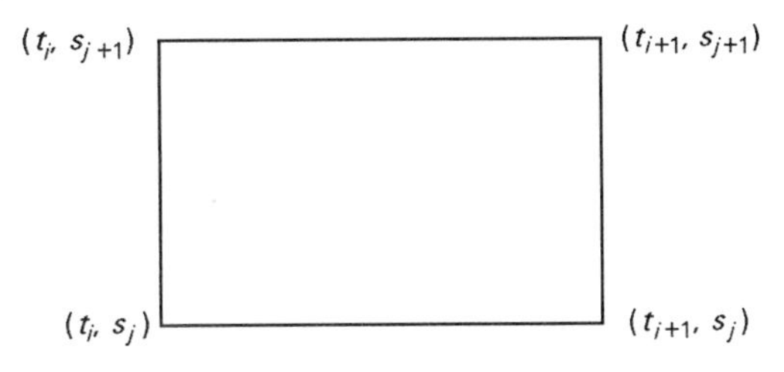

less than ε, that is we pick numbers $t_0 = 0, t_1, t_2, \ldots, t_{n-1}, t_n = 1$, $s_0 = 0, s_1, \ldots, s_{m-1}, s_m = 1$ so that each rectangle is mapped within one of U or V. Using $t_0 = 0, t_1, \ldots, t_n = 1$ as our subdivision of $[0, 1]$, we get decompositions of $w_0 = h|[0, 1] \times \{0\}$, $h|[0, 1] \times \{s_1\}$, $h|[0, 1] \times \{s_2\}$, etc. As we hope to show that the elements g_0 and $g_1 \in G$ corresponding to w_0 and w_1 are equal, it will be enough to prove that the elements corresponding to any two successive restrictions of h, say $h|[0, 1] \times \{s_j\}$ and $h|[0, 1] \times \{s_{j+1}\}$, agree as iteration will do the rest. Because of this we will assume $m = 1$, so our picture is

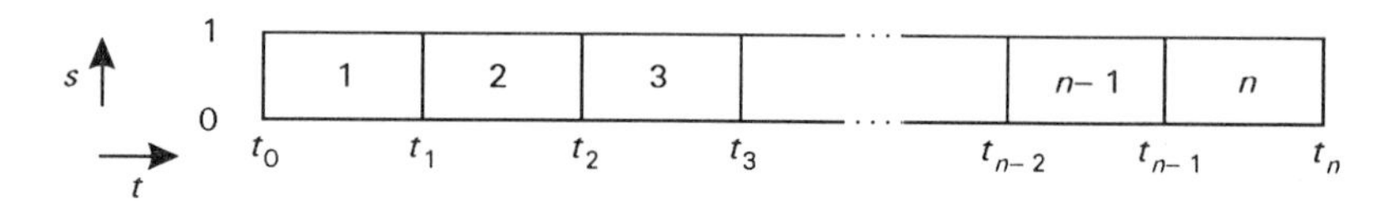

As h is a fixed-end-point homotopy the ends of the square are both mapped to x_0. We assume as always that paths v_i from x_0 to $w(t_i)$ have been chosen. We have seen that the element of G that we obtain is independent of the choice of the paths, so when we decompose w' we will use paths v_i' obtained by composing v_i with the path $w|\{t_i\} \times [0, 1] = \alpha_i$ (Fig. 10.4).

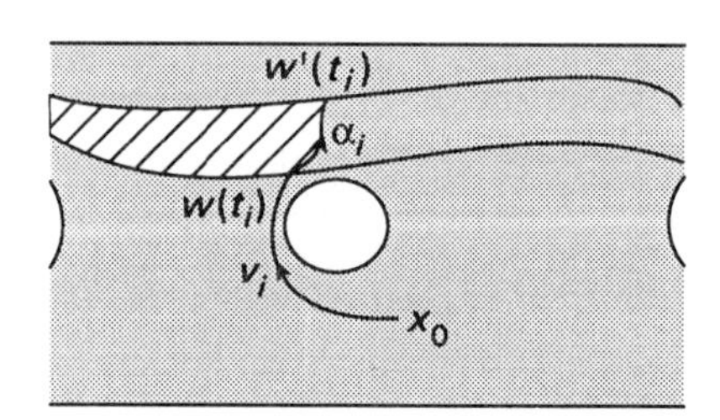

Fig. 10.4

Although we will write w' in terms of loops in U or V, it will as usual be better to do the calculation in the fundamental groupoids of X, U, V, and $U \cap V$ as follows:

$$w' \sim u_0' \cdot u_1' \cdot \ldots \cdot u_n'$$

where

$$u_i' = v_i' \cdot w_i' \cdot \bar{v}_{i+1}', \qquad i = 1, \ldots, n-1$$

$$\sim v_i \cdot \alpha_i \cdot w_i' \cdot \bar{\alpha}_{i+1} \cdot \bar{v}_{i+1}$$

within U or within V since the homotopy is just a reindexation via a linear homeomorphism of [0, 1]. The path $\alpha_i \circ w_i' \circ \bar{\alpha}_{i+1}$ is obtained by mapping three sides of the $(i+1)$st rectangle using h:

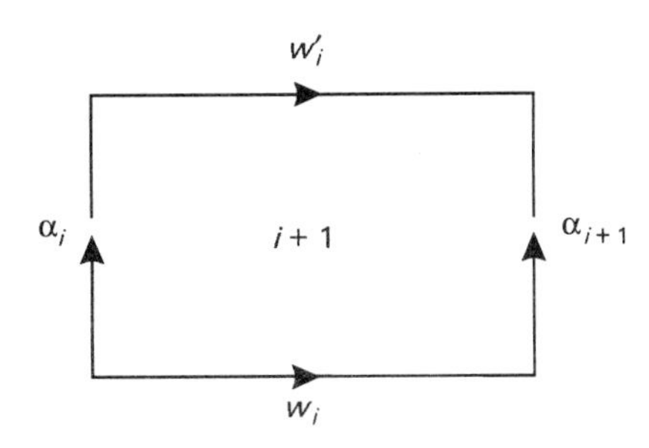

This rectangle is mapped (by h) wholly within U, or wholly within V (possibly within $U \cap V$). Thus the path $\alpha_i \circ w_i' \circ \bar{\alpha}_{i+1}$ determines in the fundamental groupoid of U (resp. of V, resp. of $U \cap V$) the same element as w_i. Composing either side with the v_i takes this back into the fundamental groups, so the elements g_0 and g_1 are equal.

Thus we have investigated all the indeterminacy linking the element g to $[w] \in \pi_1(X)$. In fact there was none. Given $[w]$, g was well defined and $\theta(g) = [w]$. We have proved more since if we compose $[w]$ and $[w']$ in $\pi_1(X)$ it is clear that by picking w and w' in these classes and composing one can pick decompositions already used for w and w' to get one for $w \circ w'$, but any decomposition defines the same element as none of the choices makes any difference. That is, the assignment of g to $[w]$ gives a homeomorphism

$$\sigma : \pi_1(X) \longrightarrow G$$

that 'splits' the epimorphism θ; in other words, $\theta\sigma$ is the identity homeomorphism on $\pi_1(X)$. The way in which σ is defined makes it clear that $\sigma\theta$ is the identity on G, so $\pi_1(X) \cong G$, the isomorphism being induced by the inclusions i_U, i_V.

We have arrived informally at the result known as Van Kampen's theorem. We state it formally in two different versions.

10.2 Van Kampen theorem (first form). *Let $X = U \cap V$ where U, V, and $U \cap V$ are non-empty, open, and pathwise connected. Pick $x_0 \in U \cap V$. Then the following diagram is a pushout square of groups.*

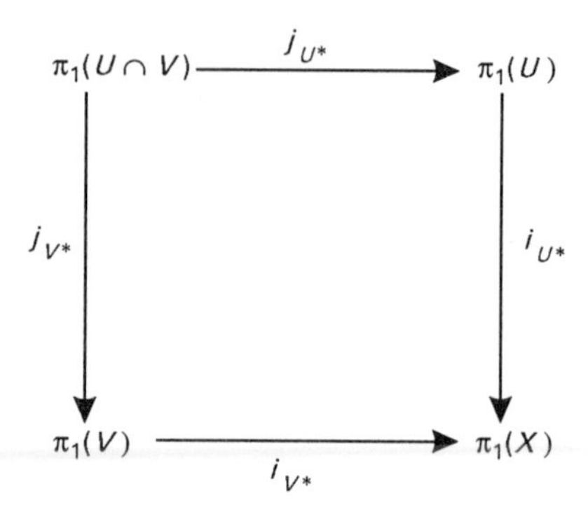

10.3 Van Kampen theorem (second form). *Let $X = U \cap V$ as before. Pick $x_0 \in U \cap V$ and suppose*

$$\pi_1(U, x_0) \text{ has a presentation } (\boldsymbol{x} : \boldsymbol{r}),$$

$$\pi_1(V, x_0) \text{ has a presentation } (\boldsymbol{y} : \boldsymbol{s}),$$

$$\pi_1(U \cap V, x_0) \text{ has a presentation } (\boldsymbol{z} : \boldsymbol{t}).$$

Then $\pi_1(X, x_0)$ has a presentation

$$(\boldsymbol{x} \cup \boldsymbol{y} : \boldsymbol{r} \cup \boldsymbol{s} \cup \{(\overline{j_{U^*}(z)})\,(\overline{j_{V^*}(z)})^{-1} : z \in \boldsymbol{z}\})$$

where $\overline{j_{U^*}(z)}$ is a word in $F(x)$ representing $j_{U^*}(z)$ and $\overline{j_{V^*}(z)}$ is a word in $F(y)$ representing $j_{V^*}(z)$.

We have given above all the necessary building blocks for constructing a formal proof of the Van Kampen theorem. We will, on purpose, not give one ourselves, but suggest the following exercises.

Exercises

1. Construct for yourself a formal proof of either form of Van Kampen's theorem. You can choose whether or not to follow the outline we have given. For instance, you may wish to verify the pushout property of the square (*) directly rather than showing that $\pi_1(X)$ is isomorphic to the algebraically constructed pushout.
2. Having attempted the previous exercise, go to your library and search out books which contain a formal proof of the Van Kampen theorem. Compare their methods with yours. Criticize both your proof and theirs. Try to improve your proof with ideas, phrases, or notation from their version. Do not think their version will necessarily be fundamentally better than yours, but

as the authors tend to have had years' more experience in the game of writing proofs, you should expect to gain something, at least in style.

3. Find a reference to Van Kampen's original paper. Try and read his proof.
4. Several times we have worked within the fundamental groupoid. Try to formulate and prove a groupoid version of the theorem (then compare with the paper by Brown (1967)).

11 Applications of the Van Kampen theorem

Having got the Van Kampen theorem, how are we to apply it? As with any result, before we try to apply it to a given example, we must check that the conditions necessary for the theorem to hold are satisfied in that example. In order to apply the Van Kampen theorem, the parts into which we divide our space must be open and path connected. As we will see, this sometimes requires an adjustment to the obvious choice of decomposition. Initially we will give nearly all the details of this necessary adjustment, but later on we will leave such details increasingly up to you.

The other thing that we will need is a starting point. The Van Kampen theorem relates the fundamental group of a space to those of its parts. If we want to calculate π_1 of a space, X, we have to find an open decomposition in terms of simpler spaces. Just as in a proof by induction, we need to start somewhere and in this situation our starting point is the proof that π_1 of the circle, S^1, is infinite cyclic, that is free on one generator, namely the class of a path going once around the circle.

We start with simple examples and get progressively to handle more complicated spaces. As we do this, our calculations will more often than not refer to earlier examples.

11.1 Calculations of fundamental groups

The *n*-leafed rose

So as not to be too ambitious to start with, we will first examine the fundamental group of a one- and a two-leafed rose.

$n = 1$. A one-leafed rose is just a circle, so π_1 (one-leafed rose) is free on one generator.

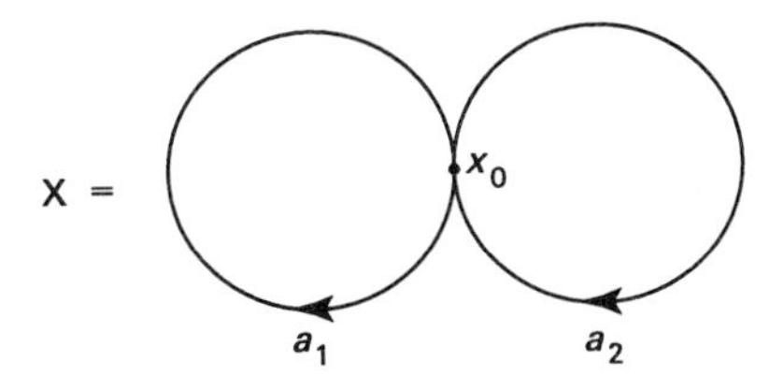

Fig. 11.1

$n = 2$. Already here the main ideas for the general case occur. A two-leafed rose (Fig. 11.1), has an 'obvious' decomposition (Fig. 11.2).

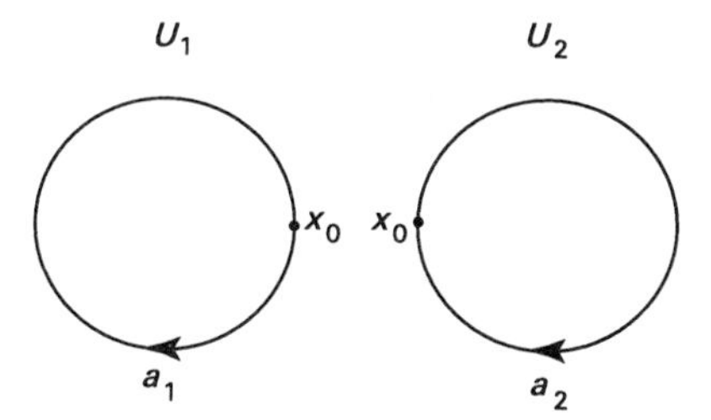

Fig. 11.2

We see that U_1, U_2 are path connected, as is $U_1 \cap U_2 = \{x_0\}$. The space $X = U_1 \cup U_2$, but U_1 and U_2 are not *open* subspaces of X, so we cannot apply the Van Kampen theorem to this present decomposition. The necessary adjustment is easily made. We let N be a small open neighbourhood of x_0 in X itself. Thus N looks as shown in Fig. 11.3.

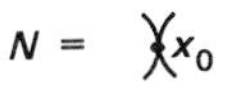

$N =$ x_0

Fig. 11.3

Now let $V_1 = U_1 \cup N$, $V_2 = U_2 \cup N$. These two subspaces *are* open and path connected; we note that $X = V_1 \cup V_2$ and $N = V_1 \cap V_2$. The final criterion is our ability to calculate $\pi_1(V_1)$, $\pi_1(V_2)$, $\pi_1(N)$, *and* the induced homeomorphisms between them.

As V_1 and V_2 are clearly homeomorphic, we need only look at one of them, say V_1 (Fig. 11.4).

This space contains U_1 and by retracting the two 'antennae', we find that U_1 is a deformation retract of V_1, so that $\pi_1(V_1) \cong \pi_1(U_1)$ and is freely generated by the class of the loop a_1. We will abuse notation and write

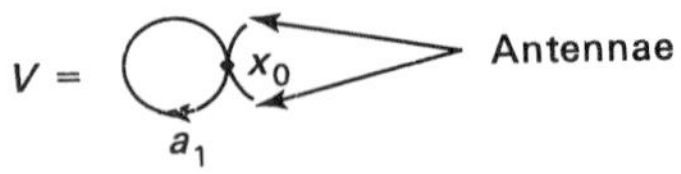

$V =$ x_0 Antennae a_1

Fig. 11.4

$$\pi_1(V_1) \cong (a_1 : \varnothing).$$

Similarly

$$\pi_1(V_2) \cong (a_2 : \varnothing).$$

Now N itself contains $\{x_0\}$ as a deformation retract, so $\pi_1(N)$ is the trivial group. This neatly absolves us from the task of calculating the induced homeomorphisms

$$\pi_1(V_1 \cap V_2) \to \pi_1(V_i), \quad i = 1, 2$$

since there is a unique homeomorphism from the trivial group to any group.

We have now collected up all the information so as to calculate $\pi_1(X)$. The Van Kampen theorem gives a presentation

$$\pi_1(X) \cong (a_1, a_2 : \varnothing),$$

that is a free group on two generators corresponding to loops around the two 'leaves' of the rose. This is as we might have guessed since a typical loop in X will clearly wander around the two leaves one after the other, perhaps going around a_1, say n_1 times, then around a_2 n_2 times, then back around a_1, and so on, corresponding to a word starting $a_1^{n_1}a_2^{n_2}a_1^{n_3}$. . . .. Thus the Van Kampen theorem provides a proof of the geometrically obvious result.

What about the general case of an n-leafed rose? The sort of intuitive reasoning just used would suggest that if X_n is an n-leafed rose with loops around the 'leaves' labelled $a_1, \ldots, a_n$, then $\pi_1(X_n)$ will be a free group on n generators corresponding to these loops. The proof by induction of this is not too hard, but as it again involves the addition of a small open neighbourhood of the basepoint, we will include a sketch of it.

We note that an n-leafed rose X_n can be written as the one-point union of X_{n-1} with a circle (Fig. 11.5). This gives $X_n = X_{n-1} \vee Y$ with Y a circle. This is not an open decomposition, so again let N be a small open neighbourhood of the basepoint x_0 in X_n. (How small is small? What were the necessary facts about the small neighbourhood N that was used in the two-leafed case? How do we know that such a neighbourhood exists? We will return to these points shortly, but will continue with the proof for the moment.)

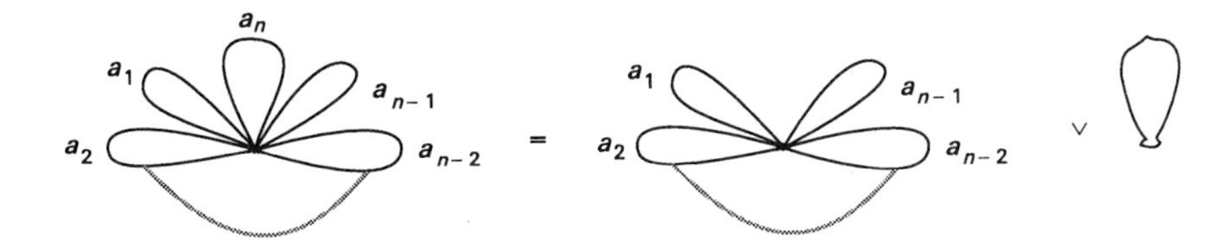

Fig. 11.5

Set $V_1 = X_{n-1} \cup N$, $V_2 = Y \cup N$; then $X_n = V_1 \cup V_2$, $V_1 \cap V_2 = N$. The basepoint is a deformation retract of N (if N has been chosen small enough!), X_{n-1} is a deformation retract of V_1, whilst the circle, Y, is a deformation retract of V_2. (Draw diagrams to illustrate this and to convince yourself it is so.) Collecting up facts, we get

1. $\pi_1(V_1) \cong \pi_1(X_{n-1}) \cong (a_1, \ldots, a_{n-1} : \varnothing)$ by our (unstated) induction hypothesis.
2. $\pi_1(V_2) \cong \pi_1(Y) \cong (a_n : \varnothing)$ since Y is a circle.
3. $\pi_1(V_1 \cap V_2) \cong (\varnothing : \varnothing)$, the trivial group.

The formula given by Van Kampen now gives

$$\pi_1(X_n) \cong (a_1, \ldots, a_n : \varnothing)$$

as expected.

What about 'small'? Clearly picking an arbitrary neighbourhood of x_0 in X_n might result in one of the leaves being within N. How can we be certain that this can be avoided? We have a lot of latitude in our way of modelling the n-leafed rose as the property that we are examining only depends on homotopy type. Hence we may assume that X_n is chosen to be a subspace of $\mathbb{R}^3$ with x_0 the origin and each leaf a vertical unit circle (Fig. 11.6).

Now let $N = \{\mathbf{x} \in X_n : \|\mathbf{x}\| < 0.1\}$. Since we have an explicit description of X_n and of N, our requirements can be checked explicitly if so

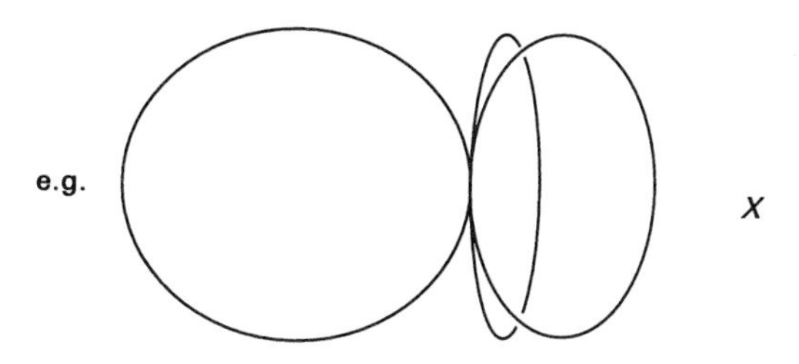

Fig. 11.6

desired. Care is required: as an example of what could go wrong consider the spaces

$$C_n = \{x \in \mathbb{R}^2 \mid \|x - (1/n, 0)\| = 1/n\}$$

so that C_n is a circle through the origin centred on the point $(1/n, 0)$ with radius $1/n$. If $m \neq n$, $C_n \cap C_m = \{\mathbf{0}\}$. Now let $X = \bigcup_{n \in \mathbb{N}} C_n$, as a subspace of the plane. If one looks at any neighbourhood, N, of $\mathbf{0}$ in X, it must contain an open set of the form $B(\mathbf{0}; \varepsilon) \cap X$ since it is an open neighbourhood of $\mathbf{0}$, but there is some $N_\varepsilon \in \mathbb{N}$ depending on ε such that for $n > N$, $2/n < \varepsilon$, that is $C_n \subset B(\mathbf{0}; \varepsilon) \cap X \subset N$ and N contains infinitely many loops of X. The question of the structure of $\pi_1(X)$ would lead us too far astray. Its structure is known, but if you feel like trying your hand at it, beware—it is not that simple.

A finite connected graph

If Γ is a finite graph with α_0 vertices and α_1 edges, than Γ is homotopy equivalent to an n-leafed rose, where $n = \alpha_1 - \alpha_0 + 1$ is the number of edges not in a maximal tree. Using the previous example, we find that $\pi_1(\Gamma)$ is free on n generators. How shall we interpret these generators?

We suppose the basepoint of Γ is chosen to be a vertex, v. Pick a maximal tree T in Γ, orient, and label the edges not in T, say $a_1, \ldots, a_n$. Now see what happens to these edges when we shrink T to a point. They become a set of generating loops of $\pi_1(X_n)$. Retracing the process back into Γ, we find that a set of free generators for $\pi_1(\Gamma)$ can be constructed using the path classes $[\bar{a}_i]$, $i = 1, \ldots, n$, where $\bar{a}_i$ is a path that goes from the basepoint, v, out in the maximal tree to the initial vertex of the oriented edge, a_i; it then crosses a_i and goes back to v in the maximal tree (Fig. 11.7).

Here $\alpha_0 = 8$, $\alpha_1 = 11$, and $n = 4$. Now $\bar{a}_1 = eba_1c^{-1}e^{-1}$, $\bar{a}_2 = eba_2c^{-1}e^{-1}$, $\bar{a}_3 = a_3g^{-1}f^{-1}$, $\bar{a}_4 = fha_4g^{-1}f^{-1}$, and $\pi_1(\Gamma) \cong \langle \bar{a}_1, \bar{a}_2, \bar{a}_3, \bar{a}_4 : \varnothing \rangle$ is the free group on four generators.

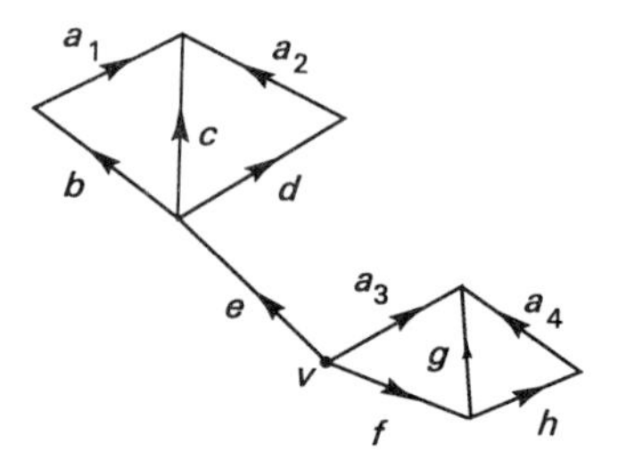

Fig. 11.7

Surfaces

Again, so as to gain some intuition of what to expect, we start with almost the simplest example possible, a torus. We have already noted that the torus can be realized as a product $S^1 \times S^1$, so by our earlier results, we know that π_1(torus) is the product of two infinite cyclic groups $C_\infty \times C_\infty$; it is therefore free Abelian on two generators. This will enable us to check our calculation independently.

The torus can be obtained by gluing a rectangle as we have seen in Chapter 4 (Fig. 11.8).

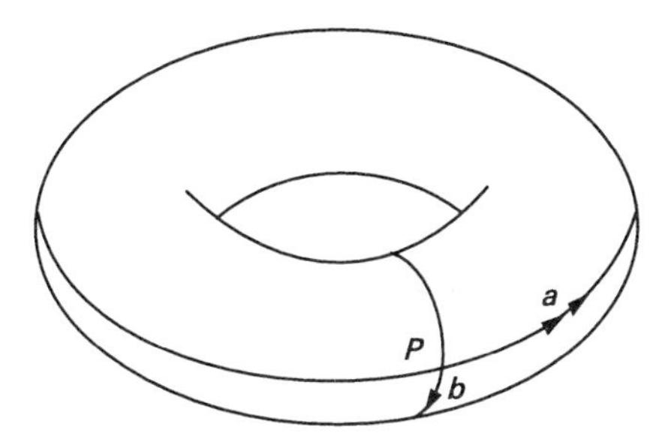

Fig. 11.8

Note that the boundary of the rectangle gives an embedded two-leafed rose. We have to look for a suitable open decomposition satisfying the conditions of the Van Kampen theorem. Such a decomposition is more easily visualized from the rectangle than from the other picture. The idea is to take one of the sets in the decomposition to be the union of the two loops a, b, and the other to be the interior of the rectangle. Of course, this does not work, as the intersection is empty and the first set is not open, but by 'thickening' the first set up to an open neighbourhood we will get around these difficulties. We do not, this time, need to be that careful with the neighbourhood.

We first pick a basepoint, the choice of which is immaterial as the space is path connected. However, a good choice of basepoint will

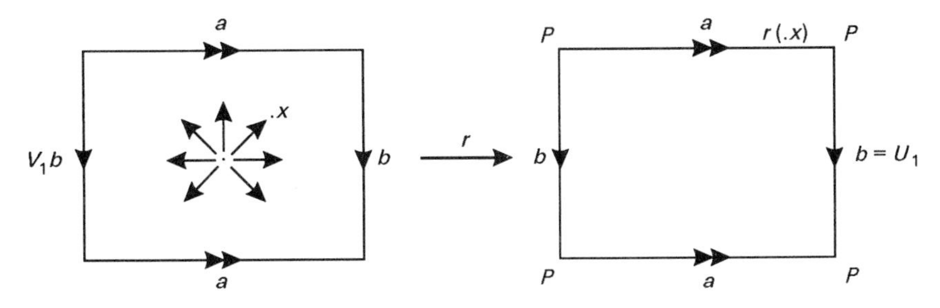

make the presentation simpler, so before choosing a basepoint, let us see how the argument will go.

We let $V_1 = X$ – centre of the rectangle, U_1 = boundary of the rectangle; then V_1 radially projects onto U_1 and U_1 is a deformation retract of V_1: (see also Fig. 11.9).

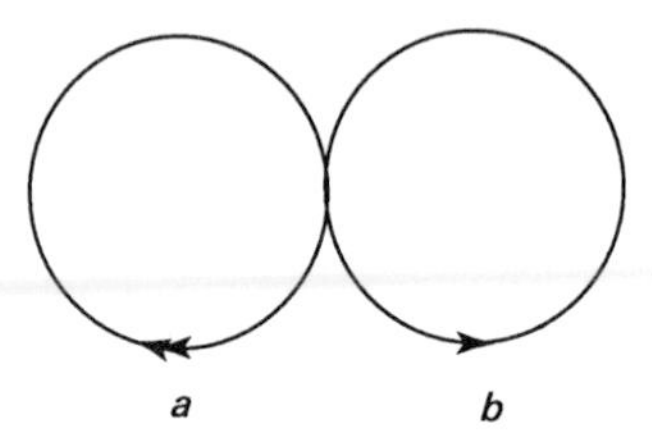

Fig. 11.9

After the identification of edges, the boundary, U_1, of the rectangle is a two-leafed rose embedded in the torus. We thus have

$$\pi_1(V_1) \cong \pi_1(U_1) \cong (a, b : \varnothing).$$

Now for V_2: for this we take the interior of the rectangle, namely $V_2 = X - U_1$. This is an open disc, and so $\pi_1(V_2) \cong (\varnothing : \varnothing)$, the trivial group.

What about $V_1 \cap V_2$? This is an open (rectangular) annulus (Fig. 11.10).

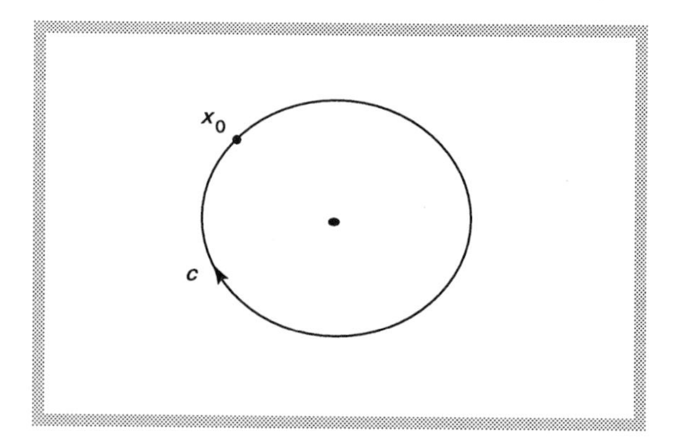

Fig. 11.10

Thus $\pi_1(V_1 \cap V_2) \cong \pi_1(S^1)$, an infinite cyclic group generated by a loop around the hole.

Looking again at the pushout diagram coming from the Van Kampen theorem, we have

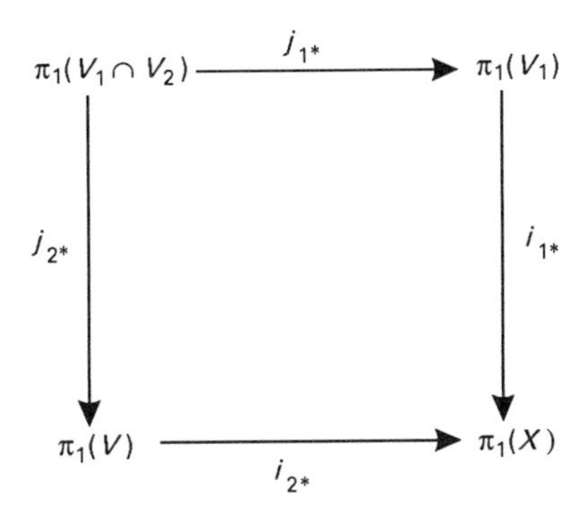

We do not need to worry about j_{2*} since $\pi_1(V_2)$ is trivial; we do, however, need to describe j_{1*}. As $\pi_1(V_1 \cap V_2)$ is infinite cyclic generated by c, it is sufficient to calculate $j_1(c)$ in terms of a and b. (As usual, we are using the same letter for a loop and the class it determines.) The radial projection r pushes the loop $j_1(c)$ onto the composite path $ab\bar{a}\bar{b}$, that is $j_{1*}(c) = aba^{-1}b^{-1}$.

At this stage, it pays to note that this neat expression for $j_{1*}(c)$ occurs only if $r(x_0)$ is the starting vertex of the boundary surface symbol. The structure of the group π_1(torus) does not depend on the choice of x_0, so x_0 is for us to choose. Choosing this basepoint so that r sends $j_1(c)$ onto $ab\bar{a}\bar{b}$, rather than $\bar{a}bab$ or $\bar{b}ab\bar{a}$ or something more complex still, pays off in the simplicity of the resulting presentation. The moral of the story is 'When a calculation does not depend on a choice but cannot be made without that choice, choose well—it will pay off later'.

We can now finish the calculation. Van Kampen's theorem gives us a presentation

$$\pi_1(X) \cong (a, b : j_{1*}(c))$$

that is $(a, b : [a, b])$. This accords with our previous calculation since if $F(a,b)$ is the free group on generators a and b, what we have here is its Abelianization which will be the free Abelian group on two generators, that is $C_\infty \times C_\infty$. Alternatively by considering the presentation $(a, b : [a, b])$ as a presentation of an Abelian group, we write everything additively to get $(a, b : \varnothing)_{ab}$, that is $C_\infty \oplus C_\infty$ (where by tradition we replace the product, $\times$, by the direct sum, $\oplus$, when writing things additively).

A general surface (without boundary curves) is not very different. We leave the details to the reader, but will give a sketch. The surface can be represented by gluing a polygon using the corresponding standard surface symbol, A, which is a word in edges $a_1, \ldots, a_n$, for some n (Fig. 11.11).

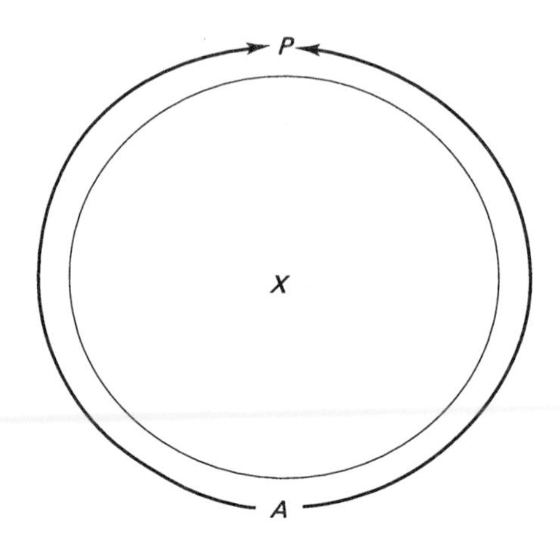

Fig. 11.11

Take U_1 = boundary of polygon (as a graph embedded in the surface) and $V_1 = X -$ centre. Choose a basepoint x_0 that projects radially onto the start vertex of A. Take $V_2 = X - U_1$. Repeating the same argument as before (details should be written down by you!), we get

$$\pi_1(X) \cong (a_1, \ldots, a_n : A).$$

Thus if X is an orientable surface, T_g, of genus g (with no boundary) then $n = 2g$ and $A = a_1 b_1 a_1^{-1} b_1^{-1} \ldots a_g b_g a_g^{-1} b_g$, so

$$\pi_1(T_g) \cong (a_1, \ldots, a_g, b_1, \ldots, b_g : [a_1, b_1] \ldots [a_g, b_g]).$$

If X is a non-orientable surface, S_g, of genus g (again with no boundary curves) then $n = g$ and $A = a_1^2 a_2^2, \ldots, a_g^2$, so

$$\pi_1(S_g) \cong (a_1, \ldots, a_g : a_1^2 \ldots a_g^2).$$

In particular, for $g = 1$, $S_1 = \mathbb{P}^2$, the projective plane, and $\pi_1(\mathbb{P}^2) \cong (a_1 : a_1^2) \cong C_2$, the cyclic group of order 2. Thus if a surface has genus 1 then its fundamental group is Abelian.

If we allow boundary curves and thus cuffs, the picture is slightly more complex. We will illustrate this with a simple example showing how to adapt the previous argument to this case.

Example: torus with two cuffs

The surface symbol is, of course, $A = aba^{-1}b^{-1}c_1 d_1 c_1^{-1} c_2 d_2 c_2^{-1}$. Everything proceeds as before until we identify $\pi_1(V_1) \cong \pi_1(U_1)$ where U_1 is the embedded graph coming from the boundary of the polygon. Instead of being simply an n-leafed rose for some n, U_1 is a slightly more complex graph (Fig. 11.12).

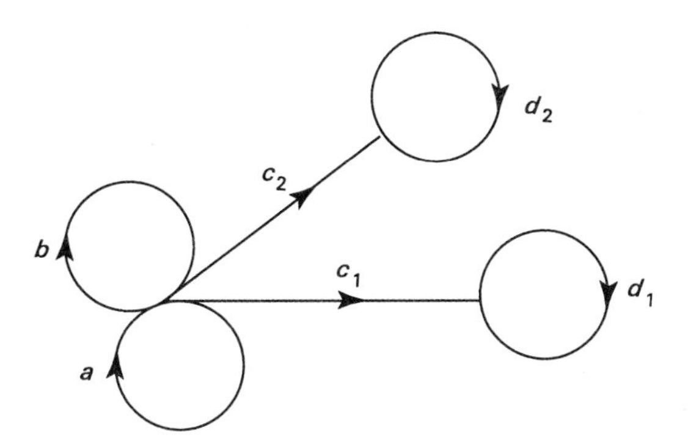

Fig. 11.12

Thus, although of course still free, $\pi_1(V_1)$ is not free on the set of edges of the polygon. We will write $d_i' = c_i d_i c_i^{-1}$, $i = 1, 2$, for the generator of $\pi_1(V_1)$ corresponding to the two cuffs. The calculation now continues more or less as before, giving us

$$\pi_1(\text{torus with two cuffs}) \cong (a, b, d_1', d_2' : [a, b]\, d_1' d_2')$$

which is a free group of rank 3, since the relation allows us to write d_2' in terms of the other generators. (Which Tietze transformations are we using here?)

We will denote by $T_{g,r}$ (resp. $S_{g,r}$) an orientable (resp. non-orientable) surface of genus g having r cuffs.

Exercises 11.1

1. Prove that if $\pi_1(T_g)$ or $\pi_1(S_g)$ is Abelian, then $g = 1$.
2. Show that $\pi_1(T_{g,r})$ is a free group on $2g + r - 1$ generators while $\pi_1(S_{g,r})$ is free on $g + r - 1$ generators.
3. Write out the proof that $\pi_1(\mathbb{P}^2)$ is cyclic of order 2. (Although this may seem a trivial exercise, it is important to gain an idea of why this 2-torsion is there as later we will be examining how to construct a space whose fundamental group is a cyclic group of given order n.)

11.2 Surface groups

We are now in a position to ask, and better still to answer completely, the obvious question: can π_1 distinguish between different surfaces? In other words, if X and Y are surfaces, and $\pi_1(X)$ and $\pi_1(Y)$ are known to be isomorphic, does this mean that X and Y are homeomorphic? If we allow surfaces with boundary the answer is clearly 'no', since both $T_{g,r}$ and $S_{2g,r}$ have fundamental groups free of rank $2g + r - 1$, yet we know that orientability is an invariant for surfaces, so $T_{g,r}$ and $S_{2g,r}$ are not homeomorphic. This suggests that we ask the question in a stronger form by requiring the surfaces to be without boundary curves. To attack this we will launch into an analysis of the *surface groups* $\pi_1(T_g)$ and $\pi_1(S_g)$.

We first look at their Abelianizations. Recall from Chapter 6 that if a group G is given by a presentation $(X : R)$, the Abelianization G_{ab} of G has a presentation $(X : R \cup \{[x, y]: x, y \in X\})$.

Examples

1. $\pi_1(T_g)_{ab} \cong (a_1, b_1, \ldots, a_g, b_g : [a_1, b_1] \ldots [a_g, b_g], [a_1, a_2], \ldots, [a_g, b_g])$.
   Thus the product of commutators $[a_1, b_1] \ldots [a_g, b_g]$ is already a consequence of the other relators and so may be deleted. (You should check the justification of this using Tietze transformations.) We have, therefore, that $\pi_1(T_g)_{ab}$ is a free Abelian group of rank $2g$.

2. $\pi_1(S_g)_{ab} \cong (a_1, \ldots, a_g : a_1^2 \ldots a_g^2, \{[a_i, a_j]: i, j = 1, 2, \ldots, g\})$.

Here it is simpler to use the alternative method of handling Abelianizations of presentations, namely by writing everything additively. This gives

$$\pi_1(S_g)_{ab} \cong (a_1, \ldots, a_g : 2a_1 + \ldots + 2a_g = 0).$$

You can now choose to attack this presentation formally with the methods from the theory of finitely generated Abelian groups or you can use additive Tietze transformations as follows.

Introduce a new generator b and set it equal to $a_1 + \ldots + a_g$; then substitute and rearrange to get

$$(a_1, \ldots, a_g, b : 2b = 0, a_g = b - a_1 - \ldots - a_{g-1}).$$

Finally, eliminating a_g gives an Abelian group of torsion free rank $g - 1$ and a single generator of order 2, that is

$$\pi_1(S_g)_{ab} \cong \underbrace{C_\infty \oplus \ldots \oplus C_\infty}_{g-1\text{-copies}} \oplus C_2.$$

Thus given a surface X, we can tell if X is orientable or not by looking for an element of order 2 in $\pi_1(X)$. Once we have determined orientability or lack of it, we can find out which surface X is from the genus which can be read off from our calculations of the torsion free rank of $\pi_1(X)_{ab}$.

Hopefully your curiosity has been aroused and you will have asked whether π_1 can detect cuffs; that is, if we know X is a surface and have some presentation (not necessarily the nice one we have given above) of $\pi_1(X)$, can we tell if X has cuffs or not? If $\pi_1(X)_{ab}$ has an element of order 2, then X must have been a non-orientable surface without cuffs, but suppose $\pi_1(X)_{ab}$ is free Abelian of rank n, n even; we could have obtained this answer from a T_g with $n = 2g$ or from a $T_{2g,r}$ with $n = 2g + r - 1$ or from an $S_{g,r}$ with $n = g + r - 1$. Thus our problem is to decide whether $\pi_1(T_g)$ is free or not. At present we only have information about its Abelianization. Might not some cunning subtle sequence of Tietze transformations applied to $(a_1, b_1, \ldots, a_g, b_g : [a_1, b_1] \ldots [a_g, b_g])$ produce a way of eliminating the relator, thus giving us a free group? Perhaps it seems unlikely, but can we be sure? There are several different methods we could use, but the easiest one for us is to use elementary ideals and other ideas from Chapter 8.

Let $r = a_1b_1a_1^{-1}b_1^{-1} \ldots a_gb_ga_g^{-1}b_g^{-1}$. We need to calculate $\partial r/\partial a_k$ and $\partial r/\partial b_k$ for each k. This gives

$$\frac{\partial r}{\partial a_k} = [a_1, b_1] \ldots [a_{k-1}, b_{k-1}](1 - a_kb_ka_k^{-1})$$

and

$$\frac{\partial r}{\partial b_k} = [a_1, b_1] \ldots [a_{k-1}, b_{k-1}]a_k(1 - b_ka_kb_k^{-1}).$$

The Abelianization of $G = \pi_1(T_g)$ was earlier seen to be free of rank $2g$. To avoid problems with the multiple use of symbols, we set s_i and t_i to be the images in $(C_\infty)^{2g}$ of a_i and b_i respectively with γ: $F(a_1, b_1, \ldots, a_g, b_g) \to G$ and $\mathcal{A}$: $G \to G_{ab}$, as in Chapter 8, the homeomorphism obtained by 'killing off' r and Abelianizing G respectively.

The images of the differentials are

$$\mathcal{A}\gamma\left(\frac{\partial r}{\partial a_k}\right) = 1 - t_k$$

and

$$\mathcal{A}\gamma\left(\frac{\partial r}{\partial a_k}\right) = s_k - 1$$

for $k = 1, \ldots, g$. This gives the following sequence of elementary ideals in $\mathbb{Z}(C_\infty)^{2g}$:

$$E_i = 0 \quad \text{for} \quad i < 2g - 1,$$

$$E_{2g-1} = (\{s_i - 1, 1 - t_i : i = 1, \ldots, g\}),$$

the ideal generated by the images of the differentials, and

$$E_{2g} = \mathbb{Z}(C_\infty)^{2g}.$$

If G had been a free group of rank $2g$ (and given our knowledge of its Abelianization, no other rank would be possible), then the elementary ideals would be

$$E_i = 0, \quad i < 2g,$$
$$E_{2g} = \mathbb{Z}(C_\infty)^{2g}.$$

Thus, as seemed likely, $\pi_1(T_g)$ is not free and its lack of freeness *is* detectable using the machinery we have at our disposal.

11.3 Spaces with finite cyclic fundamental group

We saw earlier that

$$\pi_1(\mathbb{P}^2) \cong C_2.$$

Can we find a space X which has $\pi_1(X) \cong C_3$? Clearly X cannot be a surface since our calculations showed that $\mathbb{P}^2$ was the only surface with non-trivial finite fundamental group.

A means of attacking this question is to rerun rapidly the ideas that lead to the 2-torsion in $\pi_1(\mathbb{P}^2)$. (If you have not written out this calculation in detail yet, at least write it out in 'sketch' form now.) The $a^2 = 1$ comes from the image of the loop in the annulus $V_1 \cap V_2$ when projected onto U_1, the boundary of the polygon:

$$\pi_1(V_1 \cap V_2) \to \pi_1(V_1) \cong \pi_1(U_1).$$

This suggests that we take a triangle and identify the three sides as if we had a symbol aaa (Fig. 11.13).

Now rerun the proof that $\pi_1(\mathbb{P}^2) \cong (a : a^2)$, but with $\mathbb{P}^2$ replaced by X and of course with the obvious modifications.

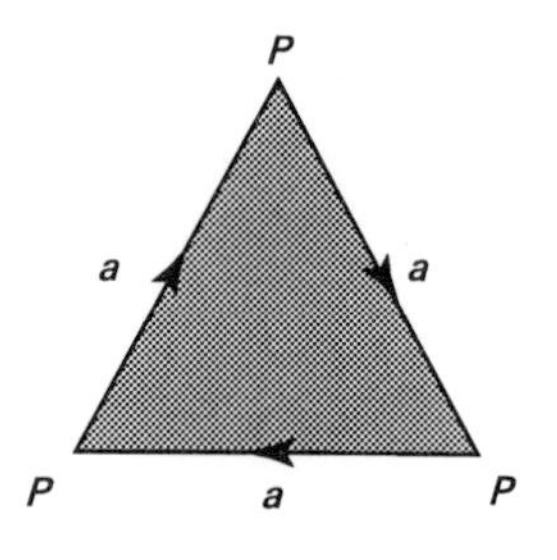

Fig. 11.13

No prizes will be given for instructions on how to construct a space with π_1 a cyclic group of order 4. Here they are (Fig. 11.14).

There are no obstacles to using a similar construction to get a space $X = X(C_n)$ with $\pi_1(X) \cong C_n$, a cyclic group of order n. Try to write down the details with a proof, using the Van Kampen theorem, that your construction does in fact give what is required.

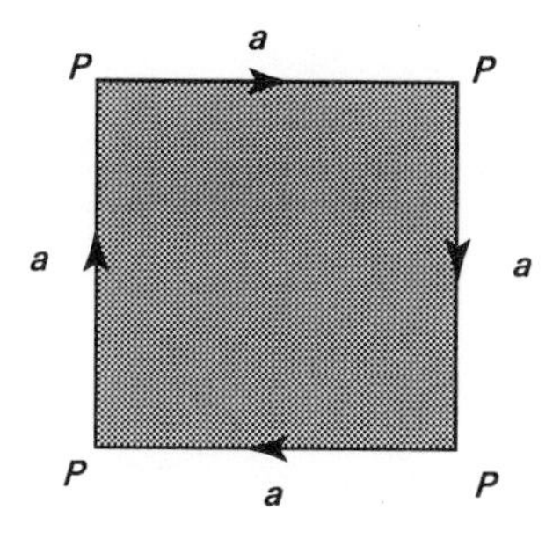

Fig. 11.14

11.4 Spaces with given fundamental group

The previous set of examples raises an interesting problem. Given a group, G, can you construct a pointed space X with $\pi_1(X) \cong G$?

To gain some insight as to how we might do this, at least for groups G with a finite presentation $(x_1, \ldots, x_n : r_1, \ldots, r_s)$, we will take another look at the construction of the space with $\pi_1(X) \cong C_3$ that we examined in the previous section. We can think of this construction as starting with a one-leafed rose into which we glue a disc D^2 via a suitable map f defined on its boundary circle, S^1. Recall that this is done by taking $U_1 \subseteq D^2$ and then gluing, using an equivalence relation, $\sim$, defined by: if $x \in S^1 \subset D^2$, $x \sim f(x)$. To get X, we use a map f that sends a generating loop $c \in \pi_1(S^1)$ into $a^3 \in \pi_1(U_1)$.

In general, if we have a space Y and an element $r \in \pi_1(Y)$, then we can choose a closed path $\alpha = \alpha_r : S^1 \to Y$ so that $[\alpha] = r$ and can use α to glue a disc onto Y to form a new space $X = Y \cup D^2$. One can show, using the Van Kampen theorem as before, that if $\pi_1(Y) \cong (x_1, \ldots, x_n : r_1, \ldots, r_s)$ then $\pi_1(X)$ has a presentation $\pi_1(X) \cong (x_1, \ldots, x_n : r_1, \ldots, r_s, r)$ in which r has been added to the relations in the presentation. Using an inductive construction it is now easy to construct a space having any (finitely presented) group G as its fundamental group. (The problems of handling non-finitely presented groups are merely ones of technique; the same method works but the justification that it works is technically slightly more demanding as it uses a more powerful form of the Van Kampen theorem than we have introduced above.)

As an example, we will see how to construct a space X having $\pi_1(X) \cong D_3$, the group of symmetries of a triangle. This group has a nice presentation, namely

$$D_3 \cong (a, b : a^3 = b^2 = (ab)^2 = 1).$$

We set $r_1 = a^3$, $r_2 = b^2$, $r_3 = (ab)^2$. Let Γ be a two-leafed rose (Fig. 11.15).

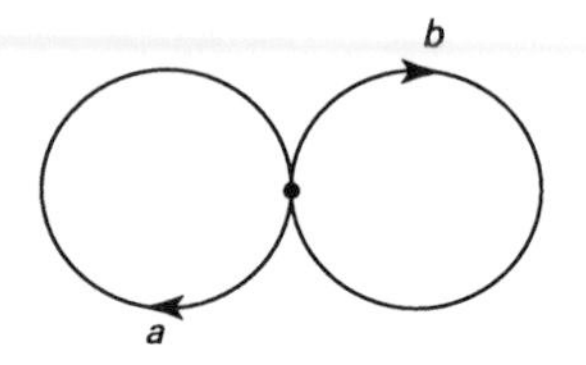

Fig. 11.15

The relator $a^3 \in \pi_1(\Gamma)$ has, as a representing loop, a map α_1 from S^1 into the left-hand circle of Γ. Gluing in a disc via this map gives a space Y_1 with $\pi_1(Y_1) \cong (a, b : a^3 = 1)$ (Fig. 11.16).

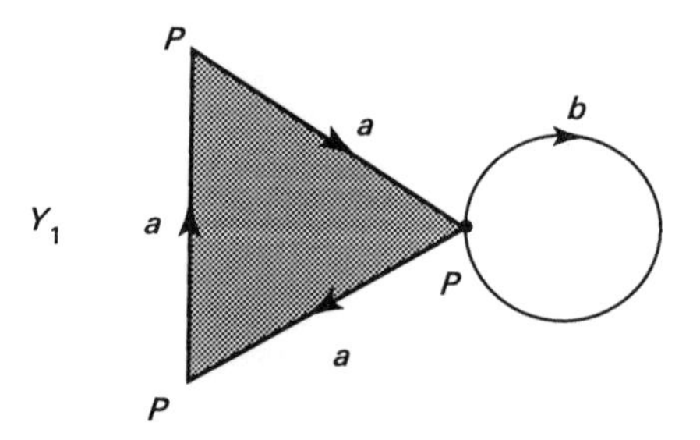

Fig. 11.16

The element b^2 in this group is represented by a loop α_2 going around the right-hand loop twice.

We use α_2 to glue in a disc to get Y_2 (Fig. 11.17).

(If we write as before $X(C_3)$ for the space we constructed earlier having $\pi_1(X(C_3)) \cong C_3$, then $Y_2 \simeq X(C_3) \vee \mathbb{P}^2$). The fundamental group of $\pi_1(Y_2)$ is $C_3 * C_2 \cong (a, b : a^3 = b^2 = 1)$. The element $r_3 = abab$ is represented by a loop α_3 going along a then along b then along a and again along b. The space $X = Y_2\ D^2$ then has $\pi_1(X) \cong D_3$. (At this point the artistic powers of the authors come up against insurmount-

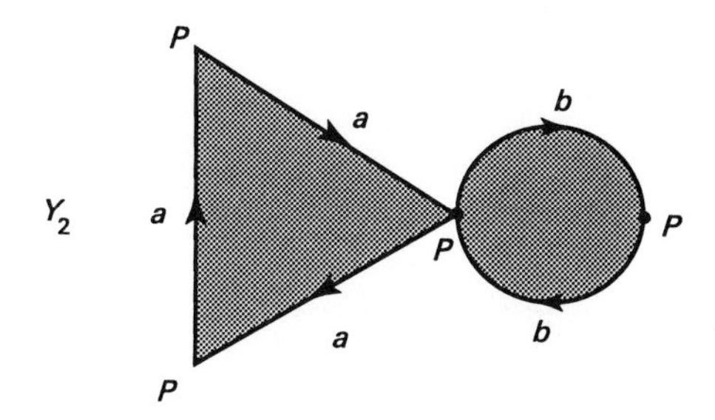

Fig. 11.17

able difficulties, so we have not attempted to draw this space even schematically.)

11.5 The group of a knot

In Chapter 6 we introduced the group of a knot and gave two different ways of obtaining a presentation of that group from a diagram of the knot. As there, we will write $G(K)$ for the group that results from a diagram of the knot, K. This notation presupposes that $G(K)$ is an invariant of the knot and not just something concocted from *one* particular diagram. We saw, however, that the Reidemeister moves left the group unchanged even though they changed the presentation, so the notation *is* justified. We also saw that the Alexander polynomial of K could be obtained from the Wirtinger presentation of $G(K)$ using the Fox derivatives and a sort of Jacobian matrix, the Alexander matrix of the presentation. So what is the geometric interpretation of this mysterious group $G(K)$? Why is it invariant? Can we see this directly? In this section we prove that $G(K)$ is, in fact, the fundamental group of the complement of K, that is $\pi_1(\mathbb{R}^3 - K)$. Moreover, the Wirtinger presentation comes from a fairly simple decomposition of $\mathbb{R}^3 - K$ after one application of the Van Kampen theorem. The decomposition will need some care in construction, but this will help later on when we come to analyse the various fundamental groups involved and the homomorphisms between them.

Suppose we are given a knot K which we think of, as usual, as being a subset of $\mathbb{R}^3$. Orient the knot and label each arc of the knot in the usual way. We shall use $\alpha_1, \alpha_2, \ldots$ as labels following on around the knot with the orientation. Thus a typical crossing would look like that shown in Fig. 11.18 with, for some n, α_{n+1} being α_1.

We next deform the knot to replace each such crossing by one which looks as shown in Fig. 11.19, with most of the knot within the

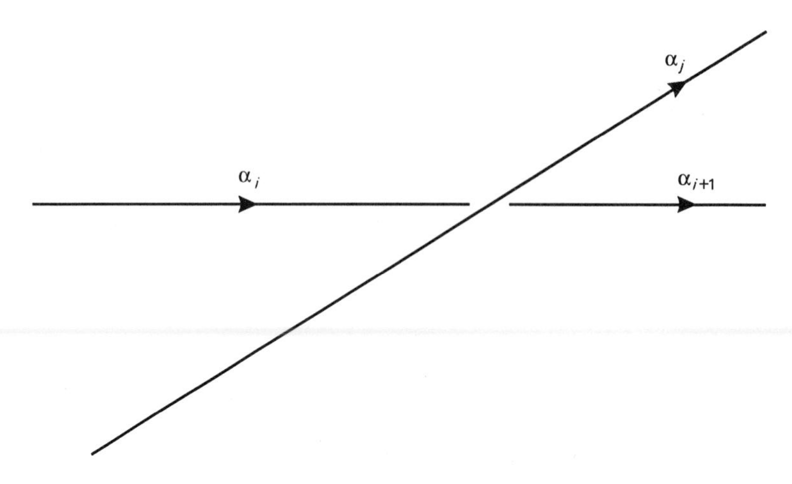

Fig. 11.18

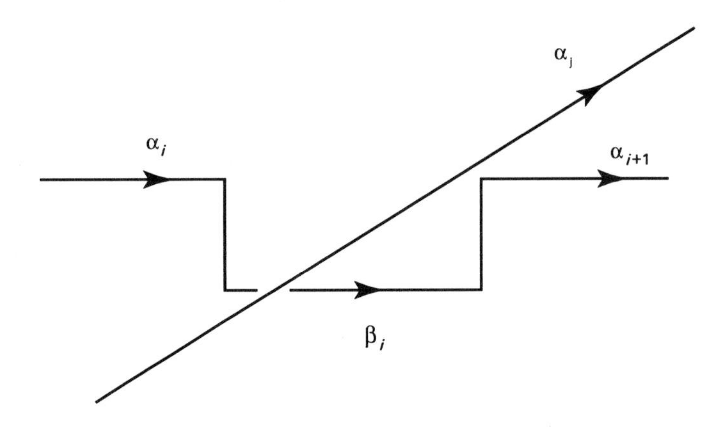

Fig. 11.19

$z = 1$ plane, but at each crossing we drop a vertical to the $z = 0$ plane followed by a segment labelled β_j in that plane from the foot of the vertical line ending α_i to the foot of that starting α_{i+1}.

Next we remove from $\mathbb{R}^3$ a 'tunnel' neighbourhood N of K. This is best thought of as follows.

Imagine that we have a closed cube of side ε fixed to the knot but able to slide along it. When it comes to a crossing section as above, it slides down the first vertical with its faces parallel to the axes in $\mathbb{R}^3$ and then it goes along β_j and back up again to the $z = 1$ part of α_{i+1}. By taking ε small enough we can ensure that $\mathbb{R}^3 - N$ is a

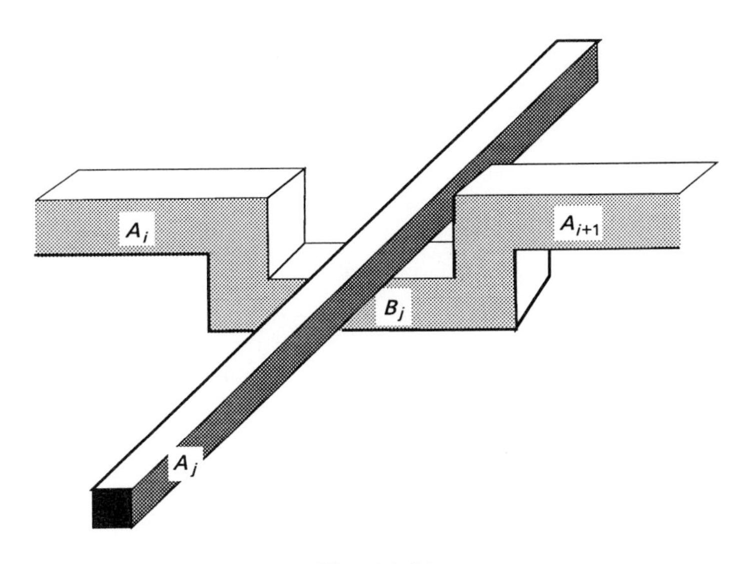

Fig. 11.20

deformation retract of $\mathbb{R}^3 - K$ so that $\pi_1(\mathbb{R}^3 - N)$ is isomorphic to $\pi_1(\mathbb{R}^3 - K)$ which we want to calculate.

That part of the tunnel containing α_i will be labelled A_i whilst that containing β_j will be called B_j (Fig. 11.20).

We are now ready to describe a decomposition of $\mathbb{R}^3 - N$ as the union of two open sets. We take

$$V_1 = \{z > 0\} - N,$$
$$V_2 = \{z < \varepsilon/2\} - N.$$

Now V_1 is half-space with some tube-shape tunnels removed. It has as a deformation retract an n-leafed rose with loops $a_1, \ldots, a_n$ where

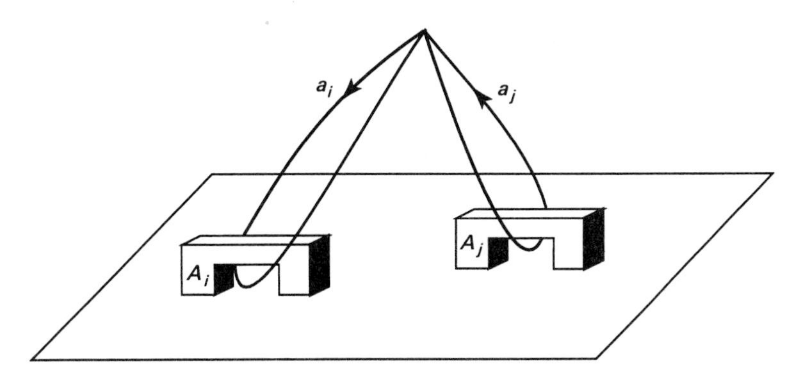

Fig. 11.21

a_i is a loop passing under the tunnel A_i that contains α_i (Fig. 11.21). Hence $\pi_1(V_1) = (a_1, \ldots, a_n : \varnothing)$.

On the other hand V_1 is an open half-space with trenches cut in it (Fig. 11.22). It is therefore contractible and $\pi_1(V_2) = \langle \varnothing : \varnothing \rangle$.

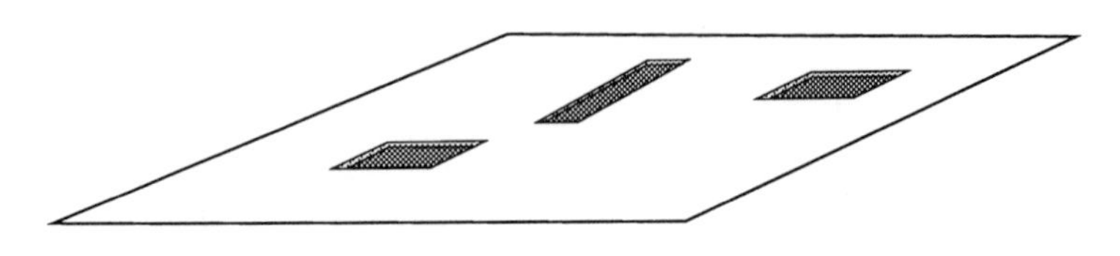

Fig. 11.22

Finally $V_1 \cap V_2 = \{0 < z < \varepsilon/2\} - N$ is an infinite plate with m holes in it, where m is the number of crossings (Fig. 11.23).

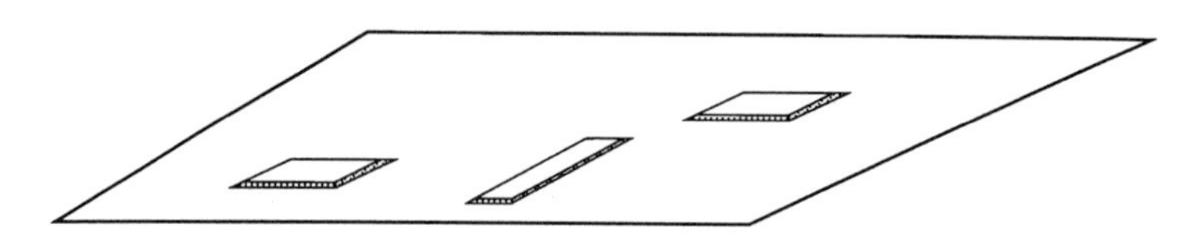

Fig. 11.23

Thus $\pi_1(V_1 \cap V_2)$ = free group of rank m. A typical generator will correspond to a circuit around a hole and so its image in $\pi_1(V_1)$ has the form $a_j a_{i+1}^{-1} a_j^{-1} a_i$ if the crossing is as shown in Fig. 11.24, or $a_j^{-1} a_{i+1}^{-1} a_j a_i$ if the crossing is as shown in Fig. 11.25.

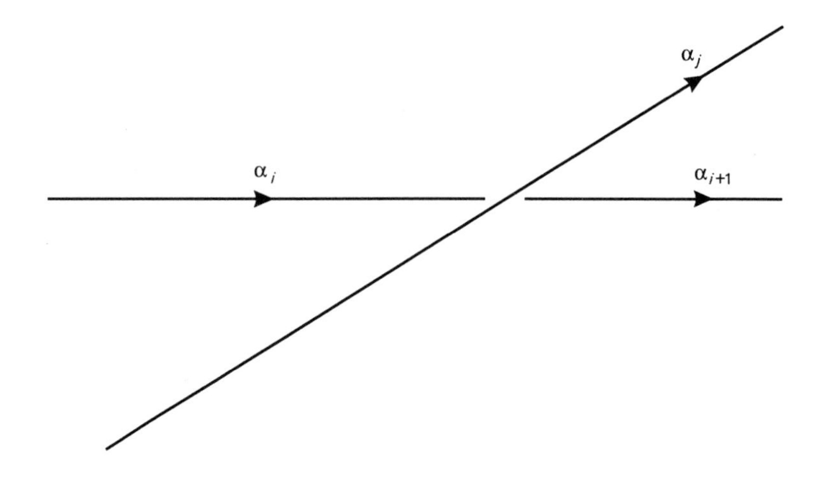

Fig. 11.24

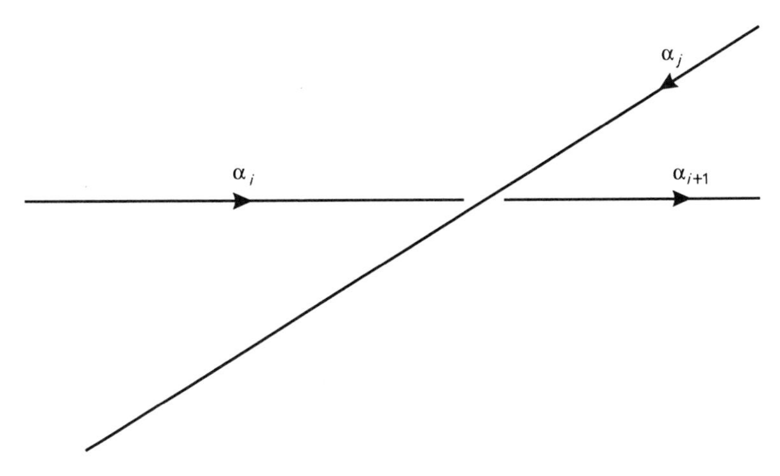

Fig. 11.25

It is a good idea to try to draw diagrams to illustrate this—a diagram drawn by you will be more use to you than one drawn by us!

Now the Van Kampen theorem can be applied and we find a presentation

$$(a_1, \ldots, a_n : r_1, \ldots, r_m)$$

where a typical r_k is as above. This is precisely the Wirtinger presentation of the group $G(K)$.

We can now see geometrically (or perhaps globally might be a better term) why $G(K)$ is an invariant of the knot K. Suppose K and L are equivalent knots. Recall that this means there is a homeomorphism φ: $\mathbb{R}^3 \to \mathbb{R}^3$ such that $\varphi K = L$. This homeomorphism therefore restricts to one from $\mathbb{R}^3 - K$ to $\mathbb{R}^3 - L$. This induces an isomorphism between $\pi_1(\mathbb{R}^3 - K)$ $(= G(K))$ and $\pi_1(\mathbb{R}^3 - L)$ $(= G(L))$ as required.

The above argument also gets around the possible objection that we picked a very special representative of the knot so as to be able to use Van Kampen's theorem. If we had wanted to we could have picked a homeomorphic image of the open decomposition $\{V_1, V_2\}$ of the complement of our original knot and could have applied exactly the same argument to obtain π_1 of that complement. Of course the description would have been a lot harder.

11.6 Torus knots

A knot K is said to be a torus knot if it can be embedded on the surface of a torus, that is we have K: $S^1 \to \mathbb{R}^3$ factors via some torus T_1 sitting in $\mathbb{R}^3$:

$$S^1 \xrightarrow{K} T_1 \hookrightarrow \mathbb{R}^3.$$

Since $\pi_1(S^1) \cong C_\infty$, and $\pi_1(T_1) \cong C_\infty \oplus C_\infty$ a torus knot K determines a pair of integers (m, n) such that writing $\pi_1(S^1) = (a : \varnothing)$ and $\pi_1(T_1) = (b, c : \varnothing)$ we have $K_*(a) = (b^m, c^n)$. The knot K thus goes around one of the two generators m times and around the other n times. As an example, it is easy to see that the trefoil is a (2,3)-torus knot as we have already mentioned (Fig. 11.26).

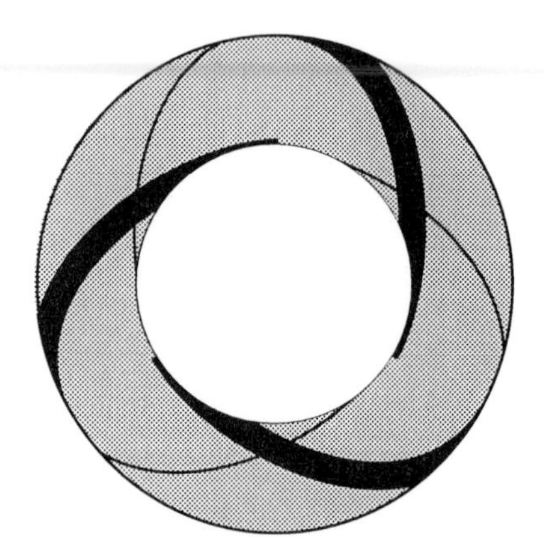

Fig. 11.26

As K is a knot one finds that m and n can have no common factor, that is they are coprime. In this case it is simple to realize K in a nice, regular way.

Consider a solid cylinder C with m line segments on its curved face equally spaced and parallel to the axis. If the ends of C are identified after a twist of $2\pi(n/m)$ then we obtain a single curve $t_{m,n}$ on the surface of a solid torus T. This $t_{m,n}$ is an (m, n)-torus knot.

We introduced these informally earlier and saw that the cinquefoil or pentoil was a (2, 5)-torus knot. In fact we examined many of the invariants of these $(2, n)$-torus knots as they formed the family able to be drawn in the form shown in Fig. 11.27.

When we calculated the knot group of these we found

$$G(t_{2,n}) \cong (a, b : a^2 = b^n).$$

This is typical of torus knots. In fact

$$G(t_{m,n}) \cong (a, b : a^m = b^n)$$

as we will see.

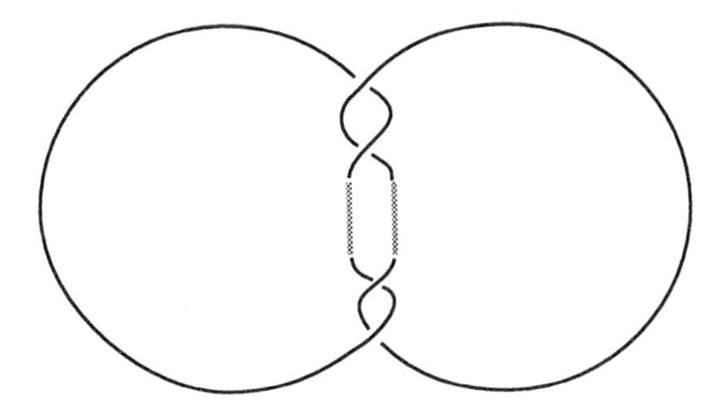

Fig. 11.27

The advantage of torus knots (and similar things hold for various others of the families of knots, e.g. pretzel knots) is that the embeddability of these knots on a torus allows a direct application of Van Kampen's theorem to $\mathbb{R}^3 - t_{m,n}$. We consider $t_{m,n}$ embedding as above on the surface of a *solid* torus T sitting in $\mathbb{R}^3$. Picking a small ε, we carve out an ε-neighbourhood N of $t_{m,n}$ in $\mathbb{R}^3$. Of course $\pi_1(\mathbb{R}^3 - N) \cong \pi_1(\mathbb{R}^3 - t_{m,n})$. The complement of N in the surface of T is an annulus, $L_{m,n}$, which results from the m parallel strips on the cylinder being joined after a twist of $2\pi(n/m)$ (Fig. 11.28).

Thus $\pi_1(L_{m,n})$ is infinite cyclic, a typical generator being the class of the centre line $\ell_{m,n}$ of $L_{m,n}$.

The space $T - N$ is a solid torus with a groove gouged in its surface and so it deforms onto a central circle (marked a in Fig. 11.28), thus $\pi_1(T - N) \cong (a : \varnothing)$.

Similarly $(\mathbb{R}^3 - T) - N$ is infinite cyclic generated by a loop b through the 'hole' in T. (In fact $\mathbb{R}^3 - T$ is homeomorphic to a solid torus with one point removed—try to see why.)

Since $\ell_{m,n}$ makes m circuits of T, the image of $\ell_{m,n}$ in $\pi_1(T - N)$ is a^m. Similarly its image in $\pi_1(\mathbb{R}^3 - T) - N)$ is b^n.

We must now adjust our decomposition to be open so as to allow ourselves to apply the Van Kampen theorem. We expand $T - N$ and $(\mathbb{R}^3 - T) - N$ slightly to get open sets V_1, V_2 with $V_1 \cap V_2$ a neighbourhood of $L_{m,n}$ V_1, deforming to $T - N$, V_2 deforming to $(\mathbb{R}^3 - T) - N$, etc. Applying Van Kampen's theorem then gives a knot group, $G(t_{m,n}) \cong \pi_1(\mathbb{R}^3 - t_{m,n}) \cong (a, b : a^m = b^{n)}$, as promised.

These groups contain much of the information about the knot. As with the surface groups, a group theoretic analysis is worth doing. We leave it to you to calculate Alexander polynomials and will instead use other methods in our analysis. (The Alexander polynomials are difficult to work with except when the smaller of m and n is 2. It is, however, easy to work out what they are.)

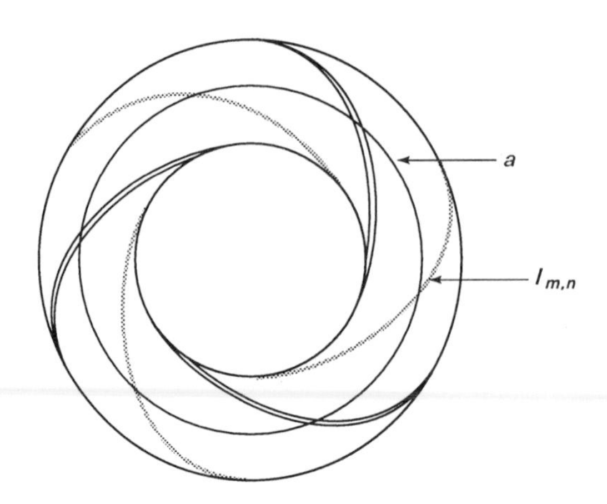

Fig. 11.28

11.7 Analysis of the knot group $G_{m,n}$

(This section uses slightly more group theory than the rest and can be omitted.) We will write $G_{m,n}$ for $G(t_{m,n})$.

We start by looking at the centre of $G_{m,n}$. Recall that if G is a group, its centre $Z(G)$ is the subgroup formed from the elements that commute with everything, that is

$$Z(G) = \{c \in G : cg = gc \quad \text{for all} \quad g \in G\}.$$

Now since $a^m = b^n$, a^m commutes with both a and b, so $a^m \in Z(G_{m,n})$ and of course, for any k, $a^{km} \in Z(G_{m,n})$ as a consequence. Let us write $N = \{a^{km} : k \in \mathbb{Z}\}$ for the subgroup generated by a^m. This subgroup N is thus in $Z(G_{m,n})$ and so is normal, and the corresponding quotient group $G_{m,n}/N$ clearly has presentation

$$G_{m,n}N = (a, b : a^m = b^n, a^m = 1)$$

and so is isomorphic to $C_m * C_n$, the free product of two cyclic groups. Taking care over the notation we will write $\bar{a} = Na$, $\bar{b} = Nb$ for the elements of $G_{m,n}/N$.

At this point, we should explain the strategy of our attack, and why we are looking at this quotient. If we are trying to write down $Z(G)$ for some group G, and we can find a non-trivial normal subgroup N in $Z(G)$, any element x in $Z(G)$ which is not in N will give a non-trivial element $\bar{x} = Nx$ in $Z(G/N)$. So we can hope to find out more about $Z(G)$ by examining $Z(G/N)$. If this is trivial then $Z(G) = N$; if it is non-trivial, we can try to repeat the process. The group G/N is in some sense smaller and simpler (even where G and G/N are infinite), so at each stage we have a better chance of obtaining complete information on the centre of the resulting quotient group.

In our case there are no problems since $Z(G_{m,n}/N)$ is trivial, so $Z(G_{m,n}) = N$. This statement seems intuitively reasonable; in the presentation of $C_m * C_n$ the generators $\bar{a}$ and $\bar{b}$ are controlled only by relations $\bar{a}^m = 1$, $\bar{b}^n = 1$, each of which involves only one generator; we have no interaction between $\bar{a}$ and $\bar{b}$, so it seems unlikely that there are any elements in $C_m * C_n$ which are central. This is not sufficient to be a proof, however, so we will have to probe a bit deeper.

In $C_m * C_n$ any element has a unique reduced form

$$\bar{a}^{x_1}\bar{b}^{y_1} \ldots \bar{a}^{x_p}\bar{b}^{y_p}$$

where

$$0 \leqslant x_1 < m,\ 0 < x_i < m \quad \text{for} \quad i > 1$$

$$0 \leqslant y_p < n,\ 0 < y_i < n \quad \text{for} \quad i < p$$

and all $x_i, y_j \in \mathbb{Z}$. Such an expression commutes with $\bar{a}$ only if it has $\bar{a}$ at both ends and commutes with $\bar{b}$ only if it has $\bar{b}$ at both ends! Hence a non-trivial element cannot commute with both $\bar{a}$ and $\bar{b}$. This completes the proof that $Z(G_{m,n}) = \langle a^m \rangle$, the subgroup generated by a^m.

We next look at elements of finite order in this group, $C_m * C_n$. Suppose $g \in C_m * C_n$ $(\cong G_{m,n}/Z(G_{m,n}))$ has normal form $\bar{a}^{x_1} \ldots \bar{b}^{y_p}$ and that it has order $r < \infty$. The word (normal form of $g)^r$ cannot be shortened unless the normal form starts and ends with the same letter, that is we must have one, but not both, of x_1 and y_p equal to zero.

If $p = 1$ this must mean that g is a power of $\bar{a}$ or $\bar{b}$. Any conjugate of a power of $\bar{a}$ or $\bar{b}$ will also have finite order. Are these all of the elements of finite order? We will try an induction on p. We have already seen that for $p = 1$ all elements of finite order are powers of $\bar{a}$ or $\bar{b}$. If p is arbitrary, assume that any g of finite order having length less than p is a conjugate of a power of $\bar{a}$ or $\bar{b}$. If $y_p = 0$, we have

$$\bar{a}^{x_p} g \bar{a}^{-x_p} = \bar{a}^{x_p + x_1} \ldots \bar{b}^{y_{p-1}} \qquad (*)$$

is of finite order and has length less than p. By the induction hypothesis this element is a conjugate of a power of $\bar{a}$ or $\bar{b}$ but then g will also be such a conjugate. If $x_1 = 0$ a similar argument applies, so we have that any element of finite order in $C_m * C_n$ is conjugate to a power of $\bar{a}$ or to a power of $\bar{b}$. In particular the elements of finite order in $G_{m,n}/Z(G_{m,n})$ cannot have order greater than $\max(m,n)$.

If we make $C_m * C_n$ Abelian, we get $C_m \oplus C_n$ which is of order mn. As $\min(m,n) = mn/\max(m, n)$ we can theoretically determine (m, n), up to order, if we are given $G_{m,n}$. Thus if $\{m, n\}$ with $m < n$ and (m', n') with $m' < n'$ are given and are distinct then $G_{m,n} \not\cong G_{m',n'}$ and $t_{m,n}$ and $t_{m',n'}$ are distinct knot types.

By combining geometric arguments with this sort of analysis, these results can be pushed further. We will not do this here but will note that Neuwirth (1961) proves the following pretty results on torus knots:

1. If K is an alternating knot and $G(K)$ has a non-trivial centre, then K is equivalent to $t_{2,2r+1}$ for some r.
2. The torus knots $t_{p,q}$ with $p > 2$ and $q > 2$ have no alternating projections.

(If you think a bit you will see that result 1 implies result 2, but result 1 itself is much harder to prove.)

Exercises 11.7

1. (a) A space Y is constructed from a torus by gluing in a disc along one of the 'vertical' sections as shown in Fig. 11.29. Use Van Kampen's theorem to calculate the fundamental group of Y.

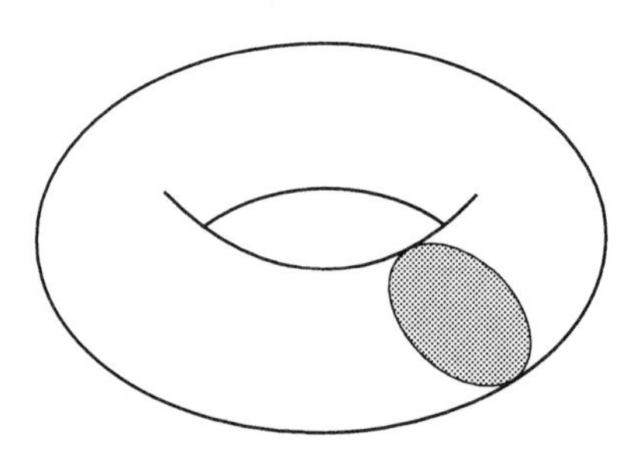

Fig. 11.29

(b) A space Z_g is constructed (rather as in (a)) by gluing a disc into a vertical circle in an orientable surface, T_g, of genus g having no boundary (thus $Y = Z_1$). Calculate $\pi_1(Z_g)$, $\pi_1(Z_g)_{ab}$, and the elementary ideals of $\pi_1(Z_g)$.

2. Find the fundamental group of $\mathbb{R}^3 - X$ when (i) X is the unlinked union of two circles, (ii) X is the linked union of two circles (Hopf link) as shown in Fig. 11.30.

Fig. 11.30 (i)

3. Let S_1 and S_2 be two surfaces and $S = S_1 \# S_2$ their connected sum. If A is a surface symbol for S_1 in edges $a_1, \ldots, a_m$ and B one for S_2 in edges $b_1, \ldots, b_n$, find a presentation for $\pi_1(S_1 \# S_2)$.

12
Covering spaces

The fundamental group $\pi_1(X, x_0)$ of a topological space X at a base-point x_0 describes the variety of possible closed paths in X at x_0. Does the algebraic structure of $\pi_1(X, x_0)$ reflect further features of the topological nature of X? In this chapter we shall see that the collection of subgroups of $\pi_1(X, x_0)$ corresponds to a collection of topological spaces that map to X: the *covering spaces* of X. A covering space preserves the *local* features of X but its overall topological structure is generally more simple. Questions about X can often be transferred into more tractable questions about a covering space. Our applications will concentrate on the covering spaces of graphs, which we use to prove two theorems about free groups: the Nielsen–Schreier theorem, which says that a subgroup of a free group is itself free; and the Schreier index formula for the number of free generators of a subgroup of a free group. Further, we show how to discover free generators for the relation subgroup of a group presentation from its Cayley quiver.

12.1 Covering maps and covering spaces

Let X be a topological space. In this chapter, all spaces are assumed to be path connected and locally path connected. A map $p: Y \to X$ is called a *covering map*, and Y a *covering space* of X, if for each point $x \in X$ there is a neighbourhood U of x such that $p^{-1}(U)$ is the disjoint union of open sets V_i of Y, so that p maps each V_i homeomorphically to U. Informally, each $x \in X$ has a neighbourhood U and Y contains a number of copies of U mapped to U by the covering map p (Fig. 12.1).

Examples

1. The identity map $X \to X$ is a covering map. More generally, if F is any space with the discrete topology, the projection map $F \times X \to X$ is a covering map. Given a neighbourhood U of $x \in X$, we have

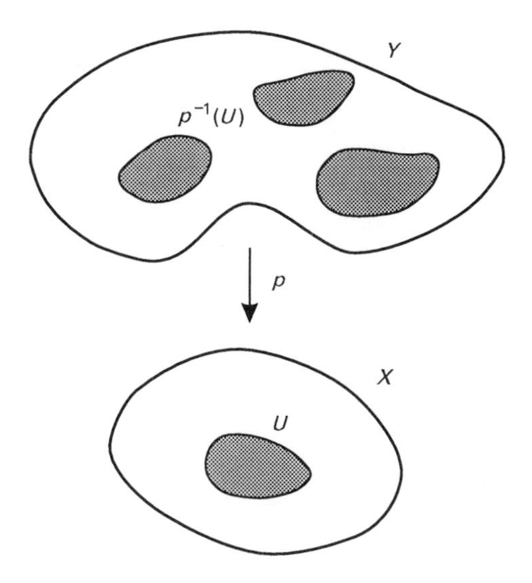

Fig. 12.1

$$p^{-1}(U) = \bigcup_{f \in F} \{f\} \times U.$$

2. The map $\mathbb{R} \to S^1$ given by $t \mapsto \exp(it)$ is a covering map. A useful picture of this map is to imagine the real line twisted into a spiral (Fig. 12.2), as we did in Chapter 9 for the computation of $\pi_1(S^1, 1)$.

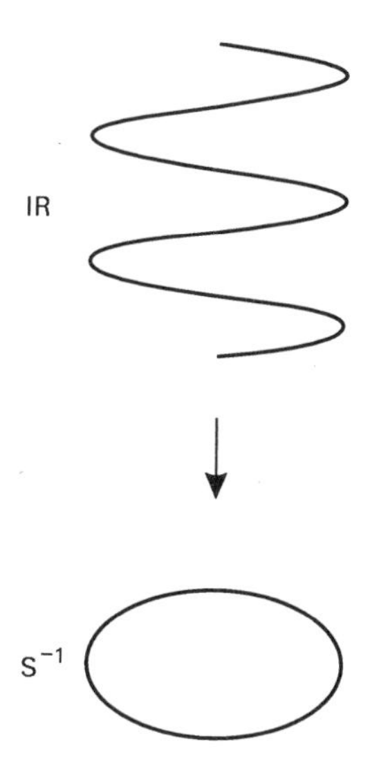

Fig. 12.2

3. The map $P: S^1 \to S^1$ given by $\exp(i\theta) \mapsto \exp(ni\theta)$ is a covering map. A small neighbourhood U of $x \in S^1$ has preimage $p^{-1}(U)$ consisting of n disjoint copies, illustrated in Fig. 12.3 here for $n = 3$.

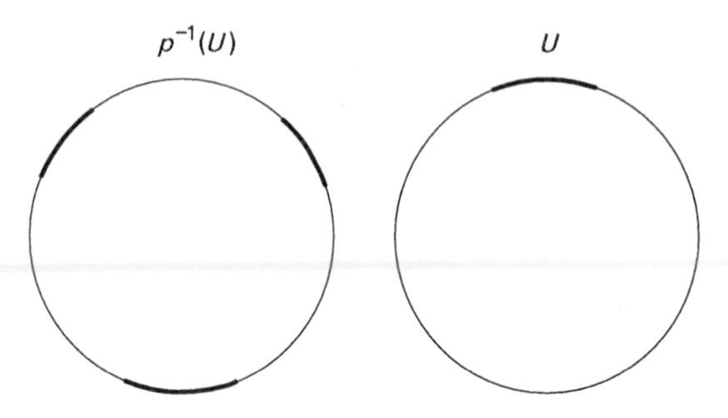

Fig. 12.3

4. $\mathbb{R}^2$ is a covering space of the torus. Divide $\mathbb{R}^2$ into squares, with the points of $\mathbb{Z} \times \mathbb{Z}$ at the corners, and map each square to the torus according to the usual gluing instructions.

5. Let X consist of two copies of S^1 joined at a single point. Then the Cayley quiver of the presentation $(x, y : \varnothing)$ of the free group with basis $\{x, y\}$ is a covering space of X. Edges coloured x are mapped around one copy of S^1, and edges coloured y around the other.

Covering spaces allow the solution of some *map-lifting* problems. Given maps $f: Z \to X$ and $p: Y \to X$, a *lift of f over p* is a map $\tilde{f}: X \to Y$ such that $p \circ \tilde{f} = f$. Lifts need not exist nor, if they exist, need they be unique. However, if $p: Y \to X$ is a covering map, many map-lifting problems admit unique solutions.

12.1 Uniqueness lemma. *Let $p: Y \to X$ be a covering map and Z a path connected space. Then two lifts of a map $f: Z \to X$ over p either are equal or differ at every point of Z.*

Proof. The details of this proof are suggested as exercises. Let us suppose that f has two lifts $\tilde{f}_1$ and $\tilde{f}_2$ over p. We want to examine the points at which $\tilde{f}_1$ and $\tilde{f}_2$ agree or differ, so let us set

$$Z_0 = \{z \in Z \mid \tilde{f}_1(z) = \tilde{f}_2(z)\}.$$

We shall show that Z_0 is both open and closed in Z.

Exercise

1. Why will this suffice to prove the uniqueness lemma? Remember that Z is path connected.

   Let $z \in Z \backslash Z_0$. Since $\tilde{f}_1(z) \neq \tilde{f}_2(z)$, but $p \circ \tilde{f}_1(z) = p \circ \tilde{f}_2(z)$, it follows that $\tilde{f}_1(z)$ and $\tilde{f}_2(z)$ lie in distinct components of $p^{-1}(U)$, where U is some neighbourhood of $f(z)$ in X.

Exercise

2. Use the continuity of f, $\tilde{f}_1$, and $\tilde{f}_2$ to deduce that there exists a neighbourhood of z contained in $Z \backslash Z_0$. Deduce that Z_0 is closed.

   Now suppose that $z \in Z_0 \cap f^{-1}(U)$; then $\tilde{f}_1(z)$ and $\tilde{f}_2(z)$ lie in the same component of $f^{-1}(U)$.

Exercise

3. Show that z has a neighbourhood contained in $Z_0 \cap f^{-1}(U)$ and deduce that Z_0 is open. ■

12.2 Path-lifting lemma. *Let X be a topological space with a chosen basepoint x_0, and let p: $Y \to X$ be a covering map. Given any path σ: $[0, 1] \to X$ with $\sigma(0) = x_0$ and $y_0 \in Y$ with $p(y_0) = x_0$, there exists a unique path $\tilde{\sigma}$: $[0, 1] \to Y$ with $\tilde{\sigma}(0) = y_0$ and $p \circ \tilde{\sigma} = \sigma$.*

Proof. The uniqueness lemma (12.1) tells us that there is at most one lift $\tilde{\sigma}$ of σ with $\tilde{\sigma}(0) = y_0$. Since p is a covering map, X is a union of open sets U_i such that $p^{-1}(U_i)$ is a disjoint union of open sets V_{ij} such that $p | V_{ij}$: $V_{ij} \to U_i$ is a homeomorphism. If the image of σ is contained in some single such U_i, then we define $\tilde{\sigma}$ by choosing some $V_{ij} \subseteq p^{-1}(U_i)$ and setting $\tilde{\sigma} = (p | V_{ij})^{-1} \circ \sigma$. In general, the image of σ need not be contained in a single U_i, but we can nevertheless write σ as a product of paths $\sigma = \sigma_1 \cdot \ldots \cdot \sigma_m$ with σ_i: $[t_i, t_{i+1}] \to X$ (where $t_1 = 0$ and $t_{m+1} = 1$) with the image of σ_i contained in a single open set as required.

Exercises

4. Show that such a decomposition $\sigma = \sigma_1 \cdot \ldots \cdot \sigma_m$ is indeed possible. To do this in detail, you will need some metric space properties of the interval $[0, 1]$. Try to generalize step one of the proof of Theorem 9.6.
5. Conclude the proof of Theorem 12.2 by showing that a lift $\tilde{\sigma}_i$ can be chosen for each σ_i such that $\sigma = \tilde{\sigma}_1 \cdot \ldots \cdot \tilde{\sigma}_m$ is a lift of σ. ■

12.3 Homotopy-lifting lemma. *Let X, x_0 and Y, y_0 be as in the path-lifting lemma. If H: $[0, 1] \times [0, 1] \to X$ is a map such that $H(0, t) = x_0 = H(1, t)$ for all $t \in [0, 1]$, then there exists a unique map $\tilde{H}$: $[0, 1] \times [0, 1] \to Y$ such that $p \circ \tilde{H} = H$ and $\tilde{H}(0, t) = y_0$ for all $t \in [0, 1]$.*

Proof. This result is a two-dimensional version of the path-lifting lemma, and to prove it we can use the same ideas. We content ourselves with a sketch. The unit square $[0, 1] \times [0, 1]$ is a compact metric space, so may be subdivided, if necessary, into smaller squares each of which is mapped by H into an open set of X on which a lift to Y is easy to define. Definitions of these lifts can be made so as to agree on common edges of the smaller squares, and so $\tilde{H}$ may be constructed. ■

The path- and homotopy-lifting lemmas indicate how the *local* similarities between X and a covering space Y allow processes occurring in X to be mimicked in Y. An important consequence of this idea now follows.

12.4 Theorem. *Let X be a space with basepoint x_0 and p: $Y \to X$ a covering map, with basepoint $y_0 \in Y$ chosen so that $p(y_0) = x_0$. Then the homeomorphism p_*: $\pi_1(Y, y_0) \to \pi_1(X, x_0)$ induced by p is injective.*

Proof. Let α be a closed path in Y at y_0 such that $p \circ \alpha$ is homotopic in X to the constant path at x_0: that is, $[\alpha]$ lies in the kernel of p_*. Let H: $[0, 1] \times [0, 1] \to X$ be a homotopy from $p \circ \alpha$ to the constant path at x_0, and let $\tilde{H}$ be its unique lift over p. Consider how $\tilde{H}$ maps the unit square into Y. The left-hand side maps to y_0, and the top and right-hand side into $p^{-1}(x_0)$, but by the preservation of connectedness (Theorem 3.6) we see that on all these three sides $\tilde{H}$ must be constant at y_0. The remaining side gives a lift of the path $p \circ \alpha$ over p and by the path-lifting lemma this must be α itself. Thus $\tilde{H}$ is a homotopy from α to the constant path at y_0, and the class of $[\alpha]$ is trivial in $\pi_1(Y, y_0)$. ■

Theorem 12.4 is the key that unlocks the structure of the collection of covering spaces of a space X, for we see that to any covering space there is associated a subgroup of $\pi_1(X, x_0)$. This association is not haphazard; it provides a neat algebraic description of the covering spaces of X as part of a profound interaction between group theory and topology. As our next step, we consider the implications of the choice of a particular basepoint $y_0 \in Y$. Suppose we pick $y_1 \in Y$ with $p(y_1) = x_0$. Since Y is path connected, there is a path σ: $[0, 1] \to Y$ with $\sigma(0) = y_0$ and $\sigma(1) = y_1$. Clearly $p \circ \sigma$ is a closed path in X at x_0 and conjugation by the class $[p \circ \sigma]$ in $\pi_1(X, x_0)$ carries $p_*(\pi_1(Y, y_1))$

to $p_*(\pi_1(Y, y_0))$. Conversely, suppose that $H \leqslant \pi_1(X, x_0)$ is conjugate to $p_*(\pi_1(Y, y_0))$; that is, suppose that there exists a class $[\alpha]$ such that $H = [\alpha]^{-1}p_*(\pi_1(Y, y_0))[\alpha]$. By the path-lifting lemma, there exists a unique path $\tilde{\alpha}$ in Y with $\tilde{\alpha}(0) = y_0$ and $p \circ \tilde{\alpha} = \alpha$. Set $y_1 = \tilde{\alpha}(1)$; then $H = p_*(\pi_1(Y, y_1))$. This argument establishes the following refinement of Theorem 12.4.

12.5 Theorem. *Let (X, x_0) be a pointed space and p: $Y \to X$ a covering map. Then the subgroups $\{p_*(\pi_1(Y, y)) \mid y \in p^{-1}(x_0)\}$ form a complete conjugacy class of subgroups of $\pi_1(X, x_0)$.* ∎

The path-lifting and homotopy-lifting lemmas were proved by exploiting a particular feature of the unit interval and unit square, namely compactness. What might we hope for as a more general map-lifting result for covering spaces, where no special features of the domain can be called on to help? Suppose that p: $Y \to X$ is a covering map and that f: $Z \to X$ has a lift $\tilde{f}$: $Z \to Y$; we assume basepoints $x_0 \in X$, $y_0 \in Y$, and $z_0 \in Z$ are fixed such that $p(y_0) = x_0$, $f(z_0) = x_0$, and $\tilde{f}(z_0) = y_0$. Then the maps p_*, f_* and $\tilde{f}_*$ fit into a commutative diagram of fundamental groups

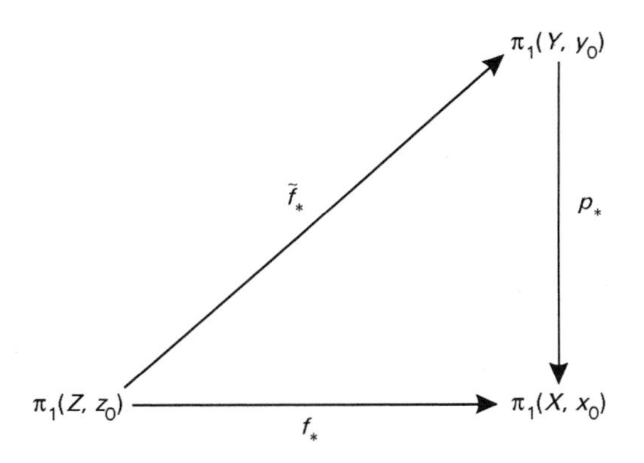

from which it follows that $f_*(\pi_1(Z, z_0)) \subseteq p_*(\pi_1(Y, y_0))$. It is one of the beauties of covering space theory that this necessary algebraic condition for the existence of a lift of f over p is also sufficient.

12.6 Map-lifting theorem. *Let p: $Y \to X$ be a covering map, with basepoints $y_0 \in Y$ and $x_0 \in X$ chosen so that $p(y_0) = x_0$. Given a map f: $Z \to X$ carrying the basepoint $z_0 \in Z$ to x_0, there exists a lift $\tilde{f}$ of f over p if and only if $f_*(\pi_1(Z, z_0)) \subseteq p_*(Y, y_0))$.*

Proof. We have seen that the condition is necessary. To establish sufficiency, we shall define $\tilde{f}$ directly. Consider an arbitrary point

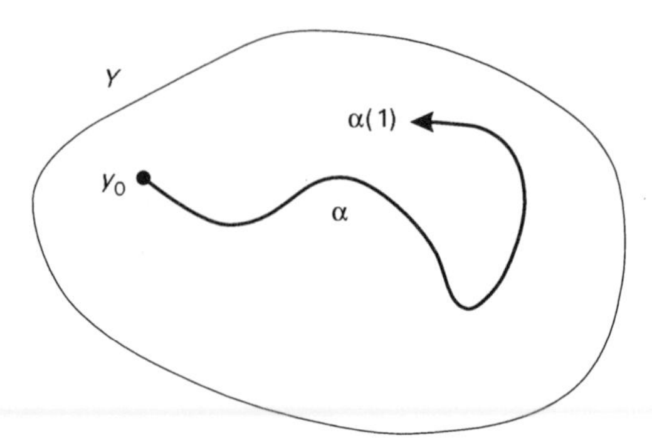

Fig. 12.4

$z \in Z$, and let σ: $[0, 1] \to Z$ be a path with $\sigma(0) = z_0$ and $\sigma(1) = z$; σ exists because of our assumption that all spaces are path connected. Now $f \circ \sigma$ is a path in X with $f \circ \sigma(0) = x_0$: by the path-lifting lemma there exists a unique path α: $[0, 1] \to Y$ such that $\sigma(0) = y_0$ and $p \circ \alpha = f \circ \sigma$. We define $\tilde{f}$ by setting $\tilde{f}(z) = \alpha(1)$ (Fig. 12.4).

We now have a candidate for $\tilde{f}$: at present we have a function $Z \to X$ that satisfies $p \circ \tilde{f} = f$ and depends on the choice of a path in Z for each $z \in Z$. We first show that the definition is in fact independent of the choice of path, and then that $\tilde{f}$ is continuous.

Given $z \in Z$, suppose that σ and τ are two paths from z_0 to z, and that τ lifts to a unique path β: $[0, 1] \to Y$ with $\beta(0) = y_0$ and $p \circ \beta = f \circ \tau$. If $\bar{\tau}$ is the reverse of τ then $\sigma \cdot \bar{\tau}$ is a closed path at $z_0 \in Z$ and so $f_*([\sigma \cdot \bar{\tau}]) \in p_*(\pi_1(Y, y_0))$. Let γ be a closed path in Y at y_0 with $[f \circ (\sigma \cdot \bar{\tau})] = [p \circ \gamma]$ and lift the homotopy $f \circ (\sigma \cdot \bar{\tau}) \simeq p \circ \gamma$ to (Y, y_0). The lifted homotopy maps the unit square into Y according to the following scheme:

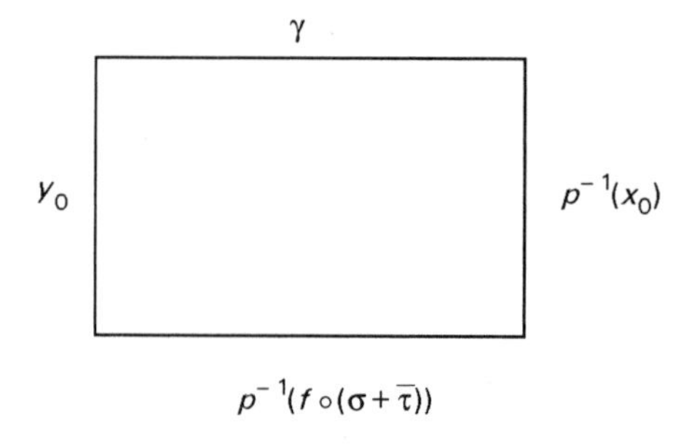

and again by the preservation of connectedness, the right-hand side must be mapped to y_0. Therefore $f \circ (\sigma \cdot \bar{\tau})$ lifts to a closed path in

(Y, y_0) at y_0 and so $\alpha(1) = \beta(1)$. This establishes that the definition of $\tilde{f}$ is independent of the choice of path.

It remains to show that $\tilde{f}$ is continuous. Let $z \in Z$ and suppose that N is a neighbourhood of $\tilde{f}(z)$ in Y. Let U be a neighbourhood of $f(z)$ in X and V an open set in Y containing $\tilde{f}(z)$ such that p maps V homeomorphically to U. Then $f^{-1}(p(N \cap V))$ is a neighbourhood of z in Z and so contains a path connected neighbourhood W of z in Z. We shall show that $\tilde{f}(W) \subseteq N$ and this will complete the proof.

Exercise

6. Why will the fact that $\tilde{f}(W) \subseteq V$ complete the proof?

Let $z' \in W$ and now let τ: $[0, 1] \to W$ be a path with $\tau(0) = z$ and $\tau(1) = z'$. Lift $f \circ (\sigma \cdot \tau)$ to a path in Y beginning at y_0. Its end point is $\tilde{f}(z')$. Now $f \circ \tau$ has its image in $p(N \cap V)$ and its lift starts at the end point of the lift of $f \circ \sigma$, which is $\tilde{f}(z)$. But $p|(N \cap V)$ is a homeomorphism, so that the end point of this lift lies in $N \cap V$. ■

It is now possible to produce a hierarchical structure for the coverings of a given space, based upon the interrelationships of their fundamental groups. Consider two covering maps p: $Y \to X$ and p': $Y' \to X$ and choose basepoints $y \in Y$ and $y' \in Y'$ with $p(y) = x_0 = p'(y')$. Let us set $H = p_*(\pi_1(Y, y))$ and $H' = p'_*(\pi_1(Y', y'))$.

12.7 Proposition. *If $H \leqslant H'$ then there is a covering map h: $Y \to Y'$ sending y to y' and such that $p \circ h = p'$.*

Proof. Apply the map-lifting theorem and lift p over p' to obtain h: $Y \to Y'$ with $p \circ h = p'$. We leave it to you as an exercise to verify that the equation $p \circ h = p'$, together with the fact that both p and p' are covering maps, suffice to ensure that h is also a covering map. ■

Of course, if $H = H'$ then Proposition 12.7 works in both directions to give covering maps g: $Y' \to Y$ and h: $Y \to Y'$ with $p' \circ g = p$ and $p \circ h = p'$. It follows that $p \circ h \circ g = p$ so that both $h \circ g$ and the identity map id lift p to a map $Y \to Y$, and so by the uniqueness lemma we must have that $h \circ g = \text{id}$. Similarly $g \circ h = \text{id}$ and so h: $Y \to Y'$ is a homeomorphism.

Exercise 12.1

1. We call two covering maps p: $Y \to X$ and p': $Y' \to X$ *equivalent* if there exists a homeomorphism h: $Y \to Y'$ such that $p \circ h = p'$. We saw in Theorem 12.5 that a covering map determines a complete conjugacy class of subgroups of $\pi_1(X, x_0)$. Show that

two covering maps are equivalent if and only if they determine the same conjugacy class of subgroups.

12.2 Deck transformations and universal covers

Let $p\colon Y \to X$ be a covering map. A *deck transformation* is a homeomorphism $h\colon Y \to Y$ such that $p \circ h = p$. A deck transformation is therefore a homeomorphism of the covering space that preserves the *fibres* of p: that is the subsets $p^{-1}(x)$ for each $x \in X$.

Exercises 12.2

1. Show that the set of all deck transformations for a fixed covering map $p\colon Y \to X$ forms a group under the operation of composition of maps.
2. Show that if h is a deck transformation and is not the identity map, then for all $y \in Y$ we have $h(y) \neq y$.

Is the group in Exercise 12.2.1 related to the fundamental group of X, or to its subgroups? Note that h is a lift of h over itself, so that by the uniqueness lemma, distinct deck transformations must differ at *every* point of Y. We may translate the conclusion of exercises 12.2.1 and 12.2.2 so as to say that the deck transformations of a covering map $p\colon Y \to X$ form a group that acts freely on Y. This idea allows us to understand the deck transformations of certain covering maps $p\colon Y \to X$ in terms of the fundamental group of X.

12.8 Theorem. *Let $p\colon Y \to X$ be a covering map and let Q be the group of deck transformations. If for some $y \in p^{-1}(x_0)$, $H = p_*(\pi_1(Y, y_0))$ is a normal subgroup of $\pi_1(X, x_0)$ then X is homeomorphic to the orbit space Y/Q and Q is isomorphic to the quotient group $\pi_1(X, x_0)/H$.*

Proof. Given $y \in Y$, let $[y]$ denote its equivalence class in Y/Q. Since $p \circ h(y) = p(y)$ for all $h \in Q$, the function $\theta\colon Y/Q \to X$ given by $[y] \mapsto p(y)$ is well defined, and is plainly surjective. For each $x \in X$ and for $y, y' \in p^{-1}(x)$, we have that $p_*(\pi_1(Y, y)) = p_*(\pi_1(Y, y'))$ since they are both conjugate, by Theorem 12.5, to $p_*(\pi_1(Y, y_0))$ which is normal. By Proposition 12.7 there is a deck transformation h with $h(y) = y'$ and it follows that θ is a bijection. We leave it to you as an exercise to complete the verification that θ is a homeomorphism.

Given a closed path α at x_0, lift α to a closed path $\tilde{\alpha}$ beginning at y_0. Then $p \circ \tilde{\alpha}(1) = x_0$ so that there exists a unique deck transformation $k_\alpha \in Q$ such that $k_\alpha(y_0) = \tilde{\alpha}(1)$. Then the map $[\alpha] \mapsto k_\alpha$ is a surjective homeomorphism $\pi_1(Y, x_0) \to Q$ with kernel H. Again we leave the verification of the details of this assertion to you. ■

A covering map p: $Y \to X$ for which $p_*(\pi_1(Y, y_0))$ is a normal subgroup of $\pi_1(X, x_0)$ is called *regular*. A covering map p: $Y \to X$ for which $\pi_1(Y, y_0) = 1$ is unique up to homeomorphism (by Proposition 12.7) and is a regular covering space of any covering space of X. For these reasons, any such space is called the *universal covering space of* X. Not every space possesses a universal covering space; see Armstrong (1982) for a further discussion. For those spaces that do possess universal covering spaces we can establish an exact correspondence between subgroups of the fundamental group $\pi_1(X, x_0)$ and covering spaces of X. This correspondence is a refinement of Theorem 12.8.

12.9 Theorem. *Let X be a space with a universal covering space $\tilde{X}$. Then $\pi_1(X, x_0)$ is the group of deck transformations of $\tilde{X}$, and if $H \subseteq \pi_1(X, x_0)$ then $\tilde{X}/H$ is a covering space of X with H as fundamental group. Thus if X is a space with a universal covering space, then there exists a covering space corresponding to any subgroup of $\pi_1(X, x_0)$.* ■

12.3 Covering graphs

We have seen how successfully the theory of covering spaces connects the topological nature of a space with the subgroup structure of its fundamental group. We shall now concentrate on a particular aspect of this theory: that of covering spaces of graphs. We have seen in Chapter 10 that the fundamental group of a graph is always a free group, so that graphs provide a topological model of free groups, and we can use the covering space theory of graphs to establish facts about free groups and their subgroups. This is the aim of the remainder of the chapter.

A choice has to be made about the type of approach that we follow. In our previous discussions, we have treated graphs essentially as combinatorial objects, their structure determined by the incidence between edges and vertices. Meanwhile our topological theory has been developed in a very general setting. Now comes the crunch. We could repeat the topological theory in a combinatorial setting, and keep everything as close to our previous discussion of graphs as possible. While the ideas may be more intuitive here, there are still subtleties to be confronted, and for a careful account of covering spaces from a combinatorial viewpoint we refer to Cohen (1989). We shall make use of the topological theory, and this choice requires us to set up a rigorous topological theory of graphs. This end could be achieved by modelling combinatorial graphs as subspaces of some $\mathbb{R}^n$, taking n as large as we need, but we face problems of ensuring that this approach is consistent and independent of the embeddings chosen. Instead we opt for a more abstract setting, and redefine a graph. The new definition will define a graph as a *one-dimensional CW complex*. CW complexes are a class of topological spaces so constructed that their homotopy theory avoids some pathologies that arise in the more general theory. Happily, most spaces encountered in everyday life can be given the structure of a CW complex. The transition from combinatorial graph to CW complex and back again will be fairly evident, and readers who prefer to keep their minds on the combinatorial side of things should try to rephrase the definitions and results according to their preference.

Let us proceed to the definition of a graph. A *graph* is a Hausdorff space Γ together with a subspace Γ^0 such that the following conditions hold:

G1: The subspace topology on Γ^0 is the discrete topology, and Γ^0 is closed;

G2: $\Gamma \backslash \Gamma^0$ is a disjoint union of open subsets e_λ each of which is homeomorphic to the open interval $(0, 1) \subseteq \mathbb{R}$;

G3: the closure $\bar{e}_\lambda$ of e_λ meets Γ^0 in either one or two points: in the first case there is a homeomorphism f_λ: $[0, 1] \to \bar{e}_\lambda$ mapping $(0,1)$ to e_λ, whilst in the second case there is a homeomorphism f_λ: $S^1 \to \bar{e}_\lambda$ mapping $S^1 \backslash \{1\}$ to e_λ;

G4: a subset $A \subseteq \Gamma$ is closed if and only if $A \cap \bar{e}_\lambda$ is closed for all λ.

Properties **G1**, **G2**, and **G3** have straightforward combinatorial reinterpretations; **G4** looks at first sight more mysterious, and indeed is redundant if there are only finitely many e_λ. Its purpose is to set up the topology on Γ when there are infinitely many e_λ. It is of course convenient to retain the obvious terminology for graphs under the new definition, so we call points of Γ^0 *vertices*, and each e_λ is an *edge*.

Exercises 12.3

1. *The Hawaiian earring*. Consider the following subset Γ of $\mathbb{R}^2$:

$$\Gamma = \bigcup_{n=1}^{\infty} \{(x, y) \mid (x - (1/n))^2 + y^2 = (1/n)^2\}.$$

We see that Γ consists of an infinite number of circles, each tangent to the y-axis at the origin. Regarding the origin as Γ^0, and each circle as an edge, show that **G1**, **G2**, and **G3** hold, but that **G4** fails. The

Hawaiian earring is also infamous for not possessing a universal covering space.

The first result in the covering space theory of graphs is that we do not need to consider any other type of space. See Massey (1967) for a proof of the following proposition.

12.10 Proposition. *Let Γ be a graph and $p\colon Y \to \Gamma$ a covering map. Then Y is a graph.* ∎

The second result of importance is that any graph possesses a universal covering space. Since this will be a graph with trivial fundamental group, it must be a tree.

The first theorem that we seek on free groups now follows easily. It was first proved by J. Nielsen in 1927 for finitely generated subgroups of free groups, and then generalized by O. Schreier.

12.11 The Nielsen–Schreier theorem. *Every subgroup of a free group is free.*

Proof. Given a subgroup G of the free group F, we construct a graph Γ such that $F \cong \pi_1 \Gamma$, the universal cover $\tilde{\Gamma}$, and a covering space $Y = \tilde{\Gamma}/G$ of Γ with $\pi_1 Y \cong G$. Then G must be a free group because it is isomorphic to the fundamental group of a graph. ∎

For subgroups of finite index in finitely generated free groups, we can be more precise and compute the number of free generators required for the subgroup. This result is another due to Schreier.

12.12 The Schreier index formula. *Let G be a subgroup of finite index i in the free group F of finite rank r. Then G is a free group of rank $i(r-1)+1$.*

Proof. We may construct a graph Γ with one vertex and with r edges whose fundamental group is F. The covering space corresponding to G will have i vertices and ir edges, and since a maximal tree must contain $i-1$ edges, there will be $ir-i+1$ edges not in a maximal tree, and this number of edges equals the number of free generators of G. ∎

Finally we show how to obtain free generators for relation subgroups in group presentations. Recall that the relation subgroup of a group presentation $(X : R)$ is the normal subgroup of the free group with basis X generated by the elements R. If there are n generators and the presentation is of a finite group G then the

Schreier index formula tells us that the relation subgroup is a free group of rank $|G|(n-1)+1$. Can we produce free generators? The Cayley quiver gives an easy method, for it is a covering graph of the rose with n leaves, and its fundamental group is isomorphic to the relation subgroup. All we have to do is find generators for the fundamental group of the Cayley quiver, and see what are their images in the fundamental group of the n-leafed rose. A detailed example will illustrate the method involved.

Consider the presentation $(a, b : a^3, b^2, (ab)^2)$ of the symmetric group of order 6. Its Cayley quiver is shown in Fig. 12.5, with 1 marked as basepoint and a maximal tree indicated.

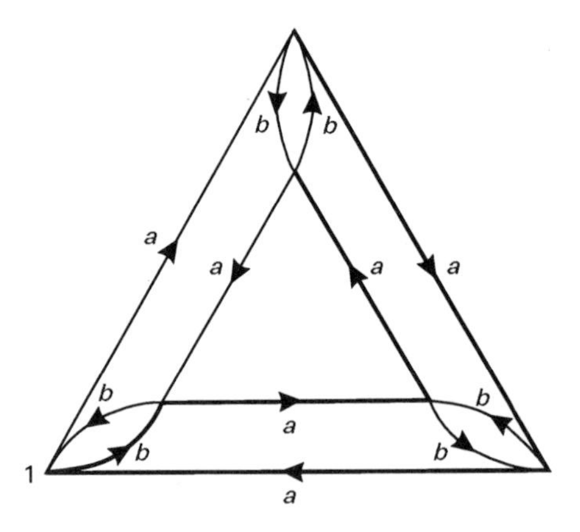

Fig. 12.5

For each edge of the quiver not in the maximal tree, we write down the label of the path that takes us from 1 to the source of the edge in the maximal tree, then along the edge, and finally back to 1 from the target vertex. In our current example there are seven such edges, and their labels are

(1) $baba$

(2) $a^{-1}ba^{-1}b^{-1}$

(3) ba^2ba^2

(4) $a^{-2}ba^{-2}b^{-1}$

(5) a^3

(6) ba^3b^{-1}

(7) b^2

These seven elements generate the relation subgroup. Each can be written as a product of conjugates of the relators, since the relation subgroup is the smallest normal subgroup of the free group with basis $\{a, b\}$ containing the relators. Note that (5), (6), and (7) are conjugates of a single relator as they stand.

Exercises 12.3 continued

2. Write the elements (1), (2), (3), and (4) in the example just given as products of conjugates of the relators a^3, b^2, and $(ab)^2$.
3. Use the Schreier index formula to compute the rank of the relation subgroup for the following presentations:

(a) $(x, y : x^{-1}yxy, y^{-1}xyx) \cong Q_8$, the quaternion group of order 8;

(b) $(x, y, z : xyz^{-1}, yzx^{-1}, zxy^{-1}) \cong Q_8$;

(c) $(x : x^{60})$, presenting the cyclic group of order 60;

(d) $(x, y, z : x^3, y^4, z^5, [x, y], [y, z], [z, x])$, which also presents the cyclic group of order 60;

(e) $(x, y : x^2, y^3, (xy)^5) \cong A_5$, the alternating group of degree 5, which again has order 60.

4. Draw Cayley quivers for examples (a) and (b) in Exercise 12.3.3, and find sets of free generators for the relation subgroups in each case.

Bibliography

Alexander, J. W. (1928). Topological invariants of knots and links. *Trans. Am. Math. Soc.*, **30**, 275–306.

Armstrong, M.A. (1983). *Basic topology.* Springer-Verlag, New York, Heidelberg, Berlin.

Bleiler, S. A. (1984). A note on unknotting number. *Math. Proc. Cambridge Philos. Soc.*, **96**, 469–71.

Brown, R. (1967). Groupoids and Van Kampen's Theorem. *Proc. London Math. Soc. (3)*, **17**, 385–401.

Brown, R. (1988). *Topology*. Ellis Horwood, Chichester.

Burde, G. and Zieschang, H. (1985). *Knots.* De Gruyter, Berlin, New York.

Cohen, D. E. (1989). *Combinatorial group theory*, London Math. Soc. Student Text 14. Cambridge University Press.

Conway, J. H. (1969). An enumeration of knots and links. In *Computational problems in abstract algebra* (ed. J. Leech). Pergamon, Oxford.

Crowell, R. H. and Fox, R. H. (1967). *Introduction to knot theory.* Springer-Verlag, New York, Heidelberg, Berlin.

Edmonds, D. (1960). A combinatorial representation for polyhedral surfaces. *Not. Am. Math. Soc.*, **7**, 646.

Freyd, P., Yetter, D., Hoste, J., Lickorish, W. B. R., Millett, K. C. and Ocneanu, A. (1985). A new polynomial invariant of knots and links. *Bull. Am. Math. Soc.*, **12**, 239–46.

Gross, J. L. and Tucker, T. W. (1987). *Topological graph theory*, Wiley Interscience, New York.

Johnson, D. L. (1990). *Presentations of groups*, London Math. Soc. Student Text 15. Cambridge University Press.

Jones, V. F. R. (1985). A polynomial invariant for knots via Von Neumann algebras. *Bull. Am. Math. Soc.*, **12**, 103–11.

Kauffman, L. H. (1987*a*). *On knots*, Ann. Math. Studies 115. Princeton University Press.

Kauffman, L. H. (1987*b*). State models and the Jones polynomial. *Topology*, **26**, 395–407.

Kauffman, L. H. (1991). *Knots and physics.* World Scientific Publishing, Singapore.

Kuratowski, K. (1930). Sur le probleme des courbes gauches en topologie. *Fund. Math.*, **15**, 271–83.

Lang, S. (1972). *Differential manifolds.* Addison-Wesley, Reading, Massachusetts.

Lickorish, W. B. R. and Millett, K. C. (1987). A polynomial invariant of oriented links. *Topology*, **26**, 107–41.

Magnus, W. and Chandler, B. (1982). *The history of combinatorial group theory*. Springer-Verlag, New York, Heidelberg, Berlin.

Massey, W. S. (1967). *Algebraic topology*. Springer-Verlag, New York, Heidelberg, Berlin.

Murasugi, K. (1987). Jones polynomials and classical conjectures in knot theory. *Topology*, **26**, 187–94.

Neuwirth, L. (1961). A note on torus knots and links determined by their groups. *Duke Math. J.*, **28**, 545–51.

Priestley, H. A. (1990). *Introduction to complex analysis*. Oxford University Press.

Reidemeister, K. (1932). *Knotentheorie*, Ergebnisse der Mathematik 1. Springer-Verlag, Berlin.

Rotman, J. (1984). *The theory of groups*. W. C. Brown.

Schubert, H. (1956). Knoten mit zwei Brücken. *Math. Z.*, **65**, 133–70.

Sutherland, W. A. (1981). *Introduction to metric and topological spaces*. Oxford University Press.

Thistlethwaite, M. B. (1987). A spanning tree expansion of the Jones polynomial. *Topology*, **26**, 297–309.

Tietze, H. (1908). Über die topologischen Invarianten mehrdimensionaler Mannigfaltigkeiten. *Monatsschr. Math. Phys.*, **19**, 1–118.

Trotter, H. F. (1964). Non-invertible knots exist. *Topology*, **2**, 275–80.

White, A. T. (1984). *Graphs, groups and surfaces*, North-Holland Mathematical Studies 8. Elsevier, Amsterdam.

Wilson, R. J. (1985). *An introduction to graph theory*. Longman, Harlow.

Index